# Photoprocesses, Photoreceptors, and Evolution

# Photoprocesses, Photoreceptors, and Evolution

**Jerome J. Wolken**
Biophysical Research Laboratory
Carnegie-Mellon University
Pittsburgh, Pennsylvania



ACADEMIC PRESS New York San Francisco London 1975
A Subsidiary of Harcourt Brace Jovanovich, Publishers

ACADEMIC PRESS, INC.
111 Fifth Avenue, New York, New York 10003

*United Kingdom Edition published by*
ACADEMIC PRESS, INC. (LONDON) LTD.
24/28 Oval Road, London NW1

Library of Congress Cataloging in Publication Data

Wolken, Jerome J
Photoprocesses, photoreceptors, and evolution.

Bibliography: p.
Includes index.
1. Photobiology. 2. Photoreceptors. 3. Evolution. I. Title.
QH515.W64 574.1'9153 75-3577
ISBN 0-12-762050-8

PRINTED IN THE UNITED STATES OF AMERICA

**The World is full of mysteries. Life is one.**

J. B. S. Haldane (1966)

# Contents

# Preface

In this monograph I have summarized a part of my experimental work on photoreceptors and photoprocesses of plants and animals. I have emphasized the important role that visible radiation has played in the development of photoreceptor systems, hence, in the evolution of life on earth.

I began these studies some two decades ago by asking questions about the effects of radiation on the behavior of living organisms: What kind of photoprocesses go on when cells are excited by light? What kind of photosensitive pigments are used in these photoprocesses? What type of photoreceptor structures evolved?

To put these studies into historical perspective, I chose as a starting point the year 1859, the publication date of Charles Darwin's "*The Origin of Species by Means of Natural Selection.*" In the more than a century which has elapsed since Darwin's publication, our ideas about the origin of life and evolution and our understanding of living processes have come a long way: from the understanding of the genetic mechanisms beginning with Mendel in 1866 to the elucidation of the genetic molecule of life, DNA, by Watson, Crick, and Wilkins in the 1950's.

To gain some feeling for this period of history, during 1970 I repeated in part Darwin's voyage on the Beagle (1831–1836) to South America and then to the Galapagos Islands where I observed firsthand the terrain and the plants and animals that continue to thrive on these islands. This experience inspired me to begin writing this monograph.

In 1971, on sabbatical leave from my university, I proceeded to University College London, where Thomas Huxley (1825–1895) lectured. Among those attending his lectures was Charles Darwin, and Huxley remained a friend and supporter of Darwin throughout his life. Darwin later lived on Gower Street, where a plaque reads, "Darwin lived here 1838–1842," and where more recently J. B. S. Haldane and J. D. Bernal speculated about the nature of life and its origin.

I then proceeded to Paris to obtain some feeling for the work of Pasteur. At the Pasteur Institute, I examined the records of his experiments which led to the refutation of the concept of the spontaneous generation of life and to his studies of crystals of biological origin. The Museum of Natural History in the Jardin des Plantes also provided an opportunity to examine the two volumes of Lamarck's "Philosophie Zoologique" (1809) on the inheritance of acquired characteristics.

While in Peru I became fascinated with the Incan sun god mythology. This led me back to South America in 1973 and to travels through Mexico to compare the ruins and temples to the Sun of the Incas to those of the Mayans, Toltecs, and Aztecs.

All of these experiences permitted me to indulge and relive a part of history.

I have set about trying to arrive at some basis as to why life on earth as we know it is possible, and whether it was the radiation from the sun and the development of photoreceptor systems that has made life possible. Therefore, the underlying theme which I have tried to develop in this monograph is that visible radiation has played a profound role in the evolution of life on earth. Whatever the environmental conditions were which shaped the molecules necessary for life, once the molecules were formed into specific structures they became ordered within a cell and became functional. Certain molecules were particularly favorable for biological processes and specifically for light absorption. Once functional, they helped determine the behavior of the organism through a variety of light-driven reactions.

Photobiologists have greatly advanced our knowledge of photoprocesses in living systems. This book was not written for these specialists, but for those students and scientists who want to obtain some idea and understanding of the photobiological world. I have emphasized directions researchers have taken to understand the phenomena, particularly my own approach. I hope the text raises questions in the minds of the readers, and so will point out many areas yet to be explored. Above all, I hope that it will present a glimpse of the beauty, the excitement, and mysteries that can be found in the world of light we live in.

**Jerome J. Wolken**

# Acknowledgments

The writing of this monograph could not have been accomplished without a sabbatical leave from Carnegie–Mellon University, 1971–1972, a grant from the American Philosophical Society, and research support from the National Aeronautical and Space Administration (NASA), the Scaife Family Charitable Trust of Pittsburgh, and the Pennsylvania Lions Eye Research Foundation. I would like to express my thanks for the hospitality extended to me at University College London by Professor J. Z. Young and Dr. P. N. Dilly, Department of Anatomy; by Professor Jacques Monod and Dr. David Perrin during my stay at Pasteur Institute, Paris; by Monsieur E. Roux, Biophysics, Atomic Energy Commission, Saclay, France; and by the Smithsonian Institute for Tropical Research, Panama.

Some of the experimental work was presented in my previously published books and in scientific journals. I would like to thank the publishers Appleton-Century-Crofts, New York; Charles Thomas Publishers, Springfield, Illinois; Van Nostrand Reinhold Company, New York; and Academic Press, New York for permission to reproduce figures and tables from my previous publications in this book.

I would like to thank Dr. Mary Ann Mogus and Dr. Christopher S. Nakagawa for the many discussions I had with them during the writing of this book. My thanks are also due to Mrs. Judith Plotz for helping to collate the first bits of material and to Ms. Paula Damsel for the interest and care in seeing that this manuscript reached its final form. I want to thank Mr. Oliver J. Bashor, Jr., for his technical assistance in many of the experiments that were described and especially for his skill in the drawing and photography of the numerous figures in the text. Also I want to thank A. Jonathan Wolken for his continuous encouragement during the writing of this text. I am grateful to all in the Biophysical Research Laboratory who have been associated with me over the years in the search to find answers about light, the phoreceptor systems, and life.

I would also like to thank and acknowledge those who have contributed photographs for use in the figures: Professor Sidney Fox, University of Miami, Miami, Florida (Figure 2.2); Dr. P. Guérin de Montgareuil, Atomic Energy Commission, Cadarche, France (Figure 2.4); Dr. K. Mühlethaler, Eidgenossische Technische Hochschule, Zurich, Switzerland (Figure 5.6); Professor David Dennison, Dartmouth College, Hanover, New Hampshire (Figure 6.2b); Professor Richard Eakin, University of California, Berkeley, California (Figure 9.3); and Dr. T. Kuwabara, National Institutes of Health, Bethesda, Maryland (Figure 10.7b). Data taken in part or adapted from other sources are acknowledged in the legends for figures and tables and in the references.

# I
# Introduction



Most cultures offer mystical and mythical accounts of the supernatural creation of life. The Judeo-Christian tradition has an account in the first book of Moses, Genesis, Chapter I, of the creation of earth in six days:

> In the beginning God created the heaven and the earth. And the earth was without form, and void; and darkness was upon the face of the deep. And the Spirit of God moved upon the face of the waters.
>
> And God said, *let there be light*: and *there was light*. And God saw the light, that it was good; and God divided the light from the darkness.

And, we are told, that beginning on the third day of creation, God brought forth living creatures: first plants, then fishes and birds, then land animals, and finally man. This biblical story dominated man's thinking on earth history from the Middle Ages to nearly the close of the eighteenth century.

Light—the sun—has from man's earliest records and biblical times been associated with life. Long before the advent of Christianity, ancient civilizations throughout the world were sun worshippers. The rituals and beliefs of ancient civilizations placed the sun and fire gods in preeminent positions. But the beliefs enshrining these gods and myths are the very beliefs that now underlie our convictions of the balance of nature and the uniqueness of life.

How remarkable that the science of today should develop on the intuitions of the past; for the deeper we delve into the origin and evolution of life, the more clearly life emerges as a light-mediated phenomenon.

Ideas about the creation of life and evolution have occurred to many civilizations throughout human history. They appear in Greek writings beginning with Aristotle in the fourth century B.C. with the view that life evolved gradually out of primeval slime.

The Kabbalists, a mystical group of Hebraic origin in the fifteenth and sixteenth centuries, held a philosophy that linked light with creation. Light was a force to bring order and structure. They believed that in addition to the direct light being absorbed there is a reflected light—a blue energy which permeates the universe and which forms a field around living beings. This reflected light can reascend through a whole series of steps very much like fluorescence. Needham (1956), a well-known biochemist and historian of science, believes it may well turn out that the "correlative thinking" of the Kabbalists had more influence on scientific minds in the dawn of modern science than has generally been credited to them.

Evolution was not a new idea when Darwin presented it in 1859. For the idea was already expressed in the middle of the eighteenth century by Diderot and in the writings of Maupertuis (1756), of Buffon (1797–1807), and of Erasmus Darwin (1794), the grandfather of Charles Darwin. Pre-Darwinian evolutionary theories presupposed a short geological time span which would not allow sufficient time for gradual evolution to operate. When geologist Sir Charles Lyell (1830–1833) amassed evidence for the long duration of earth history, he became convinced that geological phenomena could best be explained by natural forces acting over eons of time—a view that came to be termed Uniformitarianism. The first and second volumes of his "Principles of Geology" concerning the modern changes of the earth and its inhabitants appeared in 1830 and 1832, respectively. The great impact of Lyell's "Principles of Geology" on the scientific world at that time is evidenced by the fact that Darwin carried this book with him on his voyage on the *Beagle*. When Louis Agassiz (1840) showed that glaciers were a large-scale earth-moving agency, especially during the glacial ages, the Uniformitarianism school began to develop and to dominate the science of geology as well. Immanuel Velikovsky (1955) conjectured that the solar system is not immutable. He rejected the uniformitarian view of slow continuous change and advocated a thesis of cataclysmic events which interrupted the evolutionary path. He viewed the development of life on earth as a discontinuous process and produced evidence that twice in man's recorded history (1400–1500 B.C. and 700–800 B.C.) there were massive perturbations caused by the near collision between the earth and the comet Venus, and later by the near pass of Mars. Recent discoveries, for example, of volcanic ash layers interlayered within the sediment over vast areas of the sea floor, show that there have been catastrophic events in the earth's history. But these geophysical observations, especially of the earth's crust under the oceans, show that even the catastrophic events of volcanism and earthquakes result from relatively slow movements of vast segments of the earth's crust.

The first clear demonstration of the fact of evolution was made by the French naturalist, Jean-Baptiste Lamarck (1774–1829). Lamarck's work on evolution appeared in 1809, and his principal theory was elucidated between 1815 and 1822. Lamarck grasped that all life is the product of evolutionary change, that evolution resulted in the taking on of new adaptations to the environment, and that the diversity of life was the result of such adaptations.

Lamarck's view afforded a disarmingly naive understanding of the exquisitely engineered and specific adaptations to environment encoded in all living creatures. He assumed that acquired properties can be inherited. We now know they are not except under quite exceptional and limited circumstances. The unit steps of evolution are those spontaneous and predominantly fortuitous mutations which have little to do with the fitness of the individual during its lifetime. The end results are almost the same as if Lamarck were right, only the mechanism by which things happen is far more ingenious–and different–from what he thought. The Lamarckian view offered no verifiable, let alone theoretically acceptable, explanation of how an acquired trait could change the genetic blueprint. Time and again scientists motivated by misunderstandings of Darwinism or by political ideology have strived to find such proof; the consequences, notably in plant breeding, have been disastrous. And so Lamarck's ideas have generally been rejected by biologists.

Julian Huxley (1964) pointed out that the most striking simulation of Lamarckism occurs when the initial variations that inaugurated an evolutionary change are not true genetic mutations, that is, not the inheritable type. Yet, if profitable to the organism, they may be accentuated by what he calls "organic selection," and pave the way for true genetic mutations to be immediately seized upon when they turn up. Hence a new character or modification of a character, acquired by variation, and mutation, or mutation plus some selection, may arouse in the organism an activity which tends to increase the usefulness of that character. In the end, by possessing this new or changed character the individual may be caused to change its behavior toward its environment.

It was Gregor Mendel (1866) who discovered the laws governing most heredity. Although he published his results in 1866, the journal in which he published was obscure, and his work was not appreciated by the scientific community until after 1900. The science of genetics has now been refined to the point that the frontier of research deals with the molecular basis of inheritance and bears directly on the study of evolution.

The theory of evolution dates in most people's minds to the time of Charles Darwin (1808–1882), and specifically to 1859 when his "The Origin of Species by Means of Natural Selection" was first published. Darwin took a long time to bring his ideas to fruition after his voyage on the *Beagle* (1831–1836). In 1858 when he became aware that Alfred Russel Wallace (1858) had independently worked out the theory of natural selection, he hastened the long-delayed publication of his book. About the same time Henry W. Bates (1864), a collaborator of Wallace, also postulated evolution by means of natural selection and showed how natural selection accounted for mimicry in butterflies. This was an important observation, for mimicry at the morphological and behavioral level implies a molecular evolution as well. Darwin continued through 1872 to write six revisions of his book. With each new revision he tended toward a more Lamarckian point of view.

Darwin attributed evolutionary change to several forces, but the prime force was natural selection. His two basic observations were (1) all organisms exhibit variability, and (2) all organisms reproduce many more offspring than survive. Darwin saw the process of evolution as a struggle for existence among individual organisms. Since all species produce many more offspring than survive, he concluded that the environment eliminated those individuals least fit for survival. Natural selection favored features of an organism that brought it into a more efficient relationship with its environment, and this accounted for every living creature's adaptation to its specific environment.

Darwin did not explore the question of the impact of solar radiation as an evolutionary force, although photosensitivity of plants and animals interested him. He experimented with all kinds of plant movements, including phototropism, but the lack of development at that time in behavioral plant physiology provided little encouragement. These extensive observations were kept apart and published separately from his work on evolution. The evolution of the eye was also of interest to him, but the criticism of Hooker discouraged him from including any ideas about the eye in "The Origin of Species." In a letter to Asa Gray in 1860 Darwin wrote, "The eye to this day gives me a cold shudder. . . ."

Darwin provided the key to the theory of evolution. The impact of his ideas inside and outside of science was widely felt. Karl Marx was immediately taken by the philosophy of Darwin, and saw within it implications for his own ideas regarding man's struggle, in the socioeconomic system, for survival. As a result Marx dedicated "Das Kapital" (1867) to Darwin. Although they exchanged polite letters of thanks, it is doubtful that Darwin read it, for the pages remain uncut in the copy in his library at Down House, Downe.

Shortly after Darwin's publication of "The Origin of Species," Louis Pasteur showed that life could not originate on the earth's surface under existing conditions; life had to come from life (see Vallery-Radot, 1922–1939; Dubos,

1960). This dispelled, for a time, belief in the spontaneous generation of life. Pasteur did a great deal of laboratory experimentation which led him to some unique conclusions about life. For example, in his studies with crystals Pasteur was impressed by the fact that asymmetric molecules are always the products of living processes. This appeared to Pasteur to be one of the few fundamental differences between the chemistry of life and that of inanimate matter. He postulated that this peculiar asymmetry might be the manifestation of asymmetric forces in the environment (alas unknown!) acting upon the organism at the time of molecular synthesis. Pasteur (1857) wrote,

> "Life, as manifested to us, is a function of the asymmetry of the universe and of the consequences it produces. The universe is asymmetrical; for, if the whole of the bodies which comprise the solar system moving with their individual movements were placed before a glass, the image in the glass could not be superposed upon the reality. Even the movement of solar light is asymmetrical."

However, Pasteur did not rule out the spontaneous generation of life, for in 1878 (see Vallery-Radot, 1922–1939) he wrote, "I have been looking for it [spontaneous generation] for 20 years but I have not yet found it, although I do not think it is an impossibility." In this context it is interesting to note that Darwin wrote in a letter to Joseph Hooker in 1871, the frequently quoted statement,

> "It is often said that all the conditions for the first production of living organisms are now present, which could ever have been present. But if (and—oh what a big if) we could conceive in some warm little pond, with all sorts of ammonia and phosphoric salts, light, heat, electricity, etc. present, that a protein compound was chemically formed ready to undergo still more complex changes. . . ."

So we see that Darwin and Pasteur dwelled on the idea of the possibility of the spontaneous generation of life.

Claude Bernard (1866, 1878) expressed the idea that matter, including organic matter, is inheritantly inert, but organized matter is able because of its organization to respond to external stimulation, thus manifesting life-as-action. Organized matter is irritable. He classified "irritants" as "physical, chemical, and vital, e.g., sensory nerves." Irritability is repeatedly asserted by Bernard to be the fundamental characteristic of all living systems.

In the 1930's a new impetus for seeking the origins of life began to emerge, beginning with the writings of Oparin (1938, 1968), Haldane (1966), and Bernal (1951, 1967). This interest gathered great excitement with the development of outer space exploration in the 1950's. Numerous researchers are now raising the question of how life on earth originated and speculate on the existence or possibility of life in other planetary systems.

Recent experiments have been carried out in search for the origin of life. These experiments and the various resulting hypotheses have been summarized in the work and texts by Keosian (1964), Wald (1964a), Dauvillier (1965), Fox (1965a,b), Blum (1968), Calvin (1969), Kenyon and Steinman (1969), Fox and Dose (1972), and Ponnamperuma (1972), Miller and Orgel (1974). Although bits of experimental data suggestive of biochemical origins are being put together, much information bearing on the biochemical origin of life is yet to be deciphered. Crick and Orgel (1973) ask: Why is there only one stereotyped genetic code for terrestrial life? If creatures sprang to life in some "primeval soup" as many biologists suppose, it is suprising that organisms with several codes do not exist. This questioning of the origins of life results partly from uneasiness among scientists over current explanations about how life arose on earth. Nevertheless, recent research points to the fact so aptly expressed by George Wald (1970) that, "A scientist has no choice but to approach the origin of life through a hypothesis of spontaneous generation. . . ."

This brief historical review of man's thinking about the origins and evolution of life leads us back to Darwin (1859). The theory expressed in "The Origin of Species," now over a century after its publication, has profoundly affected man's way of thinking about his nature and about the universe in which he lives, for one cannot go deeply into any analysis of biological phenomena today without raising questions of evolution.

Photoprocesses are so essential for life on earth that solar energy once utilized by a living cell, for example, in photosynthesis, must have been an important force in evolution. To see how solar radiation is central to the development of photosensory systems of plants and animals becomes most vital to our thesis.

If all life is indeed made of the same molecules, then the forces, physical and chemical, that shaped these molecules will be found to be similarly structured and similarly functional at the molecular level within all living organisms.

To begin to understand the importance of solar energy as a force for life, it will be necessary to examine the spectrum of energies that impinge on the earth and to see what kinds of molecules absorb these energies and how they are structured within the cell to function as a photoreceptor system. This brings us to the molecular structure of the cell membrane, whether such a structure is common to all cellular membranes and whether the development of the photoreceptor is derivable from the cell membrane.

In these photoreceptor systems we will identify the various kinds of pigment molecules and the structures associated with photobiological phenomena. Then we will examine the relationship between the photoreceptor system and behavior, for example, phototropism and phototaxis, photosynthesis, and vision. We will then consider the photoprocesses which function in photoperiodic phenomena—"biological clocks," aging, memory, and bioluminescence.

We can now begin our search for how the photoreceptor systems evolved and explore the pathways that lead us from molecules to cells to photoreceptors and back to molecules. This is not a simple task, but life is not simple. One must, however, ask questions and strive to form a logical pattern out of the bits of experimental data which have been accumulating to better understand the world man lives in.

# II
# The Environment

## The Earth and Geological Record

What were the conditions that made life on earth possible? Our continuing exploration of the earth, moon, and other planets is supplying some answers to this question.

Our knowledge of the early history of the earth comes from geology, geophysics, and geochemistry, which have given us information and theories about the age, formation, and structure of the earth. With the new tools developed to explore geochemistry and from rocketry and satellite exploration, we are learning much about the atmosphere and extraterrestrial environment. From such data, extrapolation back to the early history of the earth is being made.

The earth is a small part of the Milky Way, which in turn is but one of the myriads of galaxies studied by astronomers. As far as we know at present, the earth may be unique in that the atmospheric and other conditions of its outer surface were suitable for the development of life.

Paleontologists have tried to put together, from fossil remains, a record of the past by carefully exploring the preserved fossil morphology. Modern techniques for dating, the use of radioactive carbon and other isotopes, have given us some idea of the age of fossilized remnants and the time when life emerged.

TABLE 2.1 GEOLOGICAL SCALE AND LIFE[a]

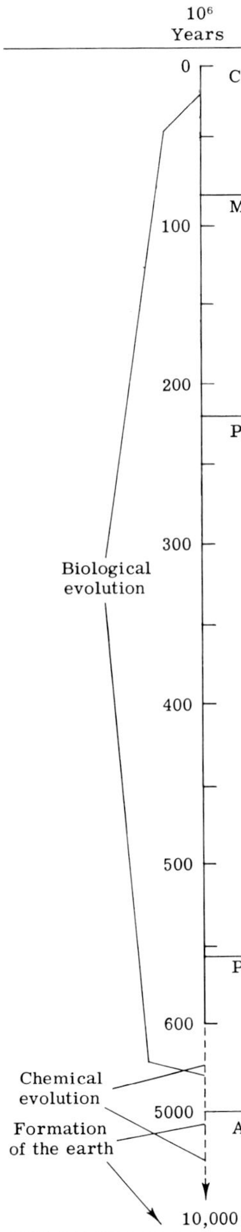


| Eras | Periods or epochs[b] | Time (in millions of years, since the beginning) | Major biological evolutionary events |
|---|---|---|---|
| Cenozoic | Pleistocene | 1 | Early man |
| | Pliocene | 10 | Large carnivores |
| | Miocene | 30 | Grazing mammals |
| | Oligocene | 40 | Large running mammals |
| | Ecocene | 55 | Whales, bats |
| | Paleocene | 65 | |
| Mesozoic | Cretaceous | 101 | Placental mammals; flowering plants |
| | Jurassic | 170 | Mammals and birds |
| | Triassic | 210 | First dinosaurs; conifers |
| Paleozoic | Permian | 253 | Gymnosperms |
| | Carboniferous[c] | | |
| | Pennsylvanian | 300 | Reptiles |
| | Mississippian | 333 | Sharks |
| | Devonian | 370 | Amphibians; fish; insects |
| | Silurian | 415 | Land plants |
| | Ordovician | 470 | Algae; fish |
| | Cambrian | 535 | All major invertebrate phyla; trilobites dominate |
| Proterozoic | Precambrian | 1000 | Fossils |
| | | 3000 | First life |
| Archeozoic | | | Origin of earth; oldest known rocks are formed |

[a] Taken in part from Calvin (1969, p. 10).

[b] In technical geological use an epoch is a subdivision of a period; the distinction is not important for these purposes. The names for the Cenozoic are technically epochs and those for the Mesozoic and Paleozoic are technically periods.

[c] American geologists call the early Carboniferous "Mississippian" and the late Carboniferous, "Pennsylvanian."

The earth was formed 5 to 6 billion years ago and hardened into its present shell. This estimate is based on evidence that the oldest reliably dated terrestrial rocks have ages in excess of 3 billion years. Assuming that the present abundance of $^{207}Pb$ resulted entirely from decay of $^{235}U$, it turns out that the age of the earth lies between 3.2 to about 5.7 billion years. The estimated age of the crust of the earth therefore is about 4.5 billion years.

Although there may be general agreement on the age of the earth, the question of the origin of the earth remains to be resolved by the cosmologists. Some hold that it gradually formed from hot gases thrown off by the sun; others believe that meteorites gradually aggregated to form one large solid planet. However they came to be evolved, the earth and the sun are believed to have a common origin. After the formation of the earth's crust, organic chemical formation began. The long span of time which included the Archeozoic and Proterozoic geological eras was the period of chemical evolution (Table 2.1). It was the period when relatively simple molecules were transformed into very complex ones (Calvin, 1969).

The study of the rocks that make up the surface layers of the earth enables us to trace an ever-changing sequence of different forms of life. Almost all pre-Cambrian rocks are severely folded, and the folding has blotted out most of the relics of life. We have fairly definite traces of the existence of calcareous seaweeds, protozoa, and marine worms. But there is no evidence of life on land. Probably the intense folding confined the seas into smaller areas then at present, and the land surface of the globe was largely covered by rainless deserts. There were, however, at least two ice ages before the Cambrian. The evidence of terrestrial rocks shows that the earth was already over a billion years old in Cambrian times.

Preserved in the Cambrian rocks are the remains of most of the main animal phyla. The only important phylum missing is that of the vertebrates (phylum Chordata). The vertebrates did not appear until the Silurian, though presumably their boneless wormlike ancestors were present in the Cambrian sea. In the upper Silurian there were already fish with fins. These fish rapidly developed paired fins from lateral folds. In the Devonian they developed bones and probably about the end of that period left the water. Most of the main animal types were thus already differentiated in their comparative anatomy, embryology, and biochemistry, which now enables us to trace their relationships to one another and to conjecture about the cause of evolution of certain forms.

### The Changing Atmosphere and Temperature

The atmosphere of the earth in its early history must have been totally different than at present. Oparin (1938) was one of the first to suggest that

TABLE 2.2 COMPOSITION OF AIR (DRY, CLEAN AIR NEAR SEA LEVEL)[a]

| *Constituent gas and formula* | *Content (% by volume)* |
|---|---|
| Nitrogen ($N_2$) | 78.084 |
| Oxygen ($O_2$) | 20.9476 |
| Argon (Ar) | 0.934 |
| Carbon dioxide ($CO_2$) | 0.0314 |
| Neon (Ne) | 0.001818 |
| Helium (He) | 0.000524 |
| Methane ($CH_4$) | 0.0002 |
| Krypton (Kr) | 0.000114 |
| Hydrogen ($H_2$) | 0.00005 |
| Nitrous oxide ($N_2O$) | 0.00005 |
| Xenon (Xe) | 0.0000087 |
| Sulfur dioxide ($SO_2$) | 0 to 0.0001 |
| Ozone ($O_3$) | 0 to 0.000007 (summer) |
| | 0 to 0.000002 (winter) |
| Nitrogen dioxide ($NO_2$) | 0 to 0.000002 |
| Iodine ($I_2$) | 0 to 0.000001 |
| Ammonia ($NH_3$) | 0 to trace |
| Carbon monoxide (CO) | 0 to trace |

[a] From "U.S. Standard Atmosphere, 1962," p. 9. Compiled by National Aeronautics and Space Administration, United States Air Force, and United States Weather Bureau, Washington, D.C.

the atmosphere of the primitive earth was not of the same oxygen–nitrogen mixture as that existing today (Table 2.2). The evidence for this was that other planetary atmospheres contain large quantities of hydrogen, methane, ammonia, and carbon dioxide. The atmosphere then had virtually none of the free oxygen it has today. Instead, it is thought that the atmosphere was reducing, that is, hydrogen was the most prevalent substance. If oxygen were present it would have immediately combined with various materials that existed then in a chemically reduced state, as well as free iron, sulfur, and other elements.

The temperature of the earth's crust would for a long time have been well above 100°C, and therefore all the water must have been present in the atmosphere as vapor until the earth cooled sufficiently to allow water to condense. The amount of hydrogen in the atmosphere must have gradually decreased as the earth cooled. There is overwhelming geological evidence that for at least 2 billion years the range of temperature on the earth's surface has been within the limits now experienced, that is, from slightly below the freezing point of water to a little above 23.5°C. If this had not been the case, life probably could not have developed.

Oxygen was probably first introduced into the atmosphere by the photodissociation of water vapor exposed to ultraviolet radiation. Hydrogen would also have been produced but would have escaped comparatively quickly from the earth's atmosphere. At first the oxygen would also have been rapidly removed as it reacted with the materials of the earth's crust, but as these became fully oxidized the level of free oxygen in the atmosphere would have started to rise. Urey (1952) has pointed out, however, that the oxygen level in the atmosphere would not have risen to anything like today's 20% as a result of photodissociation. The reason for this is that as free oxygen accumulates it tends to shield the earth from the ultraviolet radiation that is required to bring about photodissociation. Thus the process is a self-regulating one, and Urey (1952) has calculated that the level at which the atmospheric oxygen becomes limiting does not exceed about 0.1% of the present concentration.

Up to a height of about 15 km, the relative proportions of the major gases remain constant due to thorough mixing by convection currents. This region is known as the troposphere. Between 15 and 80 km above the surface, a number of distinct layers are present among which there is very little vertical circulation. This region of the atmosphere is called the stratosphere. Throughout the troposphere and the stratosphere the predominant molecular species are $N_2$ and $O_2$. The region above the stratosphere, from 80 to 100 km, is rich in ionic species and is thus named the ionosphere. The presence of these ions suggests that either high temperatures prevail in the ionosphere or that photochemical reactions, or at least photoionizations, are taking place there. The ionization in this region is induced by short-wavelength ultraviolet radiation.

At present we can divide the gases that make up the lower atmosphere into two groups. The first group, nitrogen and oxygen, makes up practically 100% of the atmosphere near the earth's surface. The other group includes water and rare gases. Water occurs in the air in the form of water vapor and can be present in concentrations up to 2–3% by volume. Dry air always contains ozone and radon. The atmosphere normally contains $10^{-6}$% ozone, the concentration increasing with the height above the earth's surface. The concentration of radon, a product of the disintegration of radioactive elements in the earth's crust, is minimal. These data are recorded in Table 2.2.

If the biosphere is the major source of the molecular oxygen in the present atmosphere, then it is possible that before the biosphere came into being there was very little molecular $O_2$ in the atmosphere. This in turn implies that the ozone screen might have been absent from the prebiological atmosphere, thus allowing the penetration of considerable amounts of short-wavelength solar ultraviolet radiation deep into the atmosphere. Some radiation of wavelengths as short as 200 nm could conceivably have reached the surface. Thus, the primitive atmosphere would have been abundantly supplied with a free energy source for photochemical reactions. These arguments require modification if there were other sources (besides the biosphere) contributing molecular oxygen to the atmosphere. Recent NASA films from a surface ultraviolet camera spectrograph left on the moon by *Apollo 16* indicate that the earth's present source of oxygen is not primarily photosynthesis but that oxygen is formed as a result of the dissociation of water vapor in the upper atmosphere caused by solar ultraviolet radiation.

There is no doubt that there was a reducing atmosphere present at the stage in the earth's history when life first arose. This may seem most unlikely in view of the fact that almost all living organisms today are dependent on oxygen for respiration. But as Oparin (1938) pointed out, the presence of a reducing atmosphere during the early stages of life was, paradoxically, a necessity. The transition from mainly anaerobic to mainly aerobic conditions on earth was brought about not by geochemical but by living forces. Since then, only minor changes such as continental shifts, temperature fluctuations, or volcanic action have harassed the living world.

## The Ocean

Despite our lack of a complete understanding of what ocean water is, we know that the oceans are a large but dilute electrolytic solution which contains all the natural elements known to man. Chlorine, sodium, magnesium, sulfate, calcium, and potassium make up 99% of the dissolved salts in seawater (Table 2.3). The ratio of any one of the major constituents to the total dissolved solids is nearly constant even though the total concentration of dissolved solids may vary from place to place. The remarkable relative constancy of the chemical composition of seawater is of great biological importance, especially since many believe that life made its appearance from the sea rather early in the 5 billion years of earth history, and most likely in the form of a bacterial cell.

If we compare the relative proportions of the different ions present in the blood sera and tissues of a number of different animals with those of seawater (Table 2.4), we find that there exists among them a great similarity, and this suggests that they had a common origin (Baldwin, 1948a,b). In this table the amounts of different ions are expressed as percentages of the total amount of sodium which is taken as 1.0. In this way it is possible to compare the relative composition of fluids which differ more or less in absolute composition. Macallum (1926) suggested that the circulating fluids of all animals originally came from the sea of millions of years ago.

But if we believe that life really originated in the sea, why does our own blood serum (Table 2.4) contain relatively more potassium and less magnesium

TABLE 2.3 COMPOSITION OF SEAWATER[a,b]

| *Elements* | *Ions* | *Content (% by weight)* |
|---|---|---|
| Sodium | $Na^+$ | 1.06 |
| Magnesium | $Mg^{2+}$ | 0.130 |
| Calcium | $Ca^{2+}$ | 0.040 |
| Potassium | $K^+$ | 0.0388 |
| Chlorine | $Cl^-$ | 1.90 |
| Sulfate | $SO_4^{2-}$ | 0.267 |
| Carbonate | $CO_3^{2-}$ | 0.007 |
| Bromine | $Br^-$ | 0.0065 |
| Borate | $H_3BO_3$ | 0.0026 |
| Other elements | | 0.09 |

[a] The large number of elements known to occur in seawater indicate that probably all of the earth's naturally occurring elements exist in the sea.

[b] Seawater is by weight 96.46% water, of which oxygen makes up 85.79% and hydrogen 10.67%.

TABLE 2.4 THE CONCENTRATION OF IONS IN THE TISSUE FLUIDS AND BLOOD OF VARIOUS ANIMALS[a,b]

| | *Percent by weight* | | | | | |
|---|---|---|---|---|---|---|
| | $Na^+$ | $K^+$ | $Ca^{2+}$ | $Mg^{2+}$ | $Cl^-$ | $SO_4^{2-}$ |
| Jellyfish | 1.0 | 0.029 | 0.022 | — | — | — |
| Squid | 1.0 | 0.082 | 0.046 | 0.129 | 1.98 | — |
| Sepia | 1.0 | 0.092 | 0.043 | 0.135 | 2.05 | 0.045 |
| Octopus | 1.0 | 0.030 | 0.027 | — | — | — |
| Conger eel | 1.0 | 0.015 | 0.030 | 0.014 | 1.32 | 0.046 |
| Dogfish | 1.0 | 0.046 | 0.027 | — | — | — |
| Hagfish | 1.0 | 0.037 | 0.025 | 0.047 | 1.78 | 0.063 |
| Frog | 1.0 | 0.024 | 0.093 | — | — | — |
| Rat | 1.0 | 0.042 | 0.021 | — | — | — |
| Man | 1.0 | 0.596 | 0.025 | 0.009 | 1.09 | 0.072 |

[a] Sodium is taken as 1.0 for comparison.

[b] Data taken in part from Young (1971, p. 27) and Baldwin (1962, p. 5).

than seawater? Macallum (1926) pointed out that according to estimates given by geologists, the primitive Jurassic ocean contained more potassium and less magnesium than is present in modern seawater. Since that time the composition of the ocean has been gradually changing. The changing composition of seawater of different geological periods is given in Table 2.5.

TABLE 2.5 COMPOSITION OF SEAWATER AT DIFFERENT PERIODS[a,b]

| | *Content (% by weight)* | | | |
|---|---|---|---|---|
| | $Na^+$ | $K^+$ | $Ca^{2+}$ | $Mg^{2+}$ |
| Modern | 1.0 | 0.0361 | 0.0391 | 0.121 |
| Eocene (Tertiary, 40–70 million years ago) | 1.0 | 0.067 | 0.031 | 0.007 |
| Jurassic (140–170 million years ago) | 1.0 | 1.0–2.5 | 0.10 | 0.001–0.0001 |

[a] From Baldwin (1948a, p. 11).
[b] Sodium is taken as 1.0 for comparison.

Although there are large differences in total ionic composition and total salinity, the relative proportions of the most important cations are remarkably similar, and far more similar to each other and to those in seawater than would have been expected. This is considered to be another piece of evidence that life must have originated in the ocean waters.

## Molecules in the Environment

The main substances of living cells are water, certain salts as found in the seas (Table 2.3), and organic compounds. The organic compounds consist for the most part of carbon, hydrogen, oxygen, and nitrogen atoms, which together constitute 99% of living matter. These atoms make up the essential constituents of life, which are the organic compounds, the carbohydrates, the lipids or fats, the proteins, and the nucleic acids (DNA and RNA).

According to Oparin (1938), Urey (1952), and Bernal (1951), the primitive atmosphere was reducing and consisted of hydrogen, methane, ammonia, nitrogen, water vapor, and possibly carbon dioxide. Oparin assumed that the first carbon compounds were hydrocarbons and not $CO_2$, that is, carbon appeared first in a reduced and not in an oxidized state, and that the primordial atmosphere contained no oxygen.

Assuming that the primordial atmosphere contained the following gaseous mixture:

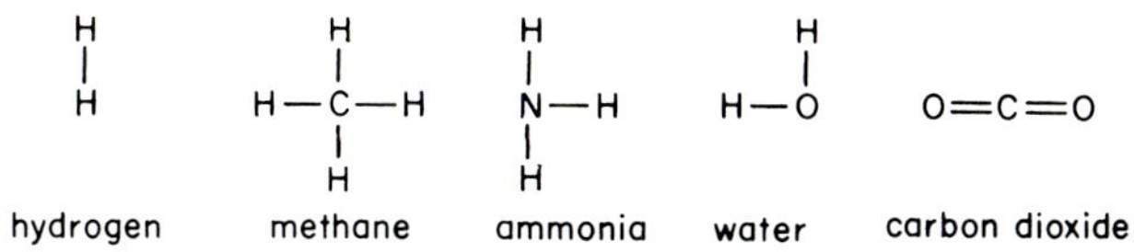

then some of the molecules that can be derived from them would be:

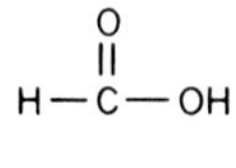
formic acid

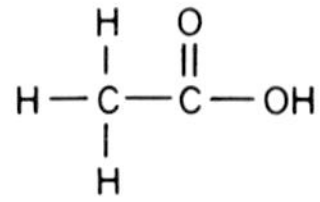
acetic acid

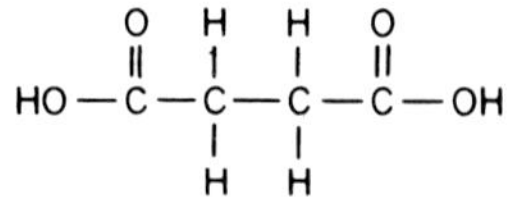
succinic acid

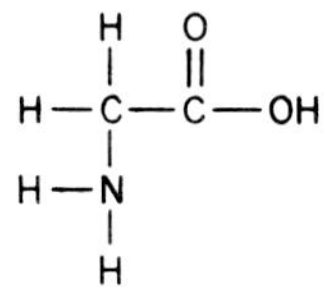
glycine

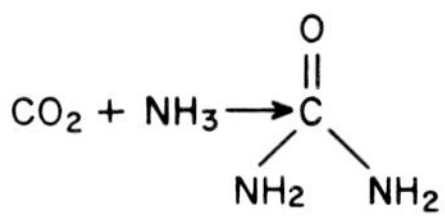
urea

From these molecules, it is easy to see how these organic compounds could be synthesized and could occur with or without the agency of living cells. Oparin (1938, 1968) envisioned that organic compounds had already formed on the earth before life arose, and that life arose from these preexisting compounds. Haldane (1928, 1954, 1966) suggested that before the origin of life, organic compounds must have accumulated until primitive oceans reached the consistancy of "hot diluted soup." He based this on the assumption that the primitive atmosphere contained $CO_2$, $NH_3$, and water vapor, but no oxygen. Haldane claimed that such a mixture exposed to ultraviolet radiation would give rise to a vast variety of organic substances. Calvin and his associates (Garrison *et al.*, 1951), in looking for ways to synthesize complex organic molecules from simple molecules in a prebiological environment, irradiated carbon dioxide and hydrogen with high energy ionizing radiation in the cyclotron. Formaldehyde, formic acid, acetic acid, and other reduced carbon compounds were obtained. These molecules could then be used for further synthesis and a variety of complex organic compounds would form. The diurnal alternations of sunlight and temperature on the earth's surface were also important factors in creating special conditions for the synthesis of organic molecules.

Probably the most exciting and informative experiments came from a series of experiments by Miller (1953, 1955, 1957). He was one of the first to synthesize amino acids in a simulation experiment. He applied electric discharge to a mixture of methane, ammonia, hydrogen, and water vapor. The amino acids he identified were glycine, alanine, $\beta$-alanine, aspartic acid, $\alpha$-amino-*n*-butyric acid, and other organic compounds, but purines and pyrimidines were absent (Figure 2.1). When similar gaseous mixtures were subject to high temperatures (80°–90°C), ultraviolet radiation, ionizing radiation, and electrical discharge, analysis of the contents of these reaction chambers showed that small quantities of a great variety of organic substances had been produced. These substances are of considerable biological importance, for they include many or-

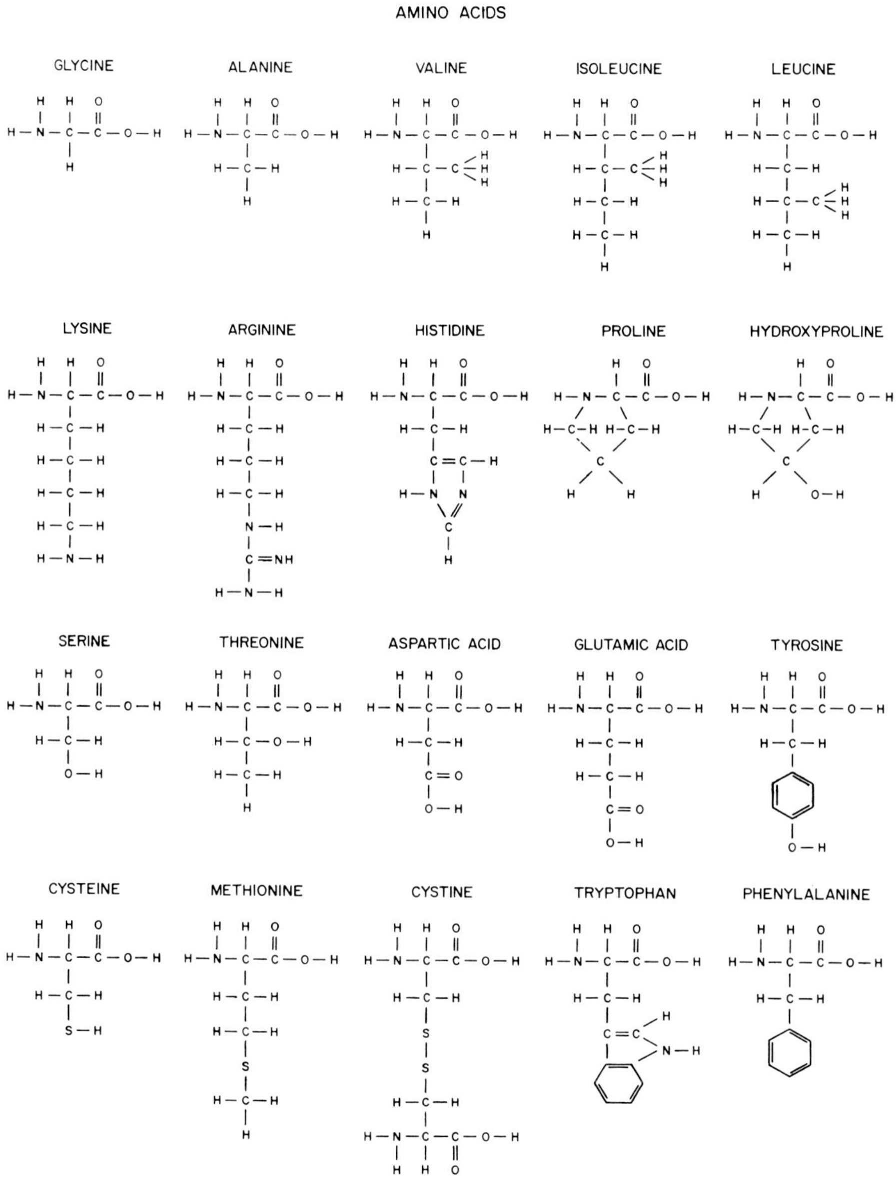


FIGURE 2.1 Amino acids: structural formulas.

ganic acids of low molecular weight, amino acids including glycine and alanine, urea and adenine, and simple sugars such as ribose.

Ponnamperuma (1965) later found that aminonitriles could be synthesized without water during electric discharge. This suggested that amino acids could be formed without aldehydes. Ultraviolet light could also be used to form amino acids or their precursors. Groth and Weyssenhoff (1960) reported the formation of glycine, alanine, and aminobutyric acid by irradiation with ultraviolet light of methane or ethane with ammonia and water. Abelson (1966) found glycinonitrile resulted when a solution of ammonium formate, ammonia, sodium cyanide, and ferrous ion was irradiated with ultraviolet light. More recent experiments of a similar nature have demonstrated the formation of simple proteins, adenosine triphosphate (ATP), and porphyrins under the same reducing conditions.

The synthesis of amino acids then is not an unusual process. Take for example the synthesis of glycine (Scheme 2.1).

$$\underset{\text{formaldehyde}}{2H_2CO} \xrightarrow{\text{dimerization}} \underset{\text{glycol aldehyde}}{CH_2OHCHO}$$

$$\underset{\text{glycol aldehyde}}{2CH_2OHCHO} + H_2O \xrightarrow[\text{oxidation-reduction}]{} \underset{\text{glycol}}{CH_2OHCH_2OH} + \underset{\text{glycolic acid}}{CH_2OHCOOH}$$

$$\underset{\text{glycolic acid}}{CH_2OHCOOH} + NH_3 \longrightarrow \underset{\text{glycine}}{CH_2NH_2COOH} + H_2O$$

SCHEME 2.1

Other molecular combinations may occur, that is, between two alcohol molecules or between two amino acid molecules (Scheme 2.2).

$$\underset{\text{glycine}}{2CH_2NH_2COOH} \xrightarrow{\text{dimerization}} \underset{\text{glycylglycine}}{CH_2NH_2CONHCH_2COOH} + H_2O$$

SCHEME 2.2

This type of reaction is then the basis of the formation of polypeptides and proteins.

An earlier finding was that of Friedrich Wöhler in 1828. He showed that urea, an organic molecule that is associated with life, could be formed by heating the inorganic salt, ammonium cyanate. Therefore it is not surprising to find that a variety of organic molecules found in living systems can be synthesized without the agency of living cells.

More recently, Fox (1965a) and his colleagues also used high temperature to bring about the synthesis of complex organic molecules. This was based on his ideas about the prebiological environment. The method also consisted of heating a mixture of amino acids to temperatures of 160°–200°C for several

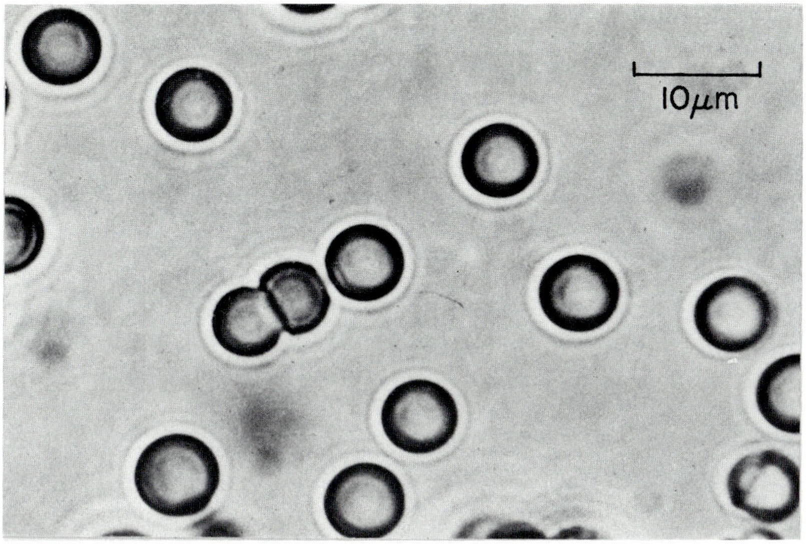


FIGURE 2.2 Proteinoid microspheres (courtesy of Professor S. Fox, University of Miami, Miami, Florida).

hours under anhydrous conditions. The mixture contained aspartic acid, glutamic acid, and lysine (Fox *et al.*, 1963). By thermal polycondensation they were able to copolymerize amino acids, which they named *proteinoids.* On analysis, 18 amino acids were identified which are found in proteins (Figure 2.1). The proteinoid molecular weights ranged from 3000 to 9000 depending on the method of preparation. Interestingly, when proteinoid is treated with hot water, microspheres separate out of the cooling clear solution. The proteinoid microspheres (Figure 2.2) vary in size from a few micrometers to 100 μm in diameter, are relatively stable, and depending on their preparation, exhibit living cell-like behavior.

Even though a variety of organic molecules found in living cells that were present in the environment have and can be synthesized in the laboratory, there is no experimental evidence to date to show that either accumulation, chemical reactions, or aggregation of these molecules resulted in the emergence of a self-reproducing cell. However, the finding of such synthesis of organic molecules and amino acids is suggestive that an environment rich in organic molecules could have formed micelles, or *proteinoids,* and through a spontaneous event given rise to a primordial cell.

The most common feature of all living organisms is that proteins play a very important part in their structure. The proteins are the largest and most complex molecules known. The units of which their structures are built consist of about 20 different amino acids (Figure 2.1). These are strung together in chains hundreds to thousands of units long, in different proportions, in all types of sequence, and with great variety of branching and folding. An infinite number of different proteins is possible for no two species of living organisms possess the same proteins.

Nucleic acid of some type is essential in all living organisms to affect the processes of reproduction, growth, and differentiation. The nucleic acids are DNA (deoxyribonucleic acid) and RNA (ribonucleic acid). Both DNA and RNA are long chains of alternating sugar and phosphate groups. In DNA the sugar is deoxyribose, that is, the carbon at the 2′ position carries simply hydrogen; and in RNA the sugar is ribose because the 2′ carbon carries a hydroxyl group. Each sugar has a nitrogen base attached to it, either a purine or a pyrimidine. The two purines, adenine and guanine, are present in both DNA and RNA. The pyrimidines commonly found in RNA are uracil and cytosine; in DNA, thymine, and cytosine. Each purine or pyrimidine with its sugar and phosphate is referred to as a nucleotide. The nucleic acids are very large structures composed of aggregates of these nucleotides. An almost endless variety of different nucleic acids is possible through variation of the nucleotide sequence, and specific differences among them are believed to be of the highest importance, for the nucleic acids are the main constituents of the genes, the bearers of hereditary information. For a more detailed discussion of the structure of nucleic acids and the genetic code, the reader is referred to Watson (1965).

The frequency of mutations in the hereditary material can be greatly increased by exposure to x-rays, and also to ultraviolet radiation. It is of interest that the wavelengths of ultraviolet radiation that produce mutations correspond to the absorption spectrum of DNA.

Another striking molecular feature that all living organisms have in common is the presence of an adenosine diphosphate (ADP) and adenosine triphosphate (ATP) system as the energy storing mechanism. This involves the synthesis of ATP from ADP and inorganic phosphate when surplus energy is being released, and the breakdown of ATP to ADP and phosphate when energy is required. The chemical energy of ATP is used to perform the chemical and mechanical work of the cell. The universal occurrence of the ADP–ATP system strongly suggests that it is the archetypical energy storing mechanism, a process that has been passed down from the very earliest forms of life.

In pressing the study of genetics to the molecular level, evolutionists have encountered a phenomenon which must surely be scored as a major triumph for the case of a common descent of all life, namely, the enormous functional, structural, and sequence similarity of molecules found in organisms of completely different morphology and complexity. Phylogenetic trees constructed from protein sequence data (Dayhoff, 1972; Fitch and Margoliash, 1967; King and Jukes, 1969) are virtually the same as those derived from other taxonomic classifications.

## Electromagnetic Energy and Photobiology

The spectrum of electromagnetic radiation extends from $\gamma$- and x-rays, less than 0.01 Å long, through ultraviolet, visible, and infrared radiation, to radio and electric waves which are kilometers long (Figure 2.3). Abundant data are now available on the spectral distribution and intensity of solar radiation incident on the surface of the earth, and this is constantly being measured (Figure 2.4). The visible part of the electromagnetic spectrum is a very narrow region

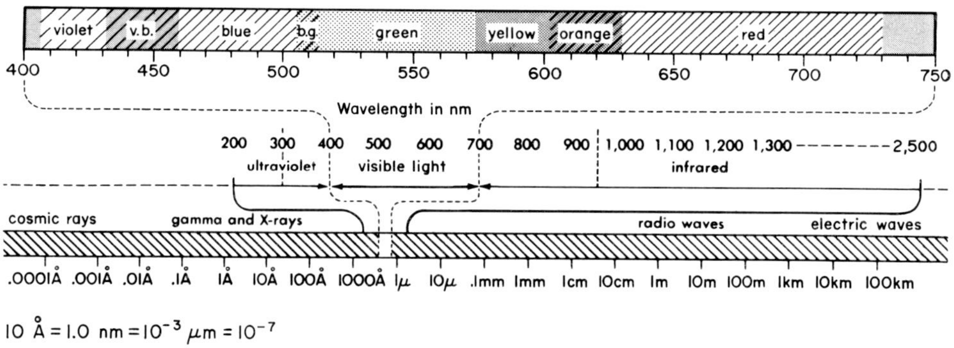


FIGURE 2.3 The electromagnetic spectrum of energy.

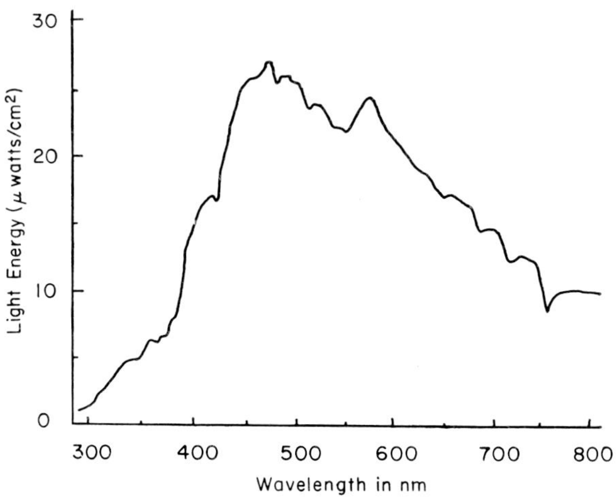


FIGURE 2.4 Solar spectrum that strikes the earth on 82 cm² taken between 1 and 2 P.M. in Cadarache, France (courtesy of Dr. P. Guérin de Montgareuil, Atomic Energy Commission, Cadarache, France).

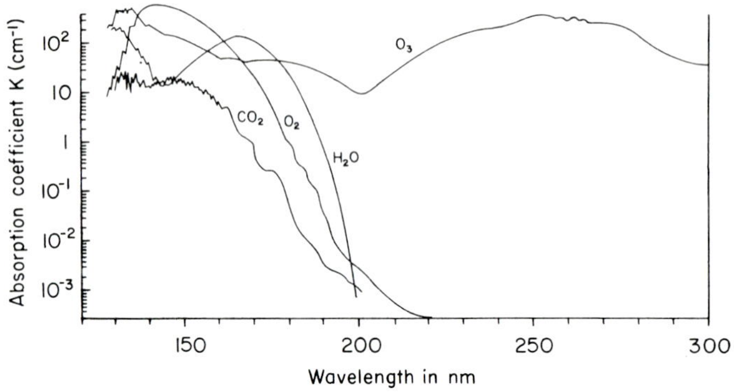


FIGURE 2.5 Ultraviolet absorption spectrum in atmospheric gases (from L. V. Berkner and L. K. Marshall, 1964; data for ozone absorption spectrum taken from K. Watanabe, 1959, and E. Vigroux, 1953, 1969).

from about 390 nm to about 760 nm (Figure 2.3). Life must have evolved by adapting to this radiation, for practically all photobiological behavior of plants and animals—photosynthesis, phototropism, phototaxis, photoperiodism, and vision—depends upon this range of radiation for excitation. The spectrum of solar radiation that reaches the surface of the earth covers this photobiological range, with a maximum around 500 nm (corresponding to 57 kcal/mole light quanta), about which these phenomena cluster. Terrestrial radiation consists mainly of infrared rays, with a maximum near 11 $\mu$m (corresponding to 2.2 kcal/mole).

Ultraviolet radiation below 300 nm is largely absorbed by ozone ($0_3$) in the upper atmosphere (Figure 2.5). Ozone is a good absorber of ultraviolet and must have helped to shield the earth from the lethal short wavelengths. The present concentration of ozone in the upper atmosphere is enough to reduce the level of ultraviolet radiation at the earth's surface by almost 200 times at 300 nm, by about $10^7$-fold at 290 nm, and to negligible levels at shorter wavelengths.

Ultraviolet radiation from 200 to 300 nm is absorbed by proteins and nucleic acids, and such absorption produces damaging effects on cells, causing mutations and even death. Before ozone filtered out most of the ultraviolet radiation in the upper atmosphere, evolving life must have been able to circumvent or repair these effects. Kelner (1949), studying bacteria, observed that ultraviolet "killing" of the bacteria could be reactivated by visible radiation. He called this phenomenon *photoreactivation.* Action spectra for photoreactivation show generally two peaks, one in the range of 300 to 380 nm and another from 430 nm to near 500 nm, depending on the organism. The process was then found to be an enzymatic photocatalysis of pyrimidine dimers (Setlow and Setlow, 1963). Photoreactivation continues to be effective and widespread in nature, for it has now been observed in algae, fungi, plants, and animals (Regan and Cook, 1969).

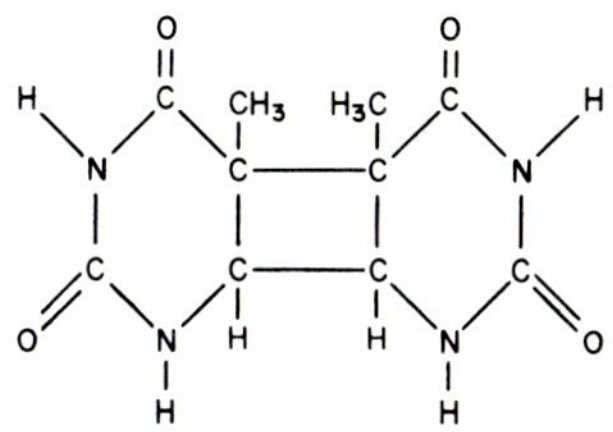

Pyrimidine dimer

Exactly what near ultraviolet radiation from about 300 nm to 400 nm plays in other photobiological processes is not clear. Action spectra for phototropism of lower plants and animals, and the spectral sensitivity for vision of most insects, show a major response near 360 nm. Since action spectra are indicative of the absorption spectrum of the molecules involved in a process, these organisms must possess molecules to absorb this energy.

Radiation in the red part of the spectrum from 600 to 700 nm is important for chlorophyll synthesis and photosynthesis. Radiation from 660 nm and into the near infrared is important for plant and animal growth, the timing of plant flowering, sexual cycles in animals, and pigment migration. The timing of flowering cycles in plants, called *photoperiodism,* is controlled in the near red part of the spectrum by the shifting of light between 660 and 730 nm. Bacterial photosynthesis takes place even further in the red, beyond 800 nm. Infrared radiation beyond 900 nm is mostly absorbed by atmospheric water vapor and by water that surrounds living cells.

Therefore, the limits of radiation effective for photobiological phenomena are less than those for photochemistry and are considered to lie between 300 nm and 950 nm (Figure 2.3). Thus, it is no accident that life on the earth has been remarkably efficient in its use of this range of energy, for living organisms are able to absorb about 83% of the total solar energy which reaches the earth.

## The Physical Nature of Light

It is important that we understand certain basic principles which are used frequently in discussions of photochemistry and photobiological phenomena. In the interactions between light and matter, electromagnetic radiation exhibits both wave and particle properties. The particles are called *photons* or *quanta* and represent the packets of energy that comprise any type of electromagnetic radiation. A particular type of radiation is characterized on the basis of either its wavelength or its energy. Max Planck discovered in 1900 the direct relationship between the frequency of electromagnetic radiation and the energy of its quanta. Albert Einstein, in 1905, extended Planck's relationship to include light. Einstein had shown—and this was a feature of Planck's derivation of the black-body spectrum as well—that the energy of a light quantum was proportional to its frequency. That is, each photon has the energy $E = hc/\lambda$, where $h$ is Planck's constant ($6.62559 \times 10^{-27}$ erg/second), $c$ is the velocity of light ($3 \times 10^{10}$ cm/second in empty space), and $\lambda$ is the wavelength of the light. The frequency of the electromagnetic radiation is $v = c/\lambda$. Then the energy of a single quantum can be calculated from $E = hv$. This equation shows that the higher the frequency of the radiation the greater the energy. For example, quanta of blue light would be more energetic than quanta of red light, and x-rays would be more energetic still since their frequencies are higher than any visible frequency. Therefore, since frequency is inversely proportional to wavelength, the shorter the wavelength the greater the energy. Einstein postulated that all the energy of a single light quantum or photon is transferred to a single electron. This one-to-one relationship between a light quantum and a particle of matter is of key importance in photochemistry. The principle that one quantum of light can bring about a direct primary photochemical change in exactly one molecule of matter is known as Einstein's Law of Photochemistry.

## The Absorption and Emission of Light

The ability of a substance to absorb light is determined by its atomic structure, that is, by the arrangement of electrons in different orbitals about the nucleus of each atom. Those electrons nearest the nucelus have relatively low energy, and those electrons in orbitals farthest away from the nucleus have a higher energy. To move an electron from an inner orbital to an outer orbital requires energy. When photons of light strike a molecule which can absorb the light, an electron in one of the orbitals may absorb the photon and gain energy sufficient to move it farther away to an outer orbital of a higher energy level. When this happens it is referred to as being in an *excited state.*

In some molecules in the excited state the high energy electrons do not escape from the molecule but return to their original low energy orbitals again, and the molecule is said to return to the *ground state.* Some of the energy may reappear as light, fluorescence, as the electron returns to the ground state.

The physical processes which illustrate the absorption of a light quantum are shown in the modified Jablonski diagram, Figure 2.6. There are four types of processes which an excited state may undergo: (1) emission of light, or a radiative transition, (2) a radiationless transition between two states without chemical reaction, (3) electronic excitation energy transfer, and (4)

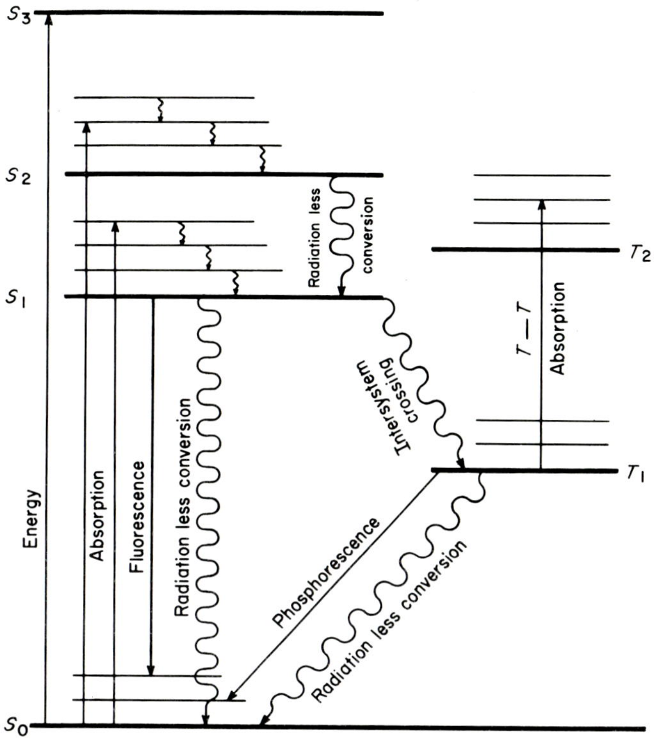


FIGURE 2.6 Energy diagram for photochemical processes [modified Jablonski diagram (Jablonski, 1935)]. Heavy horizontal bars represent the singlet energy levels $S_1$, $S_2$, and $S_3$ and ground state $S_0$, and triplet levels $T_1$ and $T_2$. The lighter bars represent the vibrational excited levels associated with each electronic state.

chemical reaction. These processes do not necessarily follow in the order given.

For molecules with an even number of electrons, in practically all cases, the photochemical behavior is describable in terms of singlet ($S = 0$) and triplet ($S = 1$) excited states. The distinction between singlet and triplet excited states of molecules that absorb light is of great importance in understanding the photochemistry of pigment molecules and therefore photobiology. The physical theory underlying these principles of molecular photobehavior is beyond the scope of this book. A more detailed discussion of the application of excited states to photobiology can be found in Clayton (1965, 1971), Seliger and McElroy (1965), and Porter (1969).

The absorption of light by a biological system and the emission of light from excited molecules are quantum phenomena, and therefore a proper description of the light either absorbed or emitted during a reaction should contain the numbers of photons per second per unit wavelength.

# III

# Photosensitivity and Pigments

We now know that all living systems from bacteria to man exhibit some form of sensitivity to solar radiation. This photosensitivity is seen in behavior, for a plant or animal will move, bend, or swim toward or away from a light source. These phenomena are described as *phototropism* and *phototaxis*. Plants evolved the ability to utilize the visible part of the sun's radiation directly in the process of *photosynthesis,* and are thus able to convert light energy to chemical energy, which the plant can use to manufacture sugars and organic matter. In animals, various kinds of photosensory cells and photoreceptors evolved, giving rise to eyes and to vision.

Many other photobiological phenomena are now recognized, such as *photoperiodism,* set by the length of day, and *photomorphogenesis,* both of which control many developmental processes such as hormonal stimulation, growth and sexual cycles of plants and animals, the timing of the flowering of plants, and the color and shade changes in the skin of animals. Also, there is the phenomena of *photodynamic action,* the photosensitization that is produced in the absorption of light by a molecule which becomes activated and causes destructive photooxidation in the cell, and *photoreactivation,* the recovery of ultraviolet damage by visible radiation.

So far we have only touched on those phenomena which depend on light absorption, but many living organisms emit visible light as well. Such production of light by bacteria, fungi, insects (as in the firefly), fish, and deep-sea animals is known as *bioluminescence*. All of these complex photophenomena fall within the exciting but complex realm of photobiology.

What then is the basis for this photosensitivity in living organisms? To comprehend these phenomena we must have an adequate means for understanding them. We have already described the environment for life, the nature of light, and the spectrum of energy that strikes the earth from the sun. We now must identify the molecules which are capable of absorbing this energy and which are associated with each of the photoprocesses. Later we will widen our discussion to include the various types of photoreceptors that contain these molecules and their molecular structure. From this information we can hope to deduce relationships between the molecules which absorb and transduce radiant energy and the photoprocesses: photosynthesis by plants, vision in animals, and other phenomena.

Since photosensitivity is so widespread in nature, is it possible to trace its evolutionary aspect at the molecular, structural, and chemical level in a way that would be helpful to our understanding of the phylogenetic progression? In an attempt to do so we will now introduce the various pigment molecules, and indicate something about their biosynthesis, structure, and properties.

One of the methods for characterizing a pigment molecule is by its absorption spectrum. Most absorption spectra are presented as optical density or absorbance versus wavelength, where the absorbance $A = \log_{10} I_0/I$. Here $I_0$ is the intensity of the entering wavelengths of light and $I$ is the intensity of the wavelengths of light transmitted through the pigment solution. The systematic investigation of absorption spectra has been of great value in establishing the identity and structure of numerous pigment molecules.

## Pyrroles and Porphyrins

The pyrrole ring once introduced into the living cell seems to show what

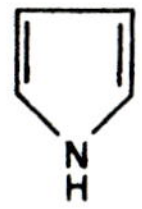

Pyrrole ring

may be loosely termed adaptive behavior on the molecular level. The existence of the pyrrole ring in living systems before it was incorporated into the porphyrins, chlorophyll, and the heme enzymes can be accounted for by a process of autocatalysis. Calvin (1962) assumed that each of these steps is catalyzed by an iron-containing compound with the end product being protoporphyrin IX (Figure 3.1). Since the end product can catalyze earlier steps in its own formation, the whole reaction series is regarded as autocatalytic. Once protoporphyrin IX was formed by a slow and random series of reactions in primitive organisms, it served to increase the probability that the earlier steps would continue to occur. That this process, or something similar to it, became incorporated into living systems relatively early in evolutionary history is suggested by the universality of pyrrole in the organic world.

The tetrapyrrole ring and the porphyrins made their appearance during the evolution of organic matter, but at a somewhat later stage than adenine or flavin derivatives. We know this since free oxygen is necessary for the synthesis of porphyrins. Thus they must have arisen during the transitional period when the earth's atmosphere began to be enriched with oxygen. Porphyrins were catalysts for chemical reactions long before they became incorporated in or were synthesized by the living cell. In living cells, the iron porphyrins and other metal porphyrins serve ideally as electron transfer catalysts. The molecule which sustains life on earth, chlorophyll, depends exclusively on the magnesium porphyrin. The biosynthetic scheme, as developed by Granick (1948, 1950, 1958) and Shemin (1948, 1955, 1956), shows certain steps in the present-day pathways for the biosynthesis of porphyrins, chlorophyll, and heme pigments (Figure 3.1).

## Chlorophyll

More than a century ago it was recognized that the pigment chlorophyll, synthesized by all green plants, is directly associated with photosynthesis. But the chemical structure of chlorophyll was not definitely established until 1913 by Willstäter and Stoll, and then later by Fischer and Stern (1940). Only recently has chlorophyll a been successfully synthesized in the laboratory by Woodward (1961).

The structure of chlorophyll was established as a cyclic tetrapyrrole molecule, which has the empirical formula $C_{55}H_{12}O_5N_4Mg$. Its greenness comes from the magnesium atom at the nucleus of the molecule (Figures 3.1 and 3.2). Chlorophyll exists in many isomeric forms. Of the various chlorophyll isomers, chlorophyll a and chlorophyll b are found in all higher plants. Chlorophyll a differs from chlorophyll b by possessing a methyl ($-CH_3$) group at the third carbon, whereas in chlorophyll b a formyl (–CHO) group occupies this position; chlorophyll b is therefore an aldehyde of chlorophyll a (Figure 3.2). Chlorophyll a and b differ in absorption spectra (Figure 3.15) as well as in solubility. For example, chlorophyll a is more soluble in petroleum ether while chlorophyll b is more soluble in methyl alcohol. These differences in solubility make it possible to separate these two chlorophylls. Chlorophyll a is present in all green plants while chlorophyll b, together with chlorophyll a, is

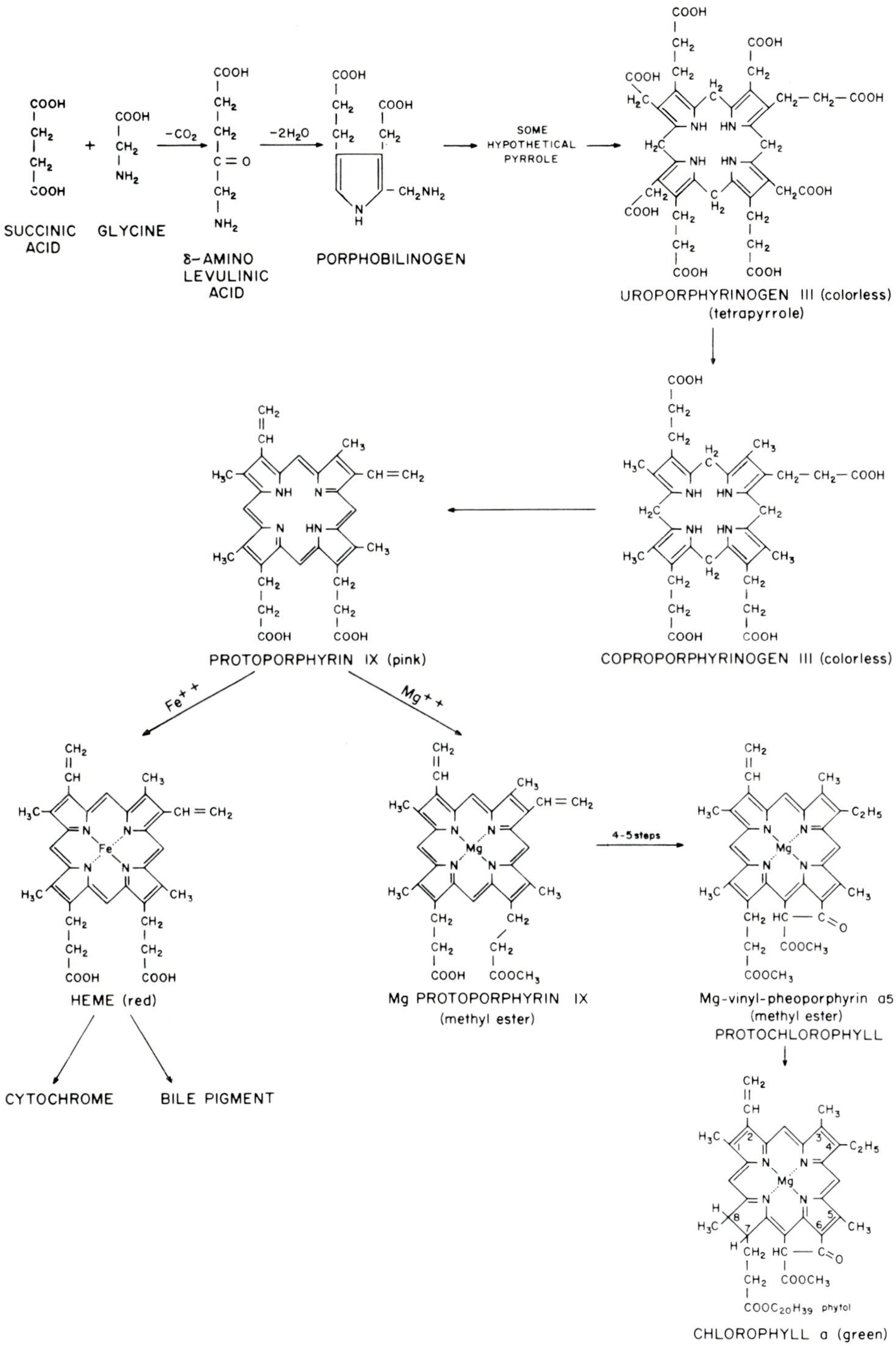


FIGURE 3.1 The biosynthesis of porphyrins, hemes, and chlorophyll (after Granick, 1950, 1958; Shemin, 1948, 1955; Calvin, 1969).

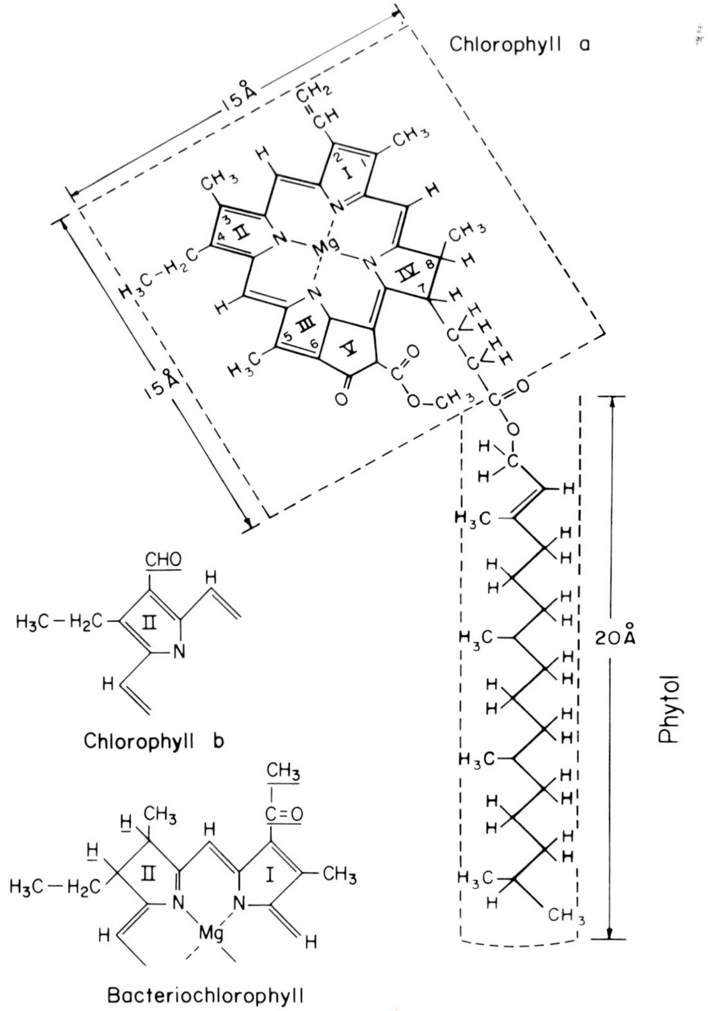


FIGURE 3.2 Structure of chlorophyll a, chlorophyll b, and bacteriochlorophyll.

found in all higher plants, ferns, mosses, green algae, and euglenoids. The other chlorophyll isomers, chlorophylls c, d, and e, are found in diatoms, brown algae, dinoflagellates, crytomonads, and crysomonads. Chlorophyll c, which lacks a phytol group, is soluble in aqueous alcohol. Chlorophyll d is believed to be an oxidation product of chlorophyll a in which the vinyl group at position 2 is oxidized to a formyl group. It is found, together with chlorophyll a, in most red algae. Chlorophyll e, together with chlorophyll a, is present in small amounts in yellow-green algae.

Pheophytin a and b are degradation products of chlorophyll a and b, respectively, from which the magnesium has been removed from the nucleus of the porphyrin part of the chlorophyll molecule. Pheophytin may be found *in vivo,* but what function it has in the cell or in photosynthesis is unknown. Pheophytin is produced by acidifying chlorophyll extracts. Its formation can be observed by a color change from green to brownish yellow. Pheophytin a in ether has absorption peaks around 408, 505, 534, and 610 nm, accompanied by the decrease in density of chlorophyll, which has a main peak in the red near 663 nm and a small peak near 667 nm. The reverse process of inserting magnesium into pheophytin to form chlorophyll is difficult, for a Grignard reagent is necessary. Therefore, it is unlikely that pheophytin is involved in the biosynthesis of chlorophyll.

Bacteriochlorophyll is found free of other chlorophylls in photosynthetic purple bacteria. It differs from chlorophyll a in that the vinyl group at position 2 is replaced by an acetyl group, and in that it contains two extra hydrogen atoms at positions 3 and 4 (Figure 3.2).

Seeds and etiolated plants (seedlings sprouted in darkness) contain no chlorophyll. However, upon exposure to light, they will turn green. The substance responsible for this reaction is *protochlorophyll,* the chlorophyll precursor (Figures 3.1 and 3.17). Protochlorophyll differs from chlorophyll in that it lacks two hydrogen atoms at positions 7 and 8 in the porphyrin part of the molecule. Thus it is an oxidation product of chlorophyll a (Figures 3.1 and 3.2).

The chlorophyll molecule in the chloroplast and its participation in photosynthesis is discussed in Chapter V.

## Heme

Heme is the iron porphyrin synthesized from the precursor protoporphyrin IX (see Figure 3.1), which is also the precursor molecule for chlorophyll and cytochrome. The heme molecule is nearly planar; the property of planarity is ascribed to the many double bonds in the molecule. The formula for the heme molecule is $C_{34}H_{32}O_4N_4Fe$. The iron atom forms bonds with the four nitrogen atoms of the tetrapyrrole (Figure 3.1). Heme with an attached oxygen is responsible for the red color of oxygenated blood. Four hemes bound to the protein globin form hemoglobin, which has a molecular weight of the order of 68,000.

Hemoglobin is the oxygen carrier of most vertebrate red blood cells and takes up oxygen to form oxyhemoglobin. The proportion of oxygen taken up by the hemoglobin depends on the concentration of the oxygen, or oxygen tension. Although hemoglobin is more characteristic of vertebrates, it has been found in a number of invertebrates as erythrocruorins. In other invertebrates,

TABLE 3.1 OXYGEN CAPACITY OF SOME BLOODS[a]

| Animal | Pigment | Color | ml $O_2$/ 100 ml blood |
|---|---|---|---|
| Mammals | Hemoglobin | Red | 25.0 |
| Birds | Hemoglobin | Red | 18.5 |
| Reptiles | Hemoglobin | Red | 9.0 |
| Fish | Hemoglobin | Red | 9.0 |
| Annelids | Hemerythrin | Red | 2.0 |
| Mollusc cephalopods | Hemocyanin (in plasma) | Blue | 8.0 |
| Crustacea | Hemocyanin (in plasma) | Blue | 3.0 |

[a] Data taken in part from Baldwin (1962, pp. 12–14).

the blood pigment is chlorocruorin, a green pigment that contains copper and is found in many crustaceans and molluscs. There are perhaps other respiratory pigments not yet identified. All have the common property of functioning as oxygen carriers (Table 3.1). The porphyrin structure then, with iron or another metallic atom, is especially well adapted to transport oxygen.

The heme pigments can be identified by their absorption spectra. In Figure 3.3 an example of the spectrum of a human red blood cell is illustrated. The spectral absorption peaks of some other animal blood cells are tabulated in Table 3.2.

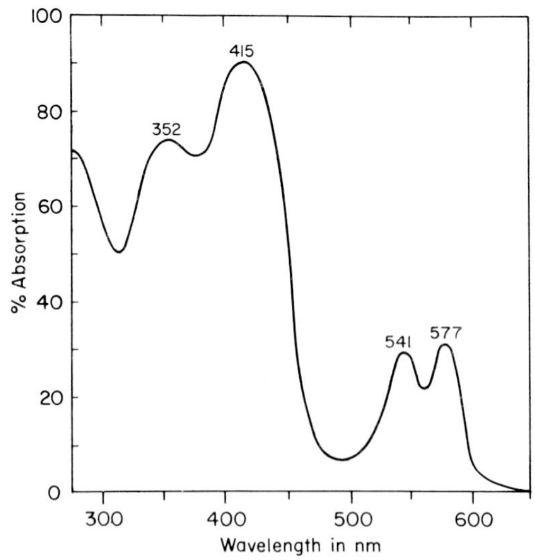


FIGURE 3.3 Absorption spectrum of a human red blood cell. (Obtained by microspectrophotometry; refer also to Table 3.2.)

TABLE 3.2 RED BLOOD CELLS
MAJOR ABSORPTION PEAKS IN nm[a]

| | | | |
|---|---|---|---|
| Human | 415 | 541 | 577 |
| Rabbit | 407 | 534 | 572 |
| Mouse | 416 | 544 | 577 |
| Chick embryo (14 days) | 413 | 543 | 576 |
| Frog (*Rana pipiens*) | 415 | 539 | 572 |
| Turtle (*Pseudemys scripta elegans*) | 408 | 539 | 574 |
| Earthworm (plasma) | 414 | 538 | 567 |
| Water flea (*Daphnia pulex*) (plasma) | 415 | 541 | 574 |

[a] Major absorption peaks obtained with the recording microspectrophotometer, M-5, of single oxygenated red blood cells. Average cross section of surface viewed 10 $\mu m^2$. From Wolken (1968c, p. 533).

For a comprehensive review of heme pigments and their functional role in animals, refer to the texts of Lemberg and Legge (1949) and Keilin (1966). For their role in discerning evolutionary patterns, the discussions in Florkin (1949), Baldwin (1964), and Allen (1966) are suggested.

## Cytochromes

The cytochromes are pigmented heme proteins that carry an iron atom in an attached chemical group (Figure 3.4). Their red color is derived from their prosthetic group or chromophore which is a derivative of iron protoporphyrin IX (Figure 3.1). The cytochromes function as electron carriers during the initial reactions of the photochemical processes. The cytochromes are designated

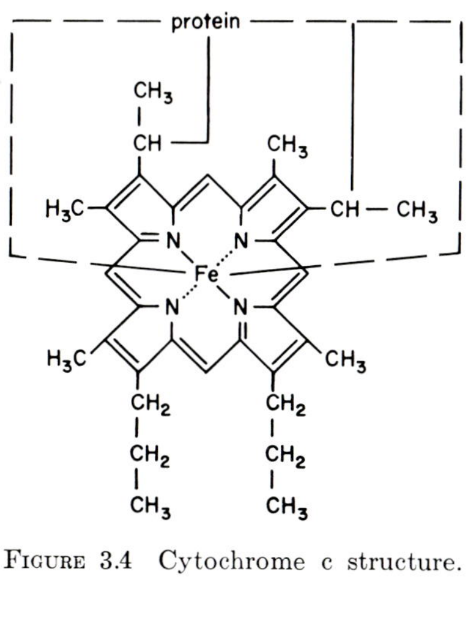


FIGURE 3.4 Cytochrome c structure.

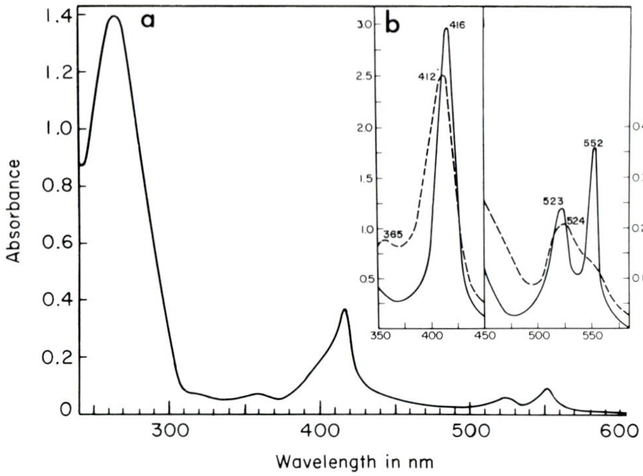


FIGURE 3.5 Cytochrome c absorption spectrum of the reduced state (a). Insert (b) expanded to show the major reduced absorption peaks (——) and oxidized absorption peaks (- - - - - -). Isolated from light grown *Euglena gracilis*.

by the letters a, b, c, and f, and are distinguished on the basis of their spectral absorption peaks in the reduced state. The spectrum of *Euglena* cytochrome-552 (a cytochrome c) in the reduced and oxidized states is shown in Figure 3.5. Note its similarity to the spectrum of hemoglobin in the red blood cell (Figure 3.3). Some comparative properties of the c-type cytochromes are listed in Table 3.3.

Cytochrome c has been isolated from animals, higher plants, algae, and bacteria. The evolutionary history of cytochrome c is well documented, and a pattern of 1.5 billion years of mutation separating yeast from man can be deduced (Fitch and Margoliash, 1967). In fact, since the rate of evolution of this molecule seems to have been relatively constant (1% change per 20 million years), there is a possibility of using cytochrome c as a paleontological clock to time the divergence of animal species from a common evolutionary pathway (Dickerson, 1971).

## The Carotenoids

The most abundant pigments found in nature are carotenoids. The carotenoids are the yellow, orange, and red fat-soluble pigments that are widely distributed in plants. In green plants carotenoids are usually found together with the chlorophylls. The carotenoids possess 40 carbon atoms in each molecule (Figure 3.6). They are named for their most familiar substance, carotene, and are divided into two main groups: The carotenes (hydrocarbons), the most

TABLE 3.3 COMPARATIVE PROPERTIES OF SOME c-TYPE CYTOCHROMES

| | *Photosynthetic bacteria*[a,b] | | *Algae*[c] *Euglena gracilis* | | *Higher plants*[b] | *Animals*[b] |
|---|---|---|---|---|---|---|
| | *Rhodospirillum rubrum* | *Chromatium* | *Light-grown* | *Dark-grown* | *Spinach* | *Beef heart* |
| Absorption maximum in nm | | | | | | |
| Oxidized | | | | | | |
| $\alpha$ | 535 | 525 | 524 | 530 | 535 | 535 |
| $\gamma$ | 409 | 410 | 412 | 412 | 412 | 410 |
| Reduce | | | | | | |
| $\alpha$ | 550 | 552 | 552 | 556 | 555 | 550 |
| $\beta$ | 521 | 523 | 523 | 525 | 526 | 521 |
| $\gamma$ | 416 | 416 | 416 | 421 | 417 | 416 |
| Isoelectric point, pH | 7.0 | 5.4 | 5.0 | <7.0 | 4.7 | 10.0 |
| $E_0'$ in volts, pH 7 | +0.32–0.365 | +0.01–0.04 | +0.35–0.40 | +0.31–0.33 | +0.365–0.38 | +0.265 |
| Sedimentation, $s_{20}$ | 2.0 | 6.0 | 1.2 | 1.4 | 6.9 | 1.8 |
| Molecular weight | 16,000 | 97,000 | 11,000 | 13,000 | 110,000 | 13,600 |

[a] Bartsch and Kamen (1960).
[b] Kamen (1956, 1960).
[c] Wolken and Gross (1963).

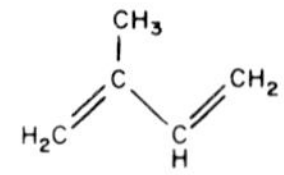

ISOPRENE

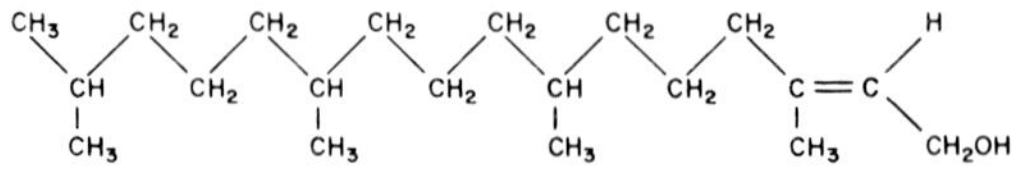

PHYTOL

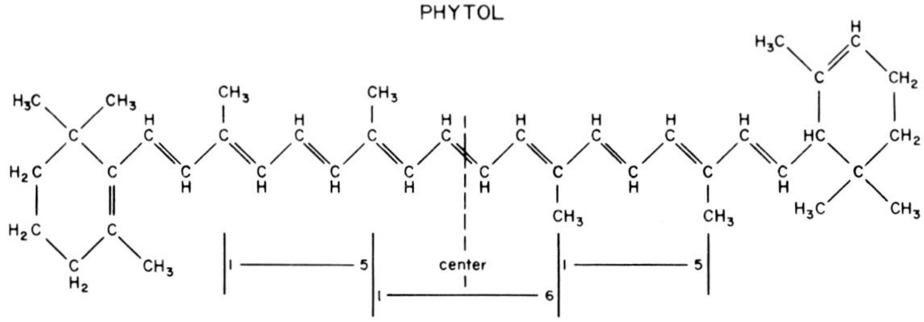

α-CAROTENE

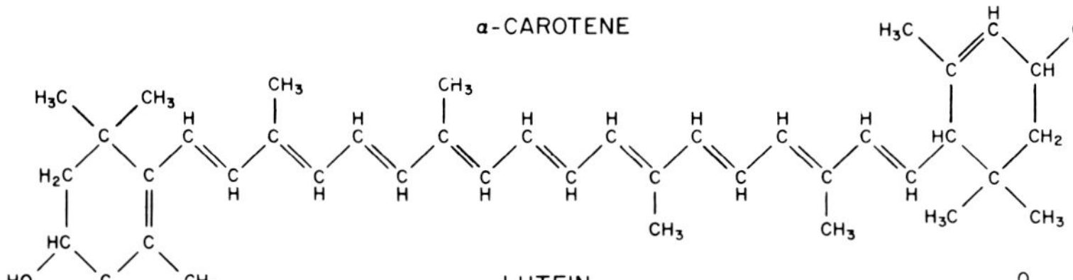

LUTEIN
(XANTHOPHYLL)

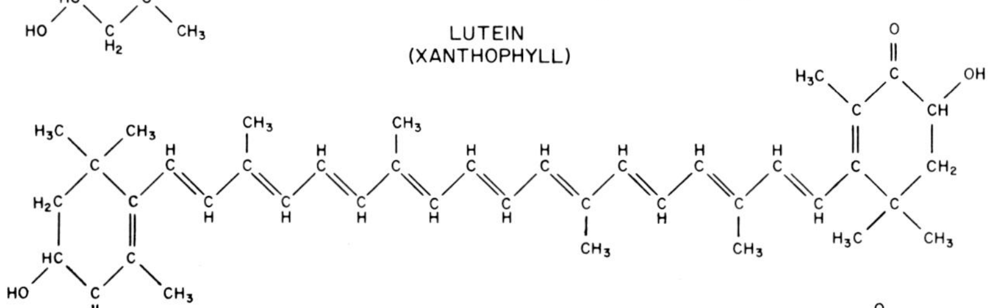

ASTAXANTHIN

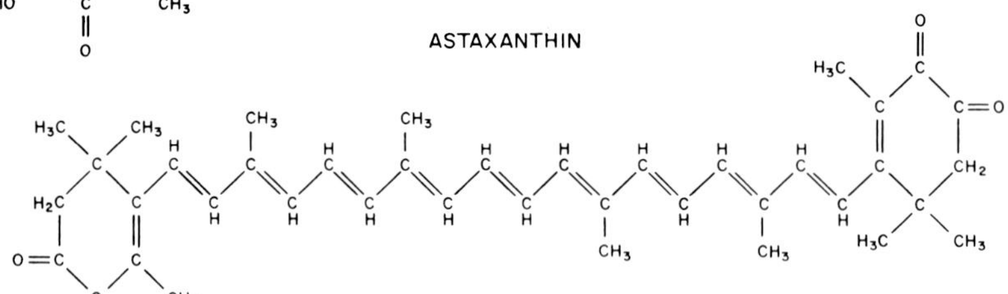

ASTACENE

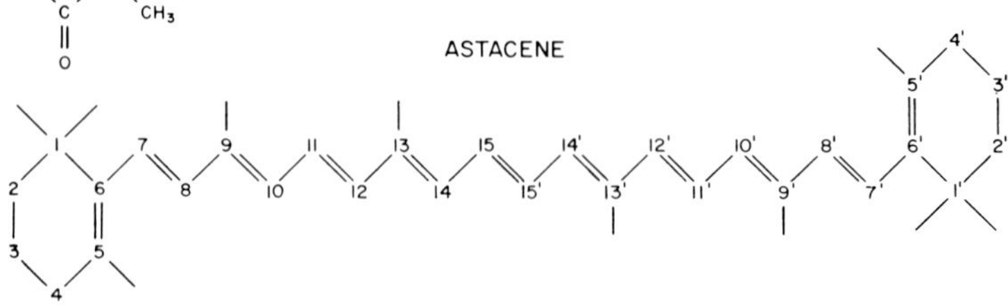

KARRER'S NUMBERING SYSTEM FOR CAROTENOIDS

FIGURE 3.6 Structures of various carotenoids.

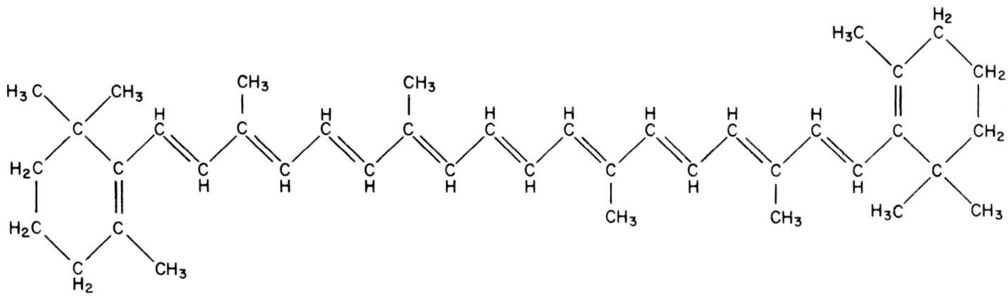

FIGURE 3.7 β-Carotene structure.

abundant of which is all-*trans*-β-carotene, $C_{40}H_{56}$ (Figure 3.7), and the xanthophylls (oxygen-containing derivatives). One of the more common xanthophylls is lutein, $C_{40}H_{56}(OH)_2$, or luteol. From their structure, carotenoids can be considered to be built from isoprene units. The linear portion of the molecule consists of four radicals of isoprene (2-methyl-1,3-butadiene) residues. The isoprene units are linked so that the two methyl groups nearest the center of the molecule are in positions 1 and 6, while all other lateral methyl groups are in positions 1 and 5 (Figure 3.6).

The carotenoid molecule, then, is made up of a chromophoric system of alternating single and double interatomic linkages—called a polyene chain of conjugated double bonds—between the carbon atoms (Figure 3.7).

The large number of these conjugated double bonds offers the possibility of either *cis* or *trans* geometric configurations. There are 20 possible geometric isomers of β-carotene, of which six *cis* isomers have been discovered in nature. The spectral characteristics, and therefore the color of the carotenoids, are largely determined by the number of conjugated double bonds in the molecule.

The biosynthesis of carotenoids is generally associated with the 20-carbon, aliphatic alcohol phytol (Figure 3.6), which is the colorless moiety of the ester-comprising chlorophyll. The striking resemblance between the carotenoid skeleton and phytol holds also for the details of spatial configuration. Mevalonic acid is a common precursor, as indicated in Figure 3.8. Geranylgeranyl pyrophosphate (GGPP), synthesized from mevalonic acid by the normal terpenoid pathway, has now been established as the immediate precursor to phytoene, the first $C_{40}$-carotene (Liaaen-Jenson and Andrews, 1972). The common precursor for α-, β-, δ-, and γ-carotene is neurosporene. The oxygenated carotenes, xanthophylls, are derivable from β-carotene.

Carotenoids are easily and abundantly synthesized by plants. However, an important change occurred during the evolutionary development of animals; namely, they became dependent on the ingestion of plants for their source of carotenoids. It is not the ingested plant carotenoids themselves but their degraded derivative, vitamin A, that is necessary for all animal life (Figure 3.9).

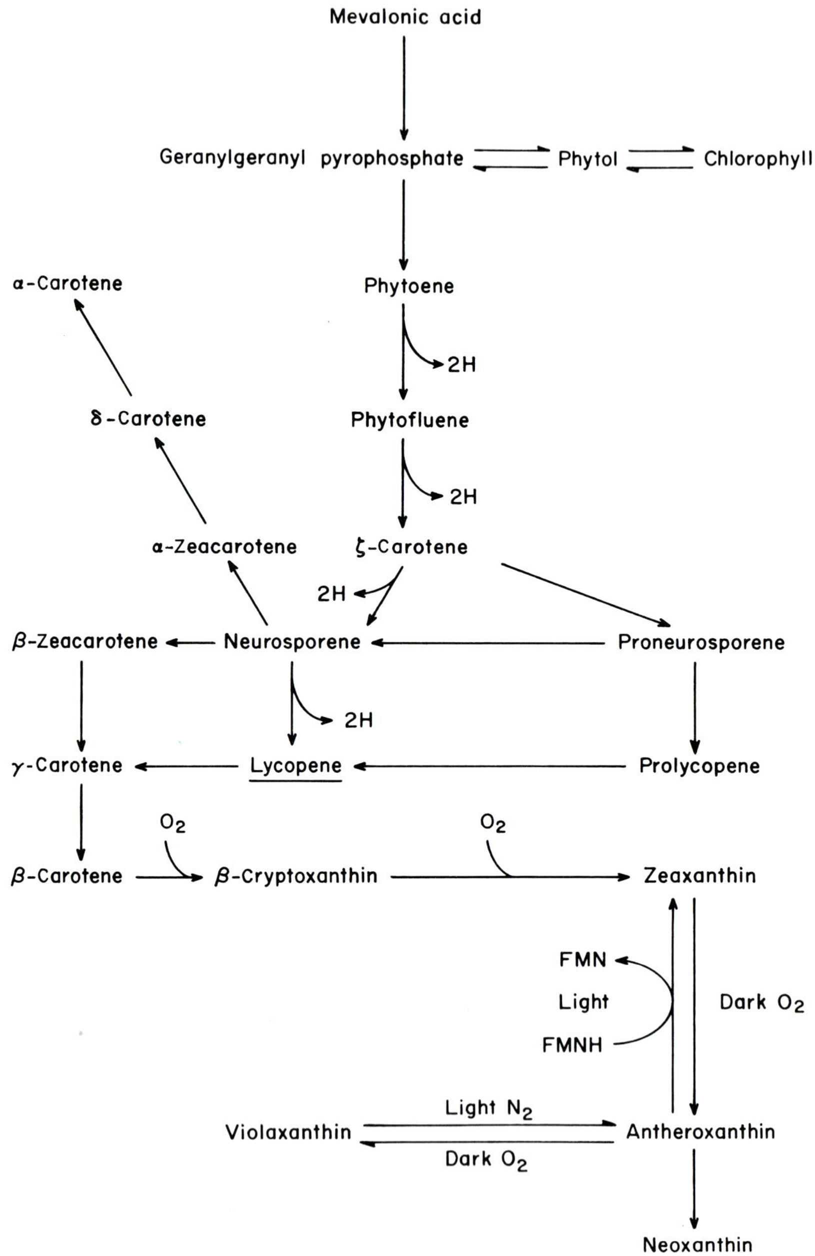


FIGURE 3.8 The biosynthesis of carotenoids.

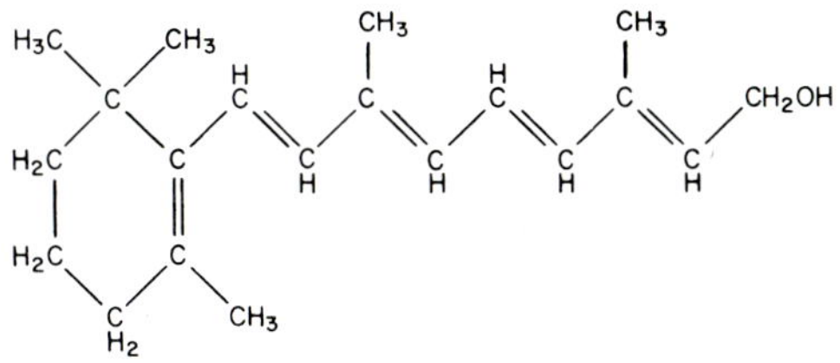

FIGURE 3.9 Vitamin $A_1$ structure.

Thus, the carotenoids can be seen to play a central role in the biochemical evolution from the plant $C_{40}$ ($\beta$-carotene) → animal $C_{20}$ (vitamin A). The importance of this to evolution and to our understanding of the visual pigment chemistry is further elaborated on in Chapter XI with regard to the visual pigments of vertebrate and invertebrate eyes. A more detailed discussion of the biosynthesis and chemical structure of the carotenoids and their distribution throughout the plant and animal kingdom can be found in Fox (1953), Goodwin (1952, 1965), Karrer and Jucker (1950), Strain (1951), and Zechmeister (1944, 1962).

## Phycobiliproteins

The bilan pigments are so named because they were first discovered in bile. They may be in fact products of metabolic transformation of hemoglobin, other hematin compounds, and chlorophyll. The name phycobilin indicates that they are derivable from algae. In red and green algae there are blue pigments, phycocyanins, and red pigments, phycoerythrins. Phycocyanins and phycoerythrins consist of a chromophore and protein. Because of the similarity of the chromophore they have been termed "phycobilins" or biliproteins. The phycobiliprotein pigments, unlike the chlorophylls and carotenoids, are water soluble. They can be identified by the peaks in their absorption spectra.

It is interesting to note that the phycobiliprotein chemical structure, like that of chlorophyll, is a tetrapyrrole, but arranged in a linear array, by the opening of the porphyrin ring (Figure 3.10). For example, the structure of phycocyanin closely resembles that of phytochrome (Figure 14.2), the pigment responsible for plant photoperiodism. Therefore its structure resembles

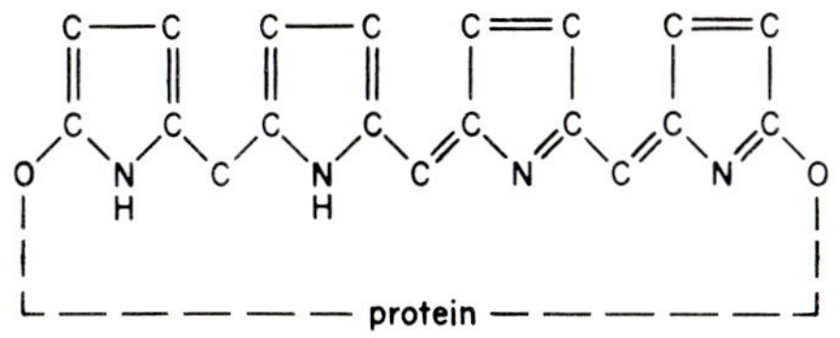

FIGURE 3.10 Phycobiliprotein structure. Note similarity to phycocyanin and phytochrome (Figure 14.2).

an open ring porphyrin (Figure 3.1), and with its system of conjugated double bonds, it resembles a carotenoid (Figure 3.7).

In the red and green algae, the biliproteins can utilize absorbed light energy by transferring it to chlorophyll a in the process of photosynthesis with an efficiency equivalent to or greater than that of chlorophyll alone. One role suggested for these pigments is to make a larger part of the spectrum of visible energy available for photosynthesis.

## Ferredoxins

Ferredoxin may have been one of the first proteins to emerge from the "primeval soup," for it is relatively simple from the point of view of biosynthesis and consequently may have played an important role in the origin of life. Ferredoxins also seem to have occupied a key position in the evolution of early primitive organisms, which may enable us to trace the evolutionary pathway from these primitive organisms to more highly evolved plants and animals.

Ferredoxin is an iron protein which is not a heme protein as are the cytochromes. Unlike the cytochromes which exhibit well-defined absorption peaks in the reduced state, ferredoxins have distinct absorption peaks in the oxidized state (Figure 3.11).

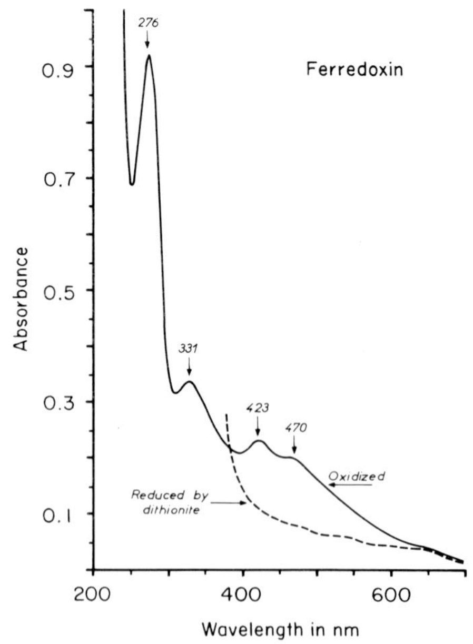


FIGURE 3.11 Absorption spectrum of ferredoxin in oxidized and reduced states (extracted from spinach).

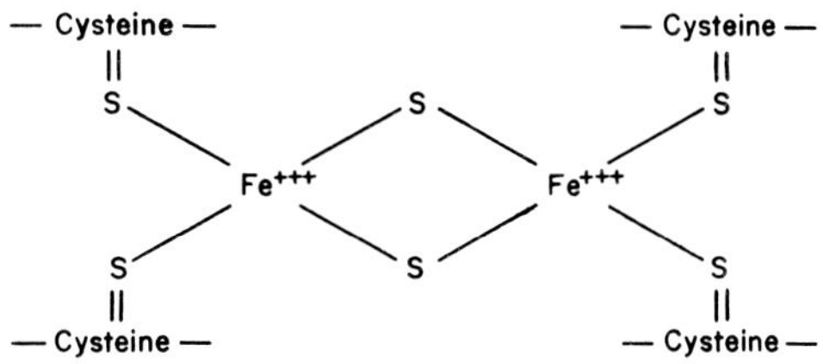


FIGURE 3.12 Basic structure of ferredoxin.

Ferredoxins are biological reducing agents; they transfer electrons, and as such they take part in metabolic processes as diverse as photosynthesis, fixation of atmospheric nitrogen, production of hydrogen, and the synthesis of steroid hormones.

The ferredoxin molecule is small in comparison to most proteins. In addition to 50 to 100 amino acids, they contain iron and inorganic sulfur (Figure 3.12). The bacterial ferredoxins are the smallest and contain 7 atoms of iron per molecule (Table 3.4).

Ferredoxins of four species of bacteria (of the genera *Clostridium* and *Micrococcus*) contain only 14 different amino acids, nine of which are common to all four organisms. These nine amino acids are identical with those that are readily synthesized in the laboratory by simulation of conditions of the primeval earth. Ferredoxins from plants have larger protein molecules, but contain only 2 atoms of iron and sulfur.

Ferredoxins from the blue-green algae are very different from those of bacteria. They resemble the ferredoxins from the more developed algae and plants.

TABLE 3.4 SOME PROPERTIES OF BACTERIA FERREDOXIN[a]

| | *Bacteria* | *Photosynthetic bacteria* | *Plants* |
|---|---|---|---|
| | *Clostridium pasteurianum* | *Chromatium* | *Spinach* |
| Iron content (atoms/molecule protein) | 7 | 3 | 2 |
| Inorganic sulfide (moles/mole protein) | 7 | 3 | 2 |
| Redox potential (volts, pH 7.55) | −0.417 | −0.490[b] | −0.432 |
| Molecular weight[b] | 6000 | 6000 | 13,000 |

[a] From Arnon (1965, p. 1464).
[b] Approximate.

There are a number of animal proteins which have similar properties to those of plant ferredoxin, for example, adrenodoxin from the adrenal glands.

Although the study of proteins such as cytochromes can give us clues about the evolution of higher organisms, only the ferredoxins seem to show developments across the evolutionary spectrum (Cammack *et al.*, 1971).

## Flavins

Riboflavin (vitamin $B_2$) is taken here as an example of the flavins and flavoproteins. Riboflavin is synthesized by a number of microorganisms and by most higher plants, particularly in growing leaves. It occurs in small concentrations in nearly all animal tissues. As one of the B vitamins it is a factor in cell respiration.

The structure of riboflavin is benzisoalloxazine-6,7,-dimethyl-9-D-ribitol (Figure 3.13). Riboflavin in solution is yellow and has spectral absorption peaks in the oxidized state around 221–227, 265–270, 365–370, and 445–460 nm (Figure 3.14). One of the more important chemical properties of riboflavin is its ability to change reversibly from the yellow-colored, oxidized form to the colorless, reduced form. Upon ultraviolet excitation it fluoresces green, 520–560 nm. Riboflavin is light sensitive in neutral or acid solution and forms lumichrome with absorption peaks near 223, 260, and 360 nm.

Riboflavin mononucleotide is riboflavin-5′-phosphate and is usually referred to as FMN. Warburg and Christian (1938,a,b,c) discovered that an enzyme catalyzing the oxidation of D-amino acids, D-amino oxidase, was a flavoprotein containing a prosthetic group distinct from FMN. The structure of this coenzyme, flavin adenine dinucleotide, is called FAD.

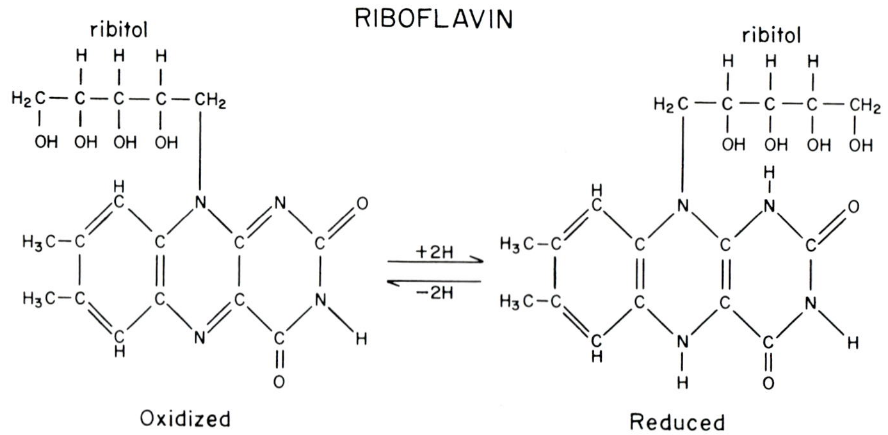


FIGURE 3.13 Riboflavin (vitamin $B_2$) structure in oxidized and reduced states.

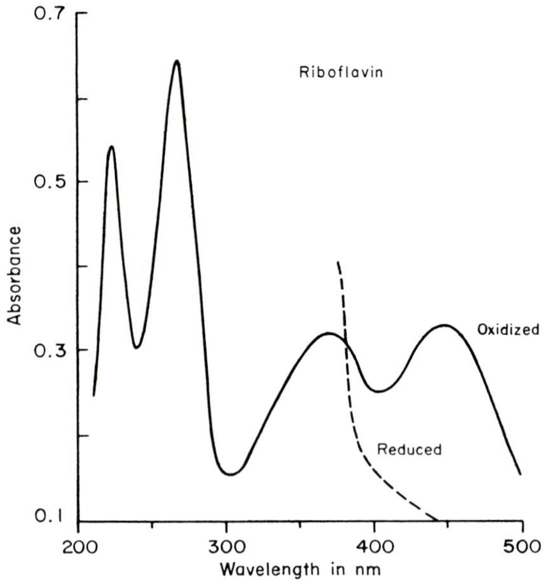


FIGURE 3.14 Riboflavin absorption spectrum in oxidized and reduced states. Reduced by dithionite.

Riboflavin and the flavoproteins may be significant molecules in photoreceptor processes associated with phototactic phenomena in algae, plants, and animals. Flavoproteins may even be involved in the biochemistry of the visual process (Fox, 1960). Riboflavin has been found in the eye, particularly in the retina of certain mammals, in the pigment epithelium of fish, and between the pigment epithelium and the choroid of frogs, rabbits, rats, and cattle.

## Melanins

Melanin pigments are black, brown, or yellow and occur in all animal phyla. The pigment is found in the hair, skin, and eyes. The main function of melanin in the skin is protection against solar ultraviolet radiation.

Melanin is a biochrome of high molecular weight and can be produced *in vitro* by oxidation of the amino acid tyrosine with the enzyme tyrosinase, a copper-containing protein. The first step in this reaction is the production of dopa (3,4-dihydroxphenylalanine; Figure 14.5) which is then oxidized enzymatically to dopa-quinone; after this stage there occurs a complicated series of further oxidations and polymerizations, leading to the formation of tyrosine-melanin. The exact structure of melanin remains to be established.

Even less is known of the structure of natural melanin produced by the chromatophores, the melanophores, and the melanocytes of cells, for the problem is complicated by the fact that melanin is always bound to a protein. However, all evidence available indicates that tyrosine is the precursor of melanin and that the biosynthesis follows the pathway of tyrosine → dopa → melanin.

Some general properties of the melanins, carotenoids, and flavins are given in Table 3.5.

TABLE 3.5 SOME PROPERTIES OF COMMON PIGMENTS[a]

| *Property* | *Carotenoids* | *Flavins* | *Melanins* |
|---|---|---|---|
| Solubility | | | |
| Water | Generally no | Yes | No |
| Organic solvents | Yes | No | No |
| Color of solution | Yellow, orange, red | Yellow–orange | Yellowish, orange, reddish, brown, gray, black |
| Absorption bands, visible | Violet, blue, blue–green | Violet | None |
| Fluorescence | Carotene and lycopene, weak yellow–green; some carotenoid acids, blue | Green | None |
| Prosthetic group bound to | Protein generally | Protein | Protein (not always) |
| Behavior toward: | | | |
| Mineral acids | Very sensitive: conc., acids give blue or green colors, finally bleaching | Stable in dilute acids | Precipitated by dilute acids |
| Alkalis | Stable | Sensitive | Dissolved (or colloidal dispersion) |
| Oxidizing agents | Very sensitive, bleaching | Stable | Variable: decolorized by $H_2O_2$ |
| Reducing agents | | Reduced by $Na_2S_2O_4$ with bleaching. reversible | Reduce ammoniacal silver |
| Biological relations | Vitamin A | Vitamin $B_2$ | Catabolite |

[a] Taken in part from Fox (1953, p. 69) and from Wolken (1971, p. 17).

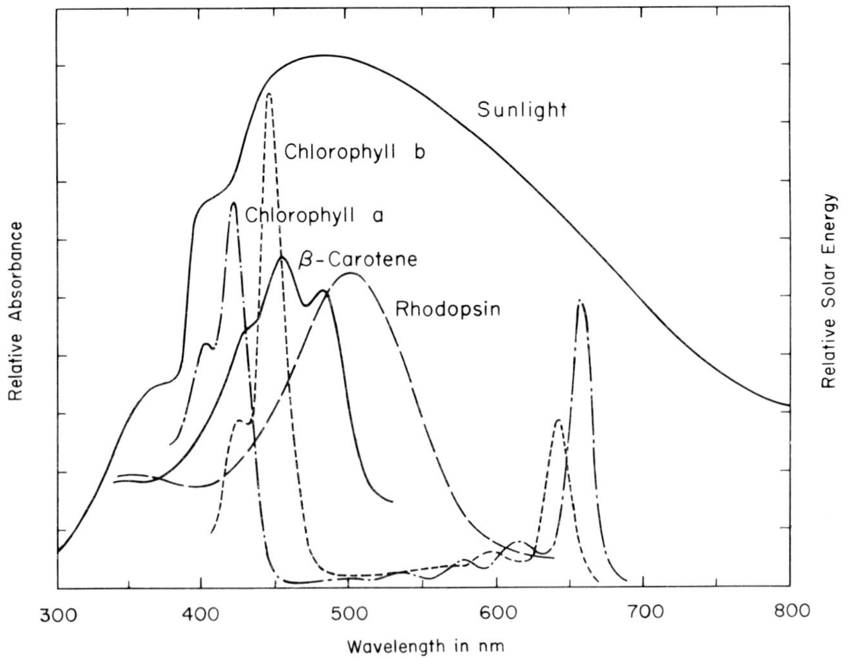


FIGURE 3.15 Chlorophyll a, chlorophyll b, β-carotene, and rhodopsin absorption spectra compared to the solar spectrum that strikes the earth.

There are many other molecules associated with photobiological processes than the ones briefly described here. These will be discussed more specifically in their relationship to particular photoprocesses.

## Solar Spectrum and Pigment Absorption Spectra

It is interesting to note that the absorption spectra of the pigments just described cover the sun's spectrum of radiation at the surface of the earth and that their absorption spectra all fall in the visible range and close to the solar energy peak around 500 nm (Figures 2.4 and 3.15).

## Action Spectra

All photobiological phenomena depend on the ability of a system to absorb light energy of the necessary wavelengths. As we have indicated, specific pigment molecules or pigment systems are required to absorb this energy.

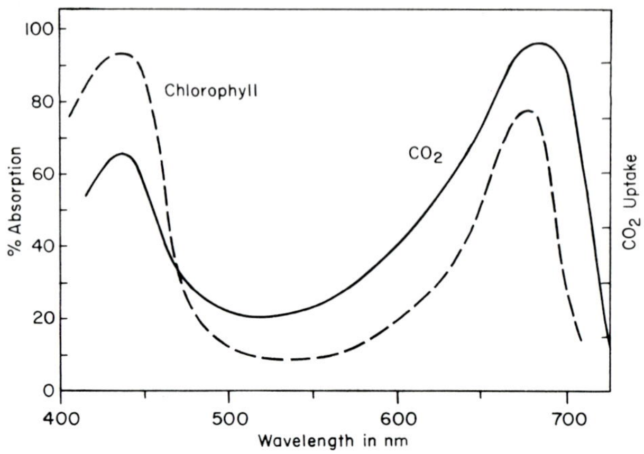


FIGURE 3.16 Action spectrum. Relative effectiveness for $CO_2$ uptake in photosynthesis compared to absorption spectrum peaks for chlorophyll.

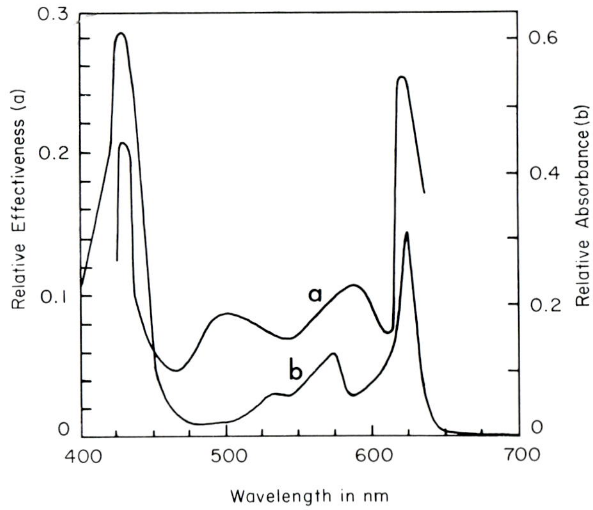


FIGURE 3.17 Action spectrum. Relative effectiveness for chlorophyll synthesis spectrum (a) compared to absorption spectrum (b) for protochlorophyll.

One of the most important means we have for identifying the kinds of molecules in a photosensitive system is by obtaining the behavioral action spectrum. We can define the term *action spectrum* as the relative response that a plant or animal makes to a light stimulus of different wavelengths. The action spectrum obtained should correspond to the absorption spectrum of the molecule or molecules responsible for this behavior.

Action spectroscopy can be an extremely powerful tool in the study of photobiological phenomena—to identify a process with a molecule. An excellent example of how we learn to associate a specific pigment can be seen in the identification of chlorophyll with photosynthesis (Figure 3.16). This same kind of association can be deduced by comparing the action spectrum for chlorophyll synthesis with the absorption spectrum of protochlorophyll, the chlorophyll precursor (Figure 3.17). Action spectra also provided some of the first evidence of the role played by nucleic acid in living cells (Jagger, 1967).

With this information about pigments, their structure, and absorption spectra, we now have some basis for discussion of photoprocesses in living cells. But first we must ask some questions about the cell, its structure, and how these pigments are related to its energetics and specific photobiological phenomena.

# IV
# The Emergence of the Cell and Its Organelles

If we believe that all forms of life as we now view it share a common ancestry and arose from a cell which gave rise to greater complexity by forming specialized organelles to carry on the life processes and that these various kinds of cells aggregated for mutual benefit to produce the whole integrated plant or animal, then we believe in evolution.

The concept of the cell as the "unit of life" and the idea that the cell could not exist without a cell membrane go back to Robert Hooke in 1665, when the microscope was in the early stages of development. It was not until the development of the compound microscope in the 1830's and its application to biological tissues that it was possible to begin to resolve the cell. The origin of the cell theory, though, is credited to Schleiden (1838) and Schwann (1839) who stated, "The cells are organisms arranged in accordance with definite laws." In the year 1858 Rudolf Virchow recognized that each cell originates from another cell. This discovery marked the beginning of the science of cellular systems.

A century of developments in optics provided new techniques for microscopy such as phase, polarization, interference, and fluorescence, which revealed detailed structural information about the cell. The development of electron microscopy in the early 1940's and the continuous introduction of newer methods for tissue preparation in the past decade have shown the cell in a structural and functional complexity hitherto unknown. Electron microscopy together with optical diffraction, x-ray diffraction, spectroscopy, and other analytical tools have now brought us closer to the molecular structure of the cell.

Cellular organisms are separated into primitive forms and advanced forms which comprise five kingdoms (Table 4.1). The number of kingdoms here is debatable, for some scientists only agree to three kingdoms, and others only to four, of the five listed here. The most difficult organisms to classify are the algae, slime molds, true fungi, and protozoa, as to whether they belong in the Protista, Fungi, or Plantae. The Monera, Protista, and Fungi are unicellular organisms which are considered primitive. Primitive cells are classified as *prokaryotes,* for they do not possess a membrane-enclosed nucleus and

TABLE 4.1 CLASSIFICATION IN KINGDOMS OF ORGANISMS

- *Primitive forms*
  - Kingdoms
    - I. Monera
      - Bacteria
      - Blue–green algae
      - Viruses (?)
    - II. Protista
      - Ciliates
      - Cryptomonads
      - Diatoms
      - Dinoflagellates
      - Euglenoids
      - Rhizopods
      - Slime molds
      - Sponges
      - Sporozoans
      - Zooflagellates
    - III. Fungi
      - True fungi
- *Advanced forms*
    - IV. Plantae
      - Algae (red, green, brown)
      - Bryophytes
      - Tracheophytes
      - Variety of well-organized multicellular types, capable of photosynthesis; grouped into two phyla
    - V. Animalia
      - Variety of well-organized multicellular types; grouped into 20 to 25 phyla

organelles, and their genetic material is dispersed in nuclei throughout their cytoplasm. This is characteristic of all species of bacteria and of blue-green algae. Eukaryotes are cells which have a membrane-enclosed nucleus where the genetic material is organized into chromosomes, and mitochondria which are membrane-bound organelles. These organelles are found in the cells of the more advanced forms of organisms (refer to Stanier, 1970, for further discussion).

The first cells to evolve were believed to be fermenting bacterial cells. Under pressures of natural selection, directly related to the depletion of their nutrients, there arose among the first fermenting bacteria many metabolic traits that are still observable among bacteria living today. These traits include the ability to ferment many different carbohydrates, to incorporate atmospheric carbon dioxide directly into metabolic compounds, and to reduce sulfate to hydrogen sulfides as a by-product of fermentation. As the ammonia available in some parts of the environment became depleted certain bacteria evolved metabolic pathways that could "fix" atmospheric nitrogen directly into amino acids. Other fermenters developed into highly motile organisms. These fermenting bacteria were "obligate anaerobes," that is, for them oxygen was a toxin. Some of the anaerobes adapted simply by retreating into the oxygen-free muds where they are still found today. Others either developed new mechanisms of oxygen detoxification or merely disappeared. Through various detoxification mechanisms the fermenters were able to dispose of the small amount of oxygen then in the environment. Many, if not all, of the various fermenting bacteria were equipped with well-developed systems for the repair of DNA. Such systems were necessary to counteract the damaging effect of ultraviolet radiation, which at that time was intense, for there was no ozone ($O_3$) in the atmosphere to filter it out. All these bacteria were *heterotrophs* needing organic substances for growth, for they had not yet evolved the means for photosynthesis.

In time, some of them developed metabolic pathways that led to the synthesis of porphyrins, bacterial chlorophyll, and hence to the evolution of bacterial photosynthesis. The process of photosynthesis requires a source of hydrogen. Bacteria can utilize inorganic substances, such as hydrogen sulfide and hydrogen gas, as well as various organic compounds in the environment. When photosynthetic bacteria became well established a second kind of photosynthesis became possible. In the second process the uptake of hydrogen was accomplished by the photolysis of water molecules. As a result, free oxygen entered the atmosphere.

The evolution of this mode of photosynthesis probably led to the appearance of the blue–green algae, which are believed to be the first organisms on the earth that were adapted to the presence of free oxygen. A direct result of the success of the blue–green algae was the evolution of new kinds of bacteria that utilized free oxygen in their metabolic processes.

TABLE 4.2 EVOLUTIONARY ERAS IN THE DEVELOPMENT OF LIFE[a]

| Era | Environment | Energy source | Outcome |
|---|---|---|---|
| I | *Anaerobic*<br>$CH_4$, $NH_3$, $H_2$ | Ultraviolet<br>High temperature | Acetate, glycine, uracil, adenine<br>"Organic soup" |
| II | *Loss of hydrogen*<br>Anaerobic; traces of $O_2$ | Ultraviolet<br>High temperature<br>Visible light | Polyphosphates, peptides, porphyrins<br>Oxidation-reductions |
| III | *Loss of ultraviolet*<br>Anaerobic; traces of $O_2$, $CO_2$ | Visible light | Surface catalysis, photochemistry<br>Synthetic reproduction cycles |
| IV | *Loss of free food*<br>Anaerobic; $CO_2$, traces of $O_2$ | Photoreduction<br>Fermentation | Multiplying metabolic cells |
| V | *Loss of anaerobiosis*<br>Aerobic; anaerobic pockets | Photosynthesis<br>Respiration | Autotrophic plants |

[a] From Gaffron (1965, p. 450).

One of the current theories on the succession toward the evolution of life, with time and environmental changes as they may have existed, was developed by Gaffron (1965) and is shown in Table 4.2.

Microorganisms offer many advantages for the study of the fundamental biological problems. As living systems they display the basic characteristics of life: metabolism, responses to environmental factors, reproduction, genetic continuity, and variability. Microbes represent life on a relatively low level of organization and complexity; they can often be studied far more effectively than more highly developed plants and animals just because they are simple.

## A Model Cell: *Euglena*

One of the more interesting unicellular organisms that has evolutionary significance and which I would like to describe here as a model to explore the structure of the cell and photobehavior is *Euglena gracilis.*

The class Euglenineae or Euglenophyta (the name derived from *Euglena*) is a natural taxonomic group within the flagellates. There is a remote, though

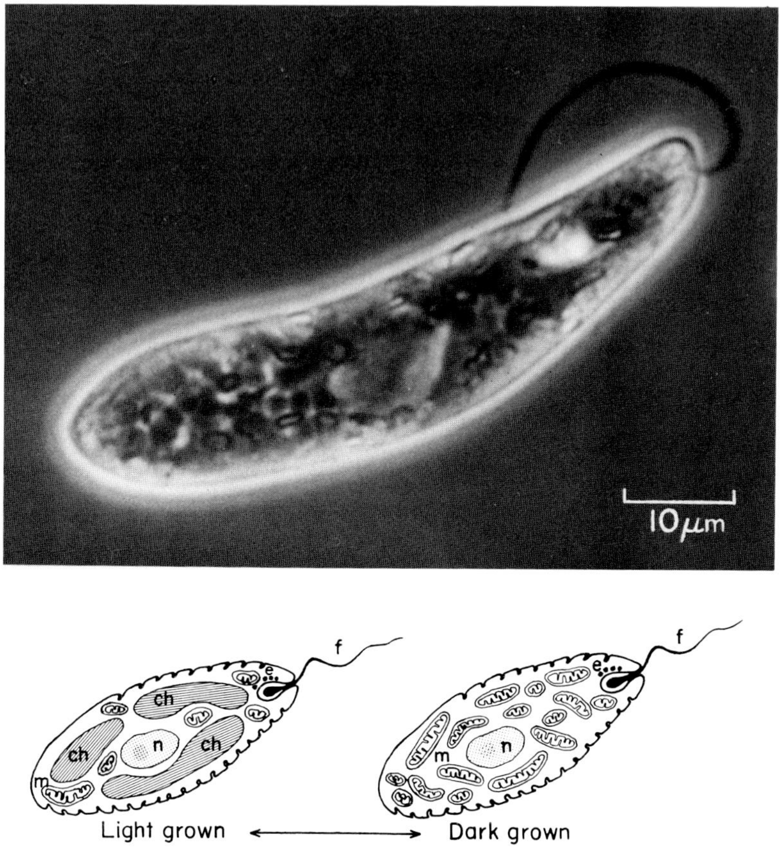


FIGURE 4.1 (a) *Euglena gracilis,* light-grown, phase contrast microscopy; (b) schematic diagram for structures in light-grown and dark-grown *Euglena*. n, Nucleus; f, flagellum; ch, chloroplast; m, mitochondrion; and e, eyespot.

definite, relationship to the green algae; the Euglenaceae and the Euglenineae are the only other organisms like the Chlorophyceae and higher plants which contain chlorophyll b in addition to chlorophyll a. Such *Euglena* are green, others are red, yellow, or brown, depending on their pigments, and some are colorless.

*Euglena* are ubiquitous in bodies of freshwater. They favor waters rich in organic materials, and some species favor damp soils, as do the closely related *Eutreptia,* common in the ocean near cities. In the sense of adaptability to laboratory conditions, *Euglena gracilis* is the weed among the photosynthetic euglenoids, and the Z strain is among the very hardiest. An active, swimming *Euglena gracilis* in the light is an elongated cell of the order of 50 by 15 μm

(Figure 4.1a,b), and some organisms reach a size of 70 by 30 $\mu$m. It has a helicoidally striated exoskeleton (pellicle) with visible ridges. The whipping flagellum protrudes from a "gullet" or cytopharnyx. The stigma, or eyespot region, consists of orange–red pigmented granules situated close to the gullet from which the flagellum protrudes. *Euglena* has many discoid chloroplasts, a single, conspicuous nucleus, numerous mitochondria, vacuoles, granules, and lipid inclusions. Many of these structures are seen in the electron micrographs, Figure 4.2a–k. *Euglena* can be profoundly altered in structure and chemistry depending upon whether they live in light or in darkness (Figure 4.2a,b). When living in the light it photosynthesizes like a plant, but in darkness it loses its chloroplasts and hence its ability to photosynthesize. It then thrives by chemosynthesis, a process typical of all animal cells. The light (photosynthesis) $\leftrightarrow$ dark (respiration) adaptation brings about changes in the chemistry and structure of the organism. This ability to adapt to light or darkness is fully reversible providing mutations do not occur. Mutations to nonphotosynthetic *Euglena* can easily be produced by changing the physical and chemical environment, for example, by growth temperatures above 33°C, by ultraviolet and ionizing radiation, and by high pressure. Chemically, mutations can be induced by drugs, for example, by streptomycin and antihistamines. The morphology, physiology, and biochemistry of the euglenoids have been reviewed by Jahn (1951) in studies summarized in Volumes I, II, and III of "Biochemistry and Physiology of Protozoa" (Lwoff, 1951; Hutner and Lwoff, 1955; Hutner, 1964), and also by Wolken (1967), Leedale (1967), and Buetow (1968).

To the taxonomist *Euglena* has presented a problem in classifying protists because it embodies features commonly and uncritically denoted as "plant-" or "animal-like." *Euglena* is a unicellular protozoan, or more accurately, an "algal flagellate"; it has the characteristics of a plant cell but shares as well some of the attributes of an animal cell. Some general distinctions between the plant and animal cells are indicated in Table 4.3. Lwoff (1951) states that it is impossible to give a satisfactory definition of protozoa. Hutner and Provasoli (1965) think of "true" protozoa as the particle-ingesting protists.

Protista is the inclusive term for all unicellular organisms. Classically, in T. H. Huxley's view, animals are denoted as those forms of life which require preformed organic molecules for their nutrition and which possess such characteristics as locomotion, flexible cell walls, and some kind of integrated "nervous" control, all in response to their reliance on ingested particulate food. Photosynthetic plants, in contrast, depend far less on externally supplied organic molecules. It is speculated that *Euglena* evolved close to the origins of the higher plants (for they do have the same photosynthetic pigments) and close to most fungi. However, the euglenoid flagellates are probably remote from the protozoan flagellates that are supposed to have given rise to the cellular animals, that is, the metazoa. The large number of question marks in Figure

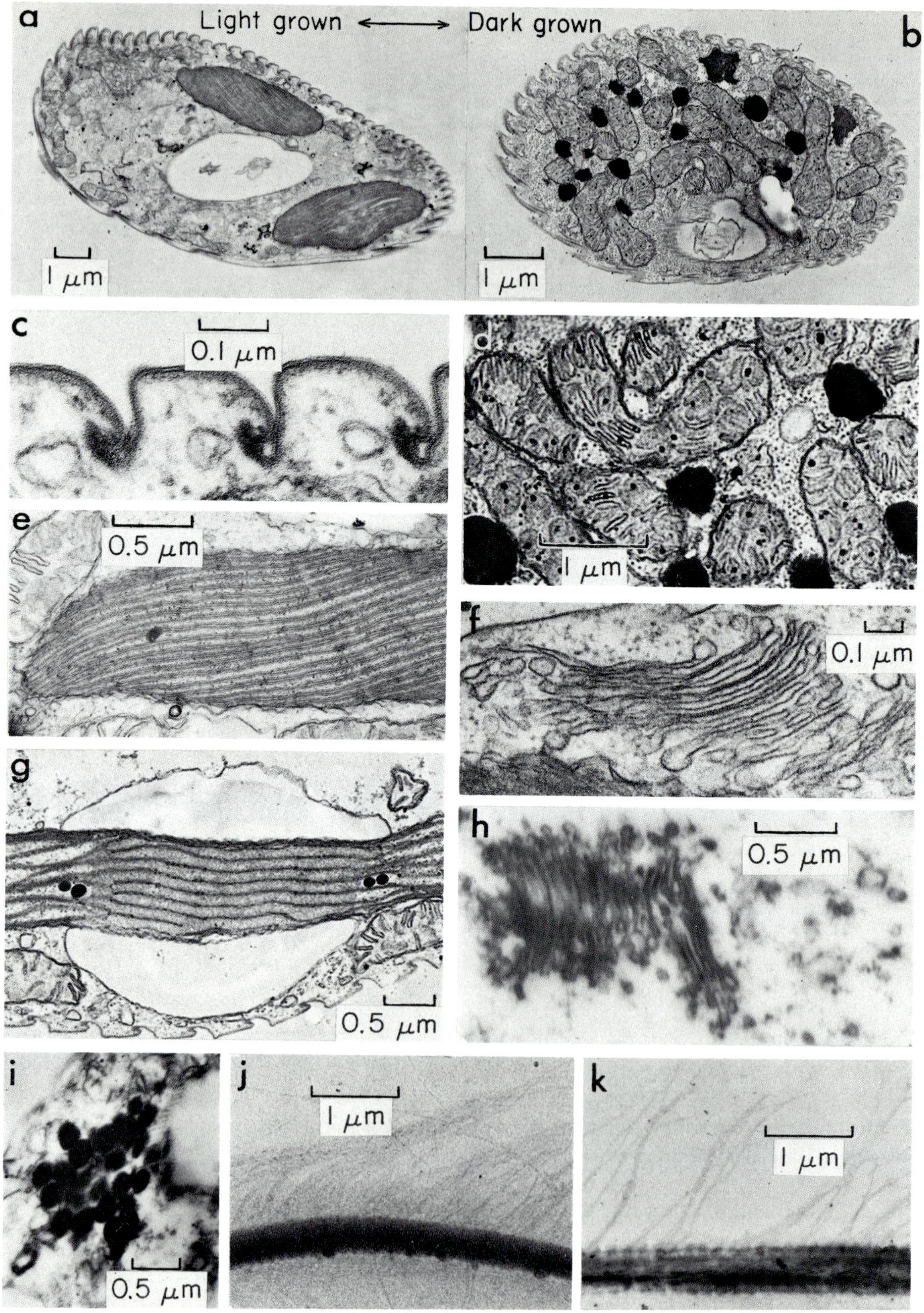
a
Light grown ⟷ Dark grown
b
1 μm
1 μm
c
0.1 μm
d
1 μm
e
0.5 μm
f
0.1 μm
g
0.5 μm
h
0.5 μm
i
0.5 μm
j
1 μm
k
1 μm

TABLE 4.3 SOME DISTINCTIONS BETWEEN PLANT AND ANIMAL CELLS[a]

| | *Plants* | *Animals* |
|---|---|---|
| Energy source | Solar radiation<br>Photosynthesis | Organic matter<br>Chemosynthesis |
| Chlorophyll | Present (except in fungi) | Absent (except in some particle-ingesting flagellates) |
| Principal reserve food | Starch, oil (in some) | Glycogen, fat |
| External foodstuff | Absorbed osmotically | Ingested |
| Active movement | Restricted (except for some flagellates and the gametes of plant cells | Usually present |
| Cell walls | Generally rigid (except in some flagellates and fungi) | Generally flexible |
| Nervous system | Generally absent (except presumably in flagellates and some fungi) | Present |

[a] From Wolken (1967, p. 3).

4.3 indicates our skeletal knowledge of the relationships among protozoan groups and their kinship to the metazoa.

From the scheme in Figure 4.3, we see that *Chlamydomonas,* from which *Euglena* is descended, is directly related to plant lines that go back to green algae. The green algae in turn go back to *Chlamydomonas,* an inference derived from the fact that the sex cells (gametes) of multicellular or filamentous green algae are very *Chlamydomonas*-like. Freshwater ponds and oceans have a wealth of *Chlamydomonas* and closely related flagellates. These green-pigmented forms differ sharply from brown-pigmented flagellates whose affinities are more to the brown seaweeds and diatoms.

*Peranema* is related to the photosynthetic *Euglena.* It is a unique example of an animal unmistakably derived from a green alga (Hutner and Provasoli, 1951, 1955, 1965). Therefore we can look upon the algal flagellates as a hetero-

FIGURE 4.2 *Euglena gracilis* structures as seen by electron microscopy, for light-grown and dark-grown. (a) Light-grown, cross section; (b) dark-grown, cross section; (c) enlargement of exoskeleton and cell membrane; (d) mitochondria; (e) chloroplast; (f) Golgi apparatus; (g) pyrenoid; (h) endoplasmic reticulum; (i) eyespot granules; (j) and (k) flagellum.

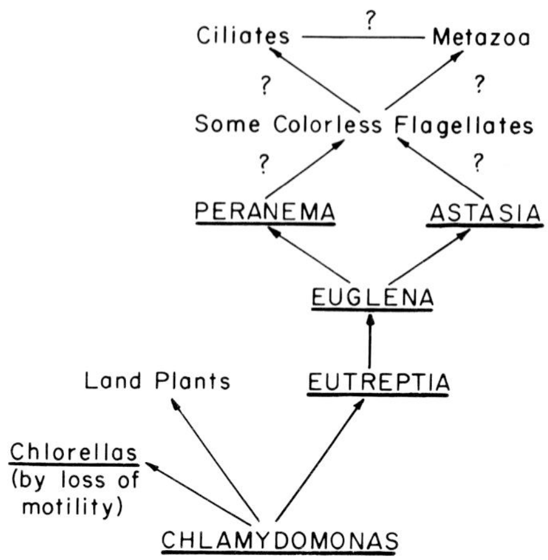


FIGURE 4.3 Affinities of the protozoan flagellates.

geneous group occupying a central position along the line of plant → animal descent.

In certain respects knowledge of the chemistry of protists is primitive, although some of the photosynthetic flagellates thrive in simple media. Unlike land plants *Euglena* cannot utilize nitrates; it will grow rapidly if given ammonia supplemented with amino acids. *Euglena* is known as one of the acetate flagellates, generally preferring acids and alcohols to sugar as its main carbon source. This is a nutritional pattern quite unlike familiar green algae, such as *Chlorella,* and the land plants whose favorite external substrate is glucose or sucrose. Since *Euglena* can grow rapidly on acetate as the sole substrate, it must have some means of making Krebs cycle intermediates from acetate. In land plants, acetate can be converted to carbohydrate by way of the "glyoxylic acid shunt." Since *Euglena* can utilize glycolic and lactic acids in media of low pH ($<4.0$), it is likely that *Euglena* may have a highly developed glyoxylic acid entry to the Krebs cycle (Hutner *et al.,* 1956).

*Euglena* diverges from land plants with respect to its vitamin needs because it requires vitamin $B_{12}$ and thiamine (vitamin $B_1$). Metabolic patterns like those of *Euglena* are common in protists; for example, the aerobic photosynthetic bacteria are nutritionally not very different, even in their need for vitamins.

## $CO_2$ Fixation

In a comparative study of the dark $CO_2$ fixation of microorganisms, Lynch and Calvin (1953) found that *Euglena gracilis* differed from other microorganisms. Photosynthetic experiments of 1- and 5-minute exposures to radioactive $CO_2$ were performed. Not only were the characteristic dark $CO_2$ fixation prod-

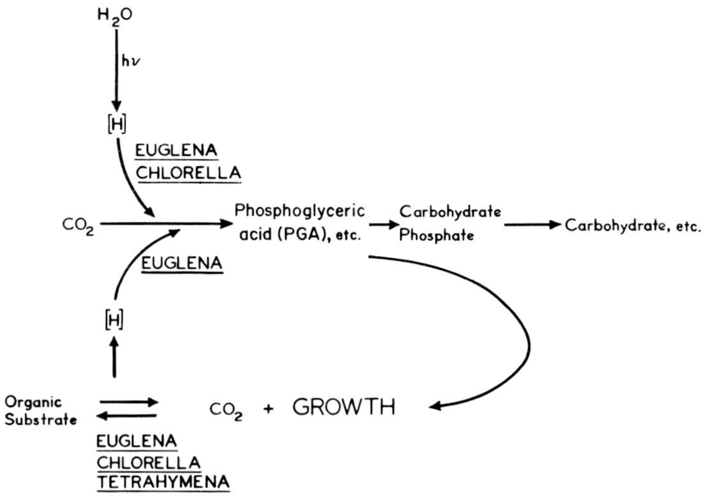


FIGURE 4.4 Diagrammatic representation of possible enzymatic interrelationships for *Euglena, Chlorella,* and *Tetrahymena* (from Lynch and Calvin, 1953).

ucts (malic, citric, aspartic, and glutamic acids) formed, but many of the phosphorylated compounds typical of $CO_2$ fixation in the light were observed. It became evident from their experiments that the coupling of energy made available by respiration and/or fermentation, with $CO_2$ reduction, existed in *Euglena* to an extent not found in any other plant or animal cell. Further proof of this close connection between dark and photosynthetic metabolic routes in *Euglena* were seen in the behavior of labeled acetate. Labeled acetate in *Euglena* in the dark found its way not only into tricarboxylic acid intermediates but into all of the compounds usually found as intermediates of photosynthetic $CO_2$ assimilation. Assuming then that the mechanism for $CO_2$ incorporation in the dark includes metabolic pathways very similar to those used in the light, they represent the relationship between *Euglena* and the plant (*Chlorella*) and the animal (*Tetrahymena*). Such relationships are indicated in Figure 4.4. Therefore, Lynch and Calvin (1953) speculated that *Euglena* represents an organism in which the separation of the plant and animal kingdoms took place. They further suggested that the evolution toward the animal kindgom involved the loss of not only the photochemical apparatus, the chloroplast, but also the ability to reduce $CO_2$ by the path of carbon dioxide fixation.

The uniqueness of *Euglena,* then, is that it has the properties of both a plant and an animal cell, and its behavior and chemistry depend upon its chemical and physical environment, particularly light. In Chapters V and VII a number of experiments with *Euglena* are described in an attempt to relate behavior, for example, phototaxis and photosynthesis, to the molecular biology of an organism.

In order to understand these experiments, let us first examine the structures

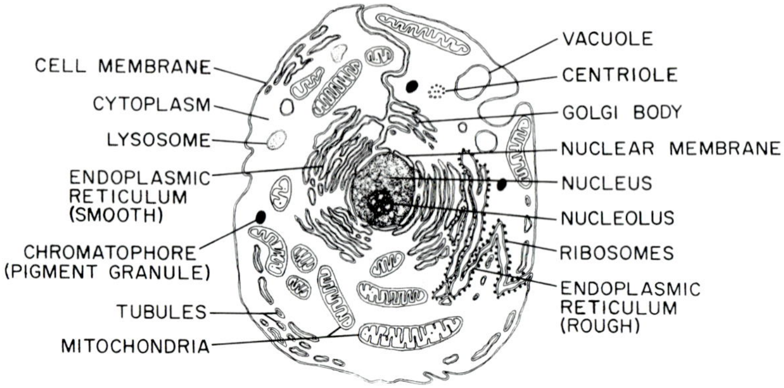


FIGURE 4.5 Structure of a generalized vertebrate eukaryote cell.

of a cell (Figure 4.5), the cell membrane, and then the membranes of its cellular organelles.

## Cell Structure

### THE CELL MEMBRANE

Since membranes are so important to the integrity of the cell and to the life processes, let us briefly examine the structure of the cell membrane before discussing in more detail specific membranes of the cellular organelles; for all photoreceptors may have evolved from the cell membrane.

The cell membrane separates the external environment from the internal environment. In order to allow for the differential diffusion of ions and the exchange of gases it has selective properties. The cell membrane was at one time envisaged as a passive barrier to diffusion and permeability. It is now thought to play an active role in chemical transport, energy transduction, and information transfer to and from the cell.

The cell membrane is described as a bilayer of lipids and proteins. The lipids are sterols, such as cholesterol, and phospholipids. Every type of membrane has a unique set of proteins to account for its function. Its molecular structure as schematized by Danielli and Davson (1935), is depicted in Figure 4.6 and is referred to as the unit membrane. They assumed that a functional protein layer was adsorbed over the membrane surfaces. Recent studies by electron microscopy and x-ray diffraction indicate that the membrane is of the order of 80 Å in thickness (Vanderkooi and Green, 1971; Robertson, 1966; Finean, 1961), each layer of the bilayer being about 40 Å.

Interesting models for the cell membrane are lipid films, which are made with phospholipids mixed with water. The lipid molecules form bimolecular

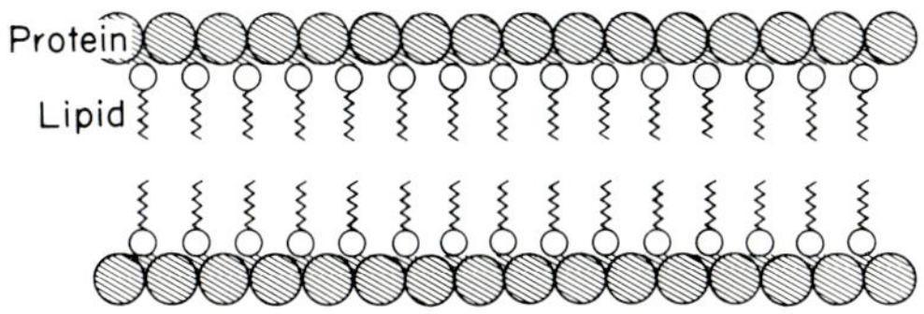


FIGURE 4.6 Schematized structure of a cell membrane as visualized by Danielli and Davson (1935).

layers attached by their nonpolar interfaces, while the polar endings are separated by layers of water. More complex models of membrane structure can be made by incorporating protein into such myelin figures. These structures have been studied by electron microscopy and x-ray diffraction and were one of the bases of the Danielli–Davson (1935) model for the cell membrane. Myelin figures are seen in a variety of cells (Figure 4.7). The myelin sheath on the other hand is a lipoprotein membrane that surrounds the axon of the

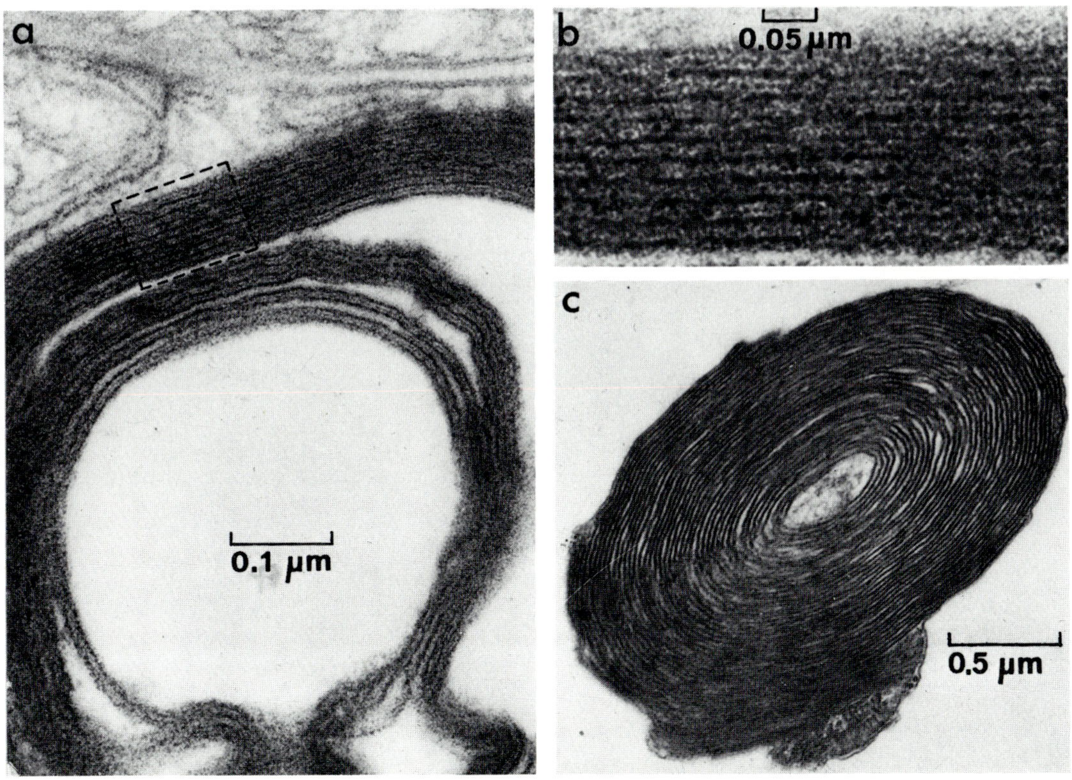


FIGURE 4.7 Myelin structures, electron micrographs (a,b,c).

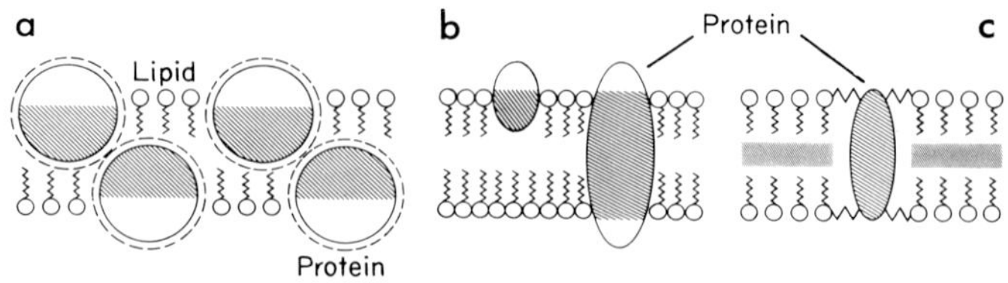


FIGURE 4.8 Possible molecular structures for a cell membrane as modified from (a) Vanderkooi and Green, 1971, (b) Singer and Nicholson, 1972, and (c) Lenard and Singer, 1966.

nerve fiber. Accordingly, the myelin sheath is formed by bimolecular layers of lipid with concentric layers of protein. Many modifications of the cell membrane have been proposed. Vanderkooi and Green (1971) suggested that some membrane proteins might not be electrostatic, but bimodal, that is, their protein molecules would possess both polar and nonpolar groups like the phospholipids. Proteins in a globular configuration would fit directly into a lipid bilayer, their hydrocarbon chains and their polar groups bonded to lipid heads (Figure 4.8a). Singer (Singer and Nicolson, 1972; Lenard and Singer, 1966) has proposed a fluid mosaic membrane model (Figure 4.8b,c). In his model the lipids are in a fluid rather than crystalline state. Oldfield (1973), however, indicates that it is possible for the membranes to contain some of their lipids in a crystalline state. If so, the cell membrane should exhibit properties of a liquid crystal, for it is temperature-dependent and concentration-dependent for its stability (Wolken, 1966a). For a recent review of the cell membrane structure reference should be made to Bretscher (1973).

The cell membrane structure then, as revealed by electron microscopy and x-ray diffraction, is a bilayer. The lipids which are in a relatively high concentration have the unique property of forming uniform (mono- or bimolecular) layers. This is due to the presence of hydrophilic or water-soluble groups at one end of the molecule and hydrophobic or fat-soluble groups at the other end of the molecule. The differences in solubilities between the water-soluble component and the water-insoluble lipid can serve to separate one part of a reaction from another.

## Cellular Organelles and Membranes

The living cell is a dynamic system concerned with the processes of energy transfer and conversion and processes vital to growth, maintenance, and reproduction. Depending upon their environment and function, cells vary in shape, size, and internal organization. They contain a variety of specialized organelles such as nucleus, mitochondria, endoplasmic reticulum, ribosomes, Golgi apparatus, chromatophores, flagella and cilia for movement, photoreceptors for photosynthesis and for visual excitation, as well as other receptors that respond to environmental stimuli. Let us now examine the structure of a few of these organelles in *Euglena* (Figure 4.2).

*Euglena* is a eukaryote and possesses a complete complement of cellular organelles: a nucleus, mitochondria, endoplasmic reticulum, chloroplasts, chromatophores, Golgi, and other structures. The nucleus is a large spherical (approximately 5 $\mu$m in diameter) membrane-enclosed body. The center of the nucleus is occupied by a large dense ovoid nucleolus and other structures referred to as *carysomes* or *endosomes* which resemble nucleoli (Leedale, 1967).

The cytoplasm contains organelles which are systems of complicated membranous structure. One such organization of membranes is the endoplasmic reticulum (Figure 4.2h) to which the ribosomes for protein synthesis are attached. This complex cytoplasmic system of vesicles and tubules was first revealed by electron microscope in cultured animal cells, which showed that its elements may be integrated in a continuous network in the endoplasm. *Euglena,* like some animal cells, seem to have both bundles of tubular elements, and associated with the endoplasmic reticulum another cluster of membranes with numerous vesicles called the *Golgi apparatus* (Figure 4.2f). A eukaryotic animal cell is schematized with its component parts as developed from electron microscopic studies in Figure 4.5.

## Mitochondria

Of the cytoplasmic organelles, the most numerous are the mitochondria (Figure 4.9). Mitochondria in *Euglena* are not particularly different in structure from the mitochondria found in all plant and animal cells. Lipids, mainly

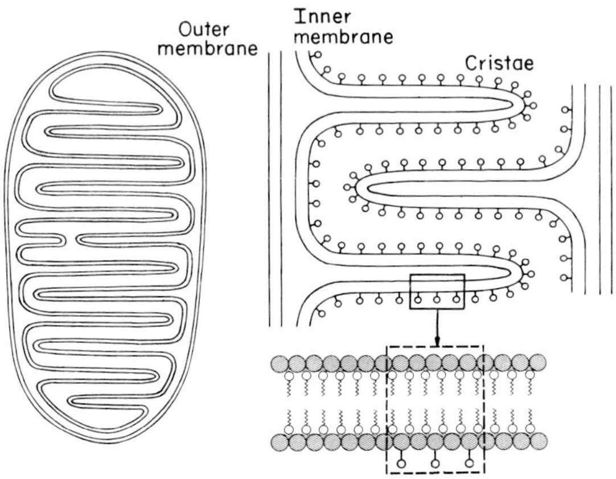


FIGURE 4.9 Schematic molecular structure of a mitochondrion (see Figure 4.2d, electron micrograph, and Figure 4.6, model for cell membrane).

phospholipids, make up 30 to 40% of the dry weight of mitochondria. Of these, chemical analysis shows that phosphatidylethanolamine, phosphatidylcholine (lecithin), and cholesterol are the major phospholipids. Although mitochondria contain the enzymatic system for the biosynthesis of fatty acids, they do not possess all the enzymes required for assembly of the complex phosphatides.

Structural studies by electron microscopy indicate that mitochondria possess two membranes, a limiting membrane and an inner membrane (Figure 4.2d). A system of ridges protrudes from the inside surface of the inner membrane. These ridges have been designated as *cristae*. It has been suggested that the oxidative enzymes of the mitochondria are built into these cristae. Mitochondria also contain DNA that is not related to the DNA of the cell nucleus. The mitochondria in animal cells apparently have enough DNA and enough associated protein-synthesizing machinery to produce only a fraction of the structural protein and enzymes needed by these organelles in order to function. Nonetheless, all the necessary machinery is there: DNA, messenger RNA, and special mitochondrial ribosomes. The presence of DNA associated with protein synthesis implies that the mitochondria have a functional genetic system.

## Cilia and Flagella

Cilia and flagella are similar structures. Their motion propels the cell through its medium, or if the cell is fixed in place the cilia move things past it. In *Euglena* the elongated flagellum is of the order of 30 $\mu$m long. In some organisms, there are more than one flagellum. The flagella arise from two "roots" at the bottom of the gullet. In cross section, the flagellum is 0.25 to 0.40 $\mu$m in diameter. In sections, the flagella appear to consist of a number of elementary filaments (axonemata) embedded in a matrix and covered by a membrane. The elementary filaments number 11, of which nine (paired microtubules) are peripherally located while the other two are found in the center of the flagellum. The pattern is known as the "9 + 2 array" and is associated with all motile cells. This arrangement is observed for a variety of plant and animal cells with flagella or cilia, from bacteria to the sperm tails of man (Pitelka, 1963, 1969).

Microtubules from any kind of eukaryotic flagella and cilia are composed of a protein called *tubulin*. Among prokaryotes the flagella or cilia are simpler; they are small, single-stranded, and consist of a closely related protein, *flagellin*. In the tissue cells of higher animals some flagella and cilia have been modified to serve sensory functions.

At the base of every eukaryotic flagellum and cilium is a distinct microtubular structure: the *basal body*. The architecture of the basal body is identical with that of the centriole, a structure found at opposite poles of the eukaryotic cell nucleus. Centrioles come into particular prominence during mitosis, the process by which the eukaryotic cells divide. Centrioles are found in nearly all animal cells, in the cells of many eukaryotic algae in most higher plants, but not in certain fungi.

The structural array of the basal body and the centriole is "9 + 0"; the central pair of microtubules is absent. In cells that possess mitotic centrioles, the centrioles left over from earlier cell divisions often grow projections that become flagella or cilia as the new cell differentiates. Thus, not only are basal bodies and centrioles identical in structural pattern, but also centrioles can become basal bodies. Moreover, the mitotic spindle, the structure that lies between the centrioles during cell division, is an array of microtubules composed of tubulin.

### Chloroplasts

The chloroplasts are differentiated cell organelles, the photoreceptors that contain the pigment chlorophyll for photosynthesis. In photosynthetic bacteria these organelles are described as *chromatophores,* in algae as *plastids* (Figures 4.2e and 5.5), and in higher plants as *grana*-containing chloroplasts, referred to as thylakoids (Figure 5.6). The chloroplast in higher plants is an ellipsoid body which is about 5 to 10 μm in diameter and length. The plastids in algae are of various shapes, but generally they are about 1 μm in diameter and from 5 to 10 μm in length. In different species and varieties of *Euglena* they may number from one to more than 20. I will refer to all of these structures as chloroplasts.

The *pyrenoid* is a differentiated region of the chloroplast. It is found in most green algae but not in higher plants. It is an organelle concerned with starch synthesis and, in some algae, with lipid storage. In the elongated chloroplasts of active *Euglena,* it is frequently found as a dense central region in which the lamellae seem to be held tightly together. The *pyrenoid* is continuous with the other chloroplast material, but because of its tightly packed structure, it appears as a condensation of the interlamellar material. The pyrenoid often protrudes at the surface of the chloroplast and the protrusion is frequently surrounded by a large vacuole (Figure 4.2g).

The chloroplasts, aside from their obvious relationship to photosynthesis, are of considerable importance in the study of hereditary mechanisms, for they possess a genetic system relatively independent from the nucleus of the cell (Gibor and Granick, 1964).

The structural and biochemical studies just alluded to indicate that the internal structure of the cell is a complex system of macromolecules which are differentiated and organized into organelles for function. These organelles consist of complicated internal membranes whose molecular structure is best represented by a bilayer of lipid and protein. Such a structure provides large surface areas for the orientation of receptor molecules and for the close interaction of substrate molecules and sites for enzymatic reactions.

We must now ask how structure controls the function of a given organelle. For example: How does the chloroplast structure at a molecular level determine its function as a light-trapping organelle in the photosynthesizing cell? In an attempt to answer this question we will now take up photosynthesis and the structure of the chloroplast.

# V
# Photosynthesis

Photosynthetic organisms have existed on earth for at least $2 \times 10^9$ years. The total amount of organic compounds formed each year by photosynthesis is about $1.6 \times 10^{17}$ gm or 100 billion tons. The total mass of organic material produced by green plants during the biological history of the earth has been estimated to be about $6 \times 10^{25}$ gm. This is an enormous weight which corresponds to 0.01, the mass of the entire earth ($6 \times 10^{27}$ gm).

A biochemical operation of this magnitude would quickly deplete the earth's atmosphere of its carbon dioxide. The fact that our atmosphere still contains $CO_2$ means that it is being returned to the atmosphere by equally large-scale processes. That is, the rate of $CO_2$ consumption in photosynthesis is just about balanced by the rate of restoration. Respiration by plants and animals is responsible for this restoration of atmospheric $CO_2$ in accordance with the equation

$$\underset{\substack{\text{Organic}\\\text{compounds}}}{(CH_2O)} + O_2 \rightarrow CO_2 + H_2O$$

supplemented by the decay of organic matter and the burning of fuels. As a result the average surface concentration of $CO_2$ in the air, about 0.03%

(Table 2.2), has remained practically constant for thousands of years. The set of processes maintaining the $CO_2$ balance, known as the *carbon cycle,* is important not only in maintaining a constant chemical environment in the atmosphere, but also in regulating the surface temperature of the earth.

Photosynthesis is also important in regulating the oxygen content of the atmosphere. It has been estimated that the current rate of photosynthesis is such to produce the entire oxygen content of the atmosphere, about 20%. As in the case of $CO_2$, the concentration of atmospheric oxygen is also believed to have been maintained at its present level for a long period of time as a result of a state of balance in the *oxygen cycle.*

Many scientists believe that there was practically no oxygen in the atmosphere until the first photosynthesizing plants evolved. If so, the emergence of photosynthesis has played a crucial role in the evolution of other forms of life.

The beginnings of our understanding of photosynthesis go back to the 1770's, when Lavoisier (1774) in France determined the composition of air, and Joseph Priestley in England discovered oxygen. Priestley (1772) then showed that oxygen was produced by green plants and algae. Jan Ingen-Housz, the Dutch physician, observed in 1779 that in the process of plant respiration, green plants at night or in the dark gave off a "dangerous air," carbon dioxide, which was purified by sunlight. Jean Senebier, a Swiss clergyman, in 1782 was able to show that the carbon dioxide was absorbed by the leaves of green plants when exposed to sunlight and that in the process oxygen was evolved. In 1798 Ingen-Housz published his theory that carbon, which was recognized by that time as important in the composition of organic molecules, was derived from carbon dioxide during photosynthesis. Then, in 1804, another Swiss scientist, Nicholas de Saussure, surmised that water might play an essential role, and the picture changed to one in which light acted on both carbon dioxide and water. In 1818, Pelletier and Caventou identified chlorophyll as the green pigment of plant tissue, and by 1837 chlorophyll was shown to be the chloroplast pigment. However, the crux of the phenomenon of photosynthesis was visualized in 1845 by Robert Mayer, a German physicist and physician, who pointed out that the photosynthetic process was the conversion of light energy to chemical energy. Once this point of view was recognized, meaningful research to understand photosynthesis had begun. By 1882, Englemann had demonstrated that the site of photosynthesis resided within the chloroplast of plant cells and that upon light absorption, oxygen was liberated. From then on the chloroplast was examined in order to discover how plants convert light energy to chemical energy.

An interesting history from the early researches in photosynthesis to the mid-1950's is developed chronologically by Rabinowitch in the three volumes of "Photosynthesis and Related Processes" (1945, 1951, 1956).

## The Photosynthetic Process

There are two distinct but related processes which occur in photosynthesis. One is the conversion of carbon dioxide to carbohydrates (indicated by Eq. 5.1) in green plants. The other is the conversion of light energy to chemical energy, which is the more difficult process to understand. Thus, photosynthesis is a series of light and dark reactions. The reduction of carbon dioxide to carbohydrates is a dark reaction and is separate from the primary light quantum conversion. This knowledge comes from the early work in 1905 of Blackman (Blackman and Matthaei, 1905). In the 1940's, Robert Hill (Hill and Scarisbrick, 1940a,b) at Cambridge University, separated the photo-induced production of molecular oxygen from the reduction of carbon dioxide by demonstrating that oxygen is evolved when chloroplasts are illuminated with a ferric ion as an oxidant in place of carbon dioxide.

We can say, then, that light energy is converted into chemical energy to form carbohydrates and oxygen. In this process the light energy is absorbed by chlorophyll and related pigments and is converted into chemical potential energy of some compounds. These compounds then react with water and produce oxygen and reducing agents as well as other cofactors containing high chemical potential energy. Finally, these reducing and energetic cofactors react with carbon dioxide and other inorganic compounds to produce organic compounds.

Melvin Calvin and his associates, in the 1940's at the University of California at Berkeley, began to study the pathway of carbon reduction during photosynthesis using $^{14}C$. They identified phosphoglyceric acid as the first stable product of carbon reduction during photosynthesis. With the development of two-dimensional paper chromatography combined with radioautography, analytical tools were available for separating minute amounts of radioactive compounds formed in the plant during photosynthesis. By these methods the intermediates in the carbon reduction cycle were found to be sugar phosphates.

Calvin confirmed Blackman's hypothesis that light was necessary for only two processes: (1) to produce ribulose diphosphate, the acceptor of $CO_2$, by phosphorylation of ribose monophosphate, and (2) to permit the reduction of carboxyl groups of phosphoglyceric acid (PGA) in the aldehyde group by the intermediate 1,3-diphosphoglyceric acid. In these reactions the donor of the phosphoryl group is ATP (adenosine triphosphate). For a summary of these studies, refer to Bassham and Calvin (1957).

## Two Photosystems in Photosynthesis

The first experimental observations which indicated that the photosynthetic process was not a simple photoreception sensitized by chlorophyll were made by Emerson and Lewis (1943). They observed that the quantum yield in photosynthesis was reasonably constant between 500 and 680 nm, but dropped dramatically beyond 680 nm. Since chlorophyll a is the major light absorber in this region of the spectrum (Figures 5.1 and 3.15), it seemed that light absorption by chlorophyll a alone was not sufficient for photosynthesis to proceed. Therefore, it was assumed that in the living cell there are forms of chlorophyll a, chlorophyll $a_1$ and chlorophyll $a_2$, which differ in the way they are aggregated or associated with proteins, lipids, and accessory pigments (Govindjee and Rabinowitch, 1960; Trosper and Sauer, 1968). To understand these experimental results, known as the Emerson "enhancement," two pigment photosystems for photosynthesis were found to be necessary (Duysens, 1964).

This phenomenon, termed the "red drop," is interpreted at present as due to a special form of "long wavelength absorbing" chlorophyll belonging to photosystem I. Emerson (1956), using *Chlorella* and monochromatic light, observed that the low efficiency of photosynthesis beyond 680 nm could be considerably improved by simultaneous illumination with a shorter wavelength (blue

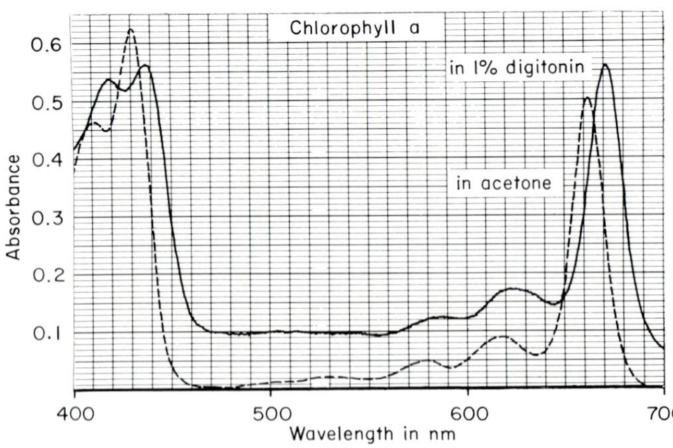


FIGURE 5.1 Absorption spectra of purified chlorophyll a (------) in acetone; chlorophyll a (——) in 1% digitonin.

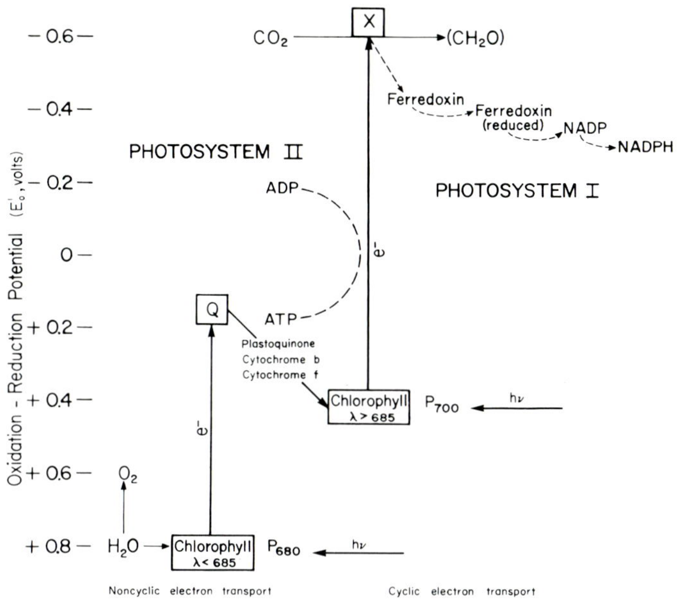


FIGURE 5.2 Scheme for the two pigment photosystems in photosynthesis, photosystem I and photosystem II.

light, 480 nm). Thus, the low efficiency of absorption of the far-red (beyond 680 nm) would require another pigment-complex absorbing below 680 nm, which has been designated as photosystem II (Emerson *et al.*, 1957).

One of the schemes to illustrate this complex process is illustrated in Figure 5.2. In examining this scheme, water is seen to serve as the electron donor in a photoreaction promoted by a chlorophyll complex, referred to as photosystem II. The electron acceptor is an unknown compound (Q). Its redox potential is around 0.0 to +0.18 *V*. The reducer (Q) transfers its electron through a series of compounds, including plastoquinone and several cytochromes, and finally to photosystem I. This complex, containing mainly chlorophyll a absorbing at longer wavelengths, is called $P_{700}$, so named because it behaves differently from a typical chlorophyll, and it has a redox potential of +0.4 *V*. Thus, an electron moving between the two photosystems loses the equivalent to about 0.2 to 0.4 *V*. This is enough energy to promote, from ADP (adenosine diphosphate) and inorganic phosphate, the formation of one to two ATP (adenosine triphosphate) molecules. In photosystem I the electron acceptor is an unknown compound (X). The light absorbed by photosystem I is then used to reduce ferredoxin. The final product of this electron transport scheme is NADP (nicotinamide adenine dinucleotide phosphate).

In summary, the two photochemical oxidation–reduction reactions are driven by two pigment systems, photosystem I and photosystem II. The reduced electron acceptor of photosystem I reduces $NADP^+$ via ferredoxin; the oxidized electron donors of photosystem II produce oxygen by the oxidation of water. Photosystem II contains chlorophyll a absorbing at relatively short wavelengths and a relatively large part of the so-called accessory pigments, chlorophyll b (in green algae and higher plants) and phycobilins (in red and blue–green algae). Therefore, photosystem I and photosystem II provide the high energy phosphates, reduced NADP and ATP, needed for the synthesis of carbohydrates and proteins from $CO_2$ and water.

## Comparative Biochemistry of Photosynthesis

Van Niel (1935, 1941, 1943) pointed out that the purple and the green photosynthetic bacteria represent remnants of an originally much wider class of organisms having a photosynthetic system simpler than that of green plants. The metabolism of purple bacteria could serve as an example for the kind of photochemistry which may have preceded that of the green plants on the evolutionary time scale. These organisms cannot evolve oxygen though many tolerate oxygen. Because they require energy-rich hydrogen donors to reduce $CO_2$ they do not contribute much to the store of free energy in the living world. Compared with the green plants and other aerobic organisms, purple bacteria seem primitive. But bacteriochlorophyll, responsible for the utilization of light energy, is similar to the chlorophyll of green plants (Figure 3.2).

Let us examine briefly what has been learned from studies on the comparative biochemistry of photosynthesis. Organisms have been classified as: *autotrophic,* those which obtain their energy for growth from sources other than organic molecules; *chemoautotrophic,* those which obtain their energy from oxidizable inorganic chemicals; and *photoautotrophic,* those which obtain their energy from light. The photoautotrophs fall into three separate groups: green plants, pigmented sulfur bacteria, and pigmented nonsulfur bacteria. Their light-driven reactions can be expressed by Eqs. 5.1–5.3.

$$\text{Green plants:}\quad CO_2 + H_2O \xrightarrow{\text{light}} \underset{\text{Organic matter}}{(CH_2O)} + O_2\uparrow \tag{5.1}$$

$$\text{Sulfur bacteria:}\quad CO_2 + H_2S \xrightarrow{\text{light}} (CH_2O) + S \tag{5.2}$$

$$\text{Nonsulfur bacteria:}\quad CO_2 + \text{succinate} \xrightarrow{\text{light}} (CH_2O) + \text{fumarate} \tag{5.3}$$

For the sulfur bacteria, $H_2S$ can be replaced by $Na_2S_2O_3$, $Na_2S_4O_7 \rightarrow H_2SO_4$, or $H_2Se$. For nonsulfur bacteria, the organic donor, succinate, can be replaced by many different organic acids that have two electrons to spare. This is schematically generalized by Eq. (5.4).

$$CO_2 + H_2A \xrightarrow{\text{light}} (CH_2O) + A \tag{5.4}$$

The characterization of the various types of photosynthesis led to a generalization by Van Niel (1941, 1949) of what is known as the "comparative biochemistry of photosynthesis." Actually the photosynthetic reaction can be schematized even further without involving carbon dioxide, for in the Hill reaction (production of $O_2$ by isolated chloroplasts in light), a quinone or ferric ion can accept the hydrogens that are activated by the light reaction. So, the general formula can be written as shown in Eq. (5.5),

$$B + H_2A \xrightarrow{\text{light}} BH_2 + A \tag{5.5}$$

or simply as a light-induced, oxidation–reduction reaction.

Many researchers attacking the mechanisms of photosynthesis hypothesized that the photochemical reaction in green plant photosynthesis is a photolysis (light-induced decomposition) of water. Indeed, experimental evidence has shown that water is photolyzed in both plant and bacterial photosynthesis. Therefore, a more acceptable equation for the bacterial systems would include both water and a general hydrogen donor. It would also exclude oxygen evolution. (See Eq. 5.6.)

$$CO_2 + 2H_2A + H_2O \xrightarrow{\text{light}} (CH_2O) + 2H_2O + 2A \tag{5.6}$$

A simple formulation for the mechanism of the overall process has yet to be conceived. Some aspects, however, are reasonably well understood. The general photosynthetic light reaction can be represented as shown in Scheme 5.1.

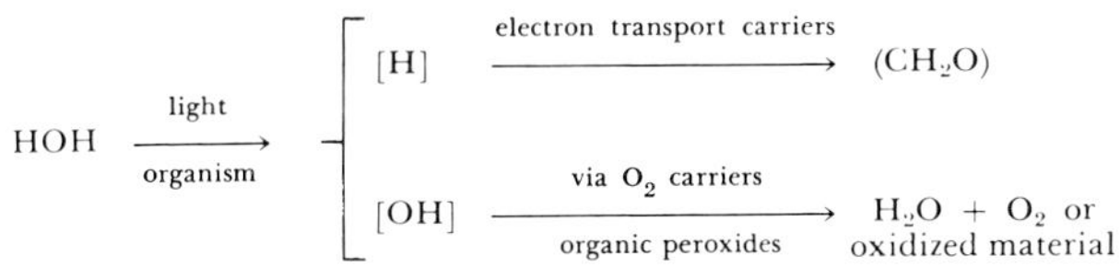


SCHEME 5.1

The photolysis of water can be regarded then as the major achievement of the chlorophyll system. Everything else follows via reaction patterns which have become well-known in general biochemistry.

## Enzymes

If the difference between the anaerobic purple bacteria and the unicellular aerobic algae signifies an evolutionary step forward—wherein does it lie? Why

Table 5.1 Cytochromes in Mitochondria and Chloroplasts[a]

| *Mitochondria and chloroplasts* | *Cytochromes* | | | |
|---|---|---|---|---|
| Yeast mitochondria | a | b | c | |
| Etiolated barley plastids (yellow) | a | $b_6$ | f | (modified c) |
| Green plant chloroplasts | — | $b_6$ | f | (modified c) |

[a] Refer to Hill and Bendall (1960).

do algae release free oxygen from water and bacteria do not? According to Gaffron (1965), one answer could be the acquisition of new enzymes or enzyme systems.

In a quest for such answers, researchers have looked for the participation of a specific enzyme system in photosynthesis for capturing electrons. Hill and Bendall (1960) experimentally demonstrated that a photosynthetic enzyme, a cytochrome, or cytochrome system, is coupled to the chlorophyll–protein complex of the chloroplast. The cytochromes then function as electron carriers during the initial reactions of the photochemical process. As Hill pointed out mitochondria and chloroplasts show some resemblance with respect to the structure-bound cytochromes. Accordingly Table 5.1 seems to indicate that the chloroplasts belong to the same category as the mitochondria, with the a cytochrome of the system being replaced by the chlorophylls. There is also the possibility that the photochemistry initiated by light absorption in photosynthesis involves the cytochrome directly. For example, in *Euglena,* two spectrally different cytochromes have been isolated. One, from the light-grown photosynthetic cells, is designated as cytochrome-552 (a cytochrome c type). In its reduced state its absorption peaks are at 552, 523, and 416 nm (Figure 3.5). The cytochrome isolated from dark-grown *Euglena* is referred to as cytochrome-556, for in the reduced state it has absorption peaks in the visible at 556, 525, and 412 nm. Its spectrum is close to that of cytochrome f which is associated with higher plants. The ratio of chlorophyll a to cytochrome-552 is of the order of 300:1.

There are similarities in the photosynthetic systems between blue–green algae and higher plants. This has led to the suggestion that the blue–green algae became a symbiont of an early cell and thus the chloroplast ferredoxins are derived from a common ancestor.

Arnon (1965) has shown that ferredoxin is also a key photochemical component of the process of photosynthesis by chloroplasts. Unlike the cytochromes, which exhibit well-defined absorption peaks in the reduced state, ferredoxins have distinct absorption peaks in the oxidized state (Figure 3.11). According to Arnon, the photoreduction of ferredoxin is coupled with oxygen evolution and with photosynthetic phosphorylation. It is interesting to note that the number of chlorophyll to ferredoxin molecules was also found to be about 300:1.

Crane (1959) suggested that plastoquinone is an important intermediate in photosynthesis. Plastoquinone was found to occur in green, red, brown, and blue–green algae and in higher plant chloroplasts. Plastoquinone is presumably part of the chloroplast membrane. It is not found in photosynthetic bacteria. The ratio of the total various quinones to chlorophyll was found to be about 150:1 (Amesz, 1973).

We can now turn to the chloroplast and examine its composition and structure and see if we can arrive at some basis for function.

## The Chloroplast

The key to understanding photosynthesis lies within the chloroplast chemistry and molecular structure. If we can unravel this chemistry and molecular structure then some insights into how the chloroplast functions will become meaningful.

Chloroplasts isolated from a variety of plant species show that the major constituents are protein (35 to 55%), lipids (18 to 37%), chlorophylls (5 to 10%), carotenoids (2%), inorganic matter (5 to 8% on dry weight basis), and the nucleic acids, RNA and DNA ($<1$ to 3%).

### Chlorophylls and Carotenoids

All chloroplasts except for bacterial chromatophores contain chlorophyll a, and all higher plants and green algae contain chlorophyll b. The chlorophylls of *Euglena* chloroplasts consist of 85% chlorophyll a and 15% chlorophyll b (Wolken, 1967). To identify and determine the concentration of chlorophyll and carotenoids in a single chloroplast, the microspectrophotometer was used (Wolken *et al.*, 1968). An absorption spectrum for the *Euglena* chloroplast is shown in Figure 5.3. It is seen that this chloroplast has major absorption peaks near 430 and 675 nm, with other absorption peaks (seen as shoulders) near 485, 580, and 630 nm. These absorption peaks are mainly due to chlorophyll a (Figures 5.1 and 5.3). The absorption peak near 485 nm is due to the carotenoids. In the ultraviolet the major absorption occurs around 265 nm and other absorption peaks, near 290, 340, and 390 nm, are associated with the proteins and lipids of the chloroplast. The number of chlorophyll molecules per chloroplast, obtained from extraction using the extinction coefficients for

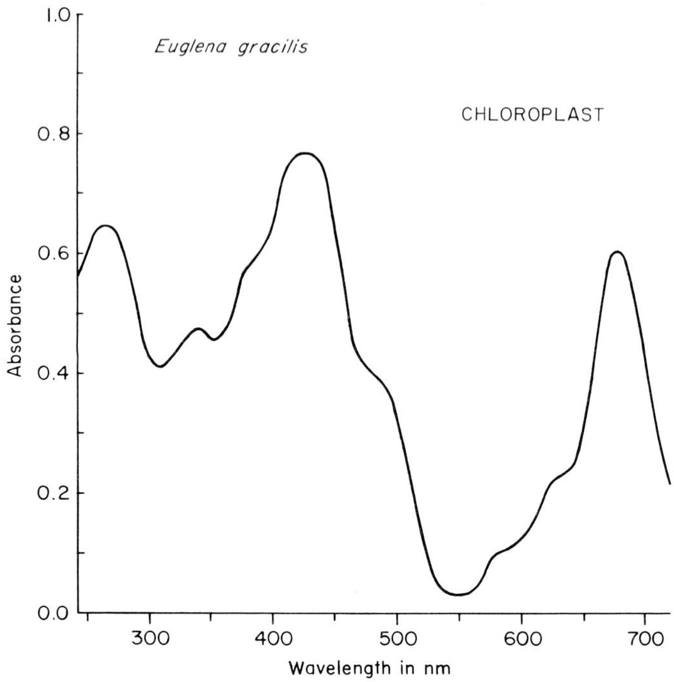


FIGURE 5.3 Absorption spectrum of *Euglena gracilis* chloroplast *in situ* (obtained by microspectrophotometry).

chlorophyll a and chlorophyll b, was found to be $1.02 \times 10^9$; and, from the absorption spectrum of the chloroplast by microspectrophotometry, the number was estimated to be $1.34 \times 10^9$. It is of interest to note that this concentration of chlorophyll molecules per chloroplast is similar to that found for a variety of photosynthetic plants (Table 5.2).

TABLE 5.2 CHLOROPHYLL CONCENTRATION IN THE CHLOROPLAST[a]

| *Organism* | *Volume of chloroplast (ml)* | *Chlorophyll molecules per chloroplast* | *Concentration of chlorophyll (moles/liter)* |
|---|---|---|---|
| *Elodea densa* | $2.8 \times 10^{-11}$ | $1.7 \times 10^9$ | 0.100 |
| *Mnium* | $4.1 \times 10^{-11}$ | $1.6 \times 10^9$ | 0.065 |
| *Euglena gracilis* | $6.6 \times 10^{-11}$ | $1.02 \times 10^9$ | 0.025 |
| *Poteriochromonas stipitata* | $1.1 \times 10^{-11}$ | $0.11 \times 10^9$ | 0.016 |

[a] From Wolken (1967, p. 67).

We must now ask what function the carotenoids play in the primary photosynthetic reaction. The photosynthetic bacterial systems are unique in being anaerobic and not evolving oxygen. The bacterium *Rhodopseudomonas spheroides* pigment mutants were studied by Stanier (1959). The blue–green mutant lacks the purple wild-type carotenoids. This mutant has been extensively studied and found to grow well photosynthetically without these carotenoids that were considered necessary. However, when the mutant cultures were exposed to light and oxygen very rapid death and bacteriochlorophyll decomposition occurred. This was not found with the purple normal wild type. But the blue–green mutant and the normal purple wild type under anaerobic conditions were found to be comparable with respect to their bacteriochlorophyll concentration and their photosynthetic rates. These experiments demonstrated that the light-trapping process was not dependent on the carotenoids if oxygen was absent and indicated that bacteriochlorophyll, and hence chlorophyll, was the only pigment necessary for the primary light reaction.

Action spectra have shown a direct relationship between the synthetic pathways of chlorophyll and carotenoid synthesis in algae (Figure 5.4) and plants (Wolken *et al.*, 1955; Ogawa *et al.*, 1973). This is not difficult to reconcile with the lack of carotenoids in the blue–green bacterial mutant. The relationship between these two classes of pigments is probably that the $C_{20}$ phytol chain on chlorophyll is derived from precursors of $C_{40}$ and $C_{20}$ carotenoids.

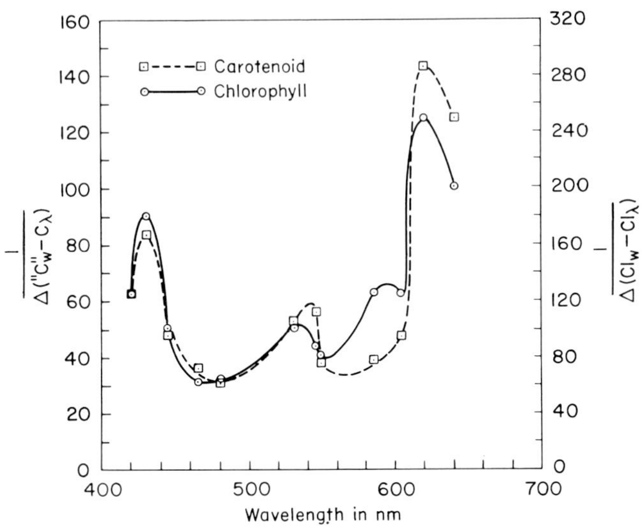


FIGURE 5.4 Reconstructed curves related to the action spectra of the substances responsible for chlorophyll and carotenoid synthesis. The left scale is used with the carotenoid curve and the right scale with the chlorophyll curve.

The mutant does have the $C_{40}$ carotenoid precursor, phytoene, which can give rise to the phytol chain via a divergent pathway. The carotenoid synthetic pathway beyond phytoene (Figure 3.8) is genetically blocked in the mutants.

Since its life and integrity are so dependent on the absence of either oxygen or light, the blue–green mutant clearly exhibits the phenomenon of photodynamic action. A thorough analysis of the mutant photosynthetic system has disclosed that the bacteriochlorophyll itself is the photosensitized compound, and that it is also photooxidized. The carotenoids in the wild-type strain somehow protect the light-sensitive bacteriochlorophyll from self-destructive oxidation. These experimental results would indicate that the carotenoids are not essential, except in catalytic amounts, for green plant photosynthesis, but are necessary for protection against photodynamic destruction.

Other investigations give support to the view that the carotenoids participate directly in the primary act of photosynthesis. One hypothesis is that the carotenoids combine with the oxidized portion of the photosynthetically split water molecule by forming epoxides across the numerous double bonds, with one or more epoxide groups resulting.

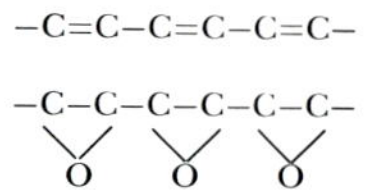

Such epoxide formation has been demonstrated in leaves and *in vitro*. Therefore the carotenoids can act as a filter to screen the light and prevent photooxidation (photodestruction) of chlorophyll at high light intensities or could function as an accessory pigment molecule in energy transfer (Lundegardh, 1967).

## Nucleic Acids

The presence of DNA in chloroplasts indicates that they may possess an autonomous genetic system similar to that of the nucleus. There is, however, no evidence concerning the function of the chloroplast DNA. The amount of DNA in chloroplasts is about the same as in *Escherichia coli* and could possess sufficient genetic information for a large number of physiological functions. Chloroplast DNA has a nucleotide composition sharply different from that of the nuclear DNA (Brawerman and Eisenstadt, 1964). The purified chloroplast DNA contains a major component with a density of 1.684 gm/cm$^3$. The density of the nuclear DNA is 1.708 gm/cm$^3$. Ray and Hanawalt (1964) have shown that the 1.684 component is absent from *Euglena* which have lost their ability to photosynthesize. Edelman *et al.* (1965) found three species of DNA—a DNA with a density of 1.707 gm/cm$^3$ which probably comes from the nucleus, and two DNA satellites with densities 1.686 gm/cm$^3$ and 1.691 gm/cm$^3$.

The chloroplasts also contain messenger RNA in sufficient quantity for maximum activity of their protein-synthesizing system. A mechanism could be postulated by assuming that the messenger RNA molecules for the structural proteins of the chloroplast are generated *in situ* by the chloroplast DNA. The replication and turnover of chloroplast DNA of *Euglena* have been shown to be more rapid than those of nuclear DNA (Manning and Richards, 1972). This information raises many interesting questions regarding organelle origins and evolution. These data and other considerations related to the origin of chloroplasts and evolution are discussed by Cohen (1970, 1973), Sager (1972), and Margulis (1970).

## Lipids

Lipids function universally as structural components of cytoplasm and intracellular membranes. Lipids and phospholipids have been isolated from all cell fractions that are regarded as membranes. When the lipids are extracted from chloroplasts their photochemistry is inhibited, suggesting that the lipids may be more than a structural part of the chloroplasts.

We have already indicated that *Euglena* is different in structure and chemistry depending upon whether it is grown in light or in darkness (Wolken, 1967). These light ↔ dark adaptations bring about changes in the chemistry of the cell, as illustrated by the analysis of lipids, proteins, and carbohydrates in Table 5.3. An interesting point in this analysis of *Euglena* is the change in total lipid concentration from 13.7% for light-grown to 9.1% for dark-grown, and to even lower concentrations of 7.8% for the heat-bleached (HB) and 6.4% for the $Mg^{2+}$-deficient mutants. Is this change in lipids related to the synthesis of chlorophyll and to the development of the chloroplast?

The major phospholipid of photosynthetic *Euglena* is phosphatidylserine, whereas dark-grown cells contain primarily phosphatidylethanolamine. Studies of the fatty acids and lipids of *Euglena* indicate at least 51 fatty acids (Korn, 1964). This analysis shows that *Euglena* is capable of fatty acid biosynthesis characteristic of plant cells when light-grown and of animal cells when dark-grown. Aaronson and Baker (1961) showed that the quantity of nonsaponifiable lipids was 13.2% when *Euglena* were light-grown and 22% when they were dark-grown. Rosenberg and Pecker (1963) demonstrated that sulfolipid accumulated before measurable amounts of chlorophyll accumulated when etiolated cells were subjected to constant illumination. The $C_{13}$ and $C_{14}$ fatty acids of the well-nourished etiolated *Euglena* disappeared, and the photosynthetic cells synthesized unsaturated fatty acids, mainly of the $C_{16}$ and $C_{18}$ series. It is known from the studies of Benson (1964) that mono- and digal-

TABLE 5.3 SYNTHESIS OF VARIOUS PRODUCTS BY LIGHT-GROWN AND DARK-GROWN *Euglena* AND *Chlorella*[a]

| | *E. gracilis* v. *bacillaris* | | | | *Chlorella* | |
|---|---|---|---|---|---|---|
| *Product* | *Light-grown* (%) | *Dark-grown* (%) | *$Mg^{2+}$ deficient light-grown* (%) | *Heat-bleached*[b] *light-grown* (%) | *Full light* (%) | *Weak light* (%) |
| Protein | 69.3 | 36.3 | 73.9 | 48.1 | 70.0 | 46.0 |
| Carbohydrates | 7.0 | 46.6 | 7.6 | 36.6 | 8.0 | 35.5 |
| Lipids | 13.7 | 9.1 | 6.4 | 7.8 | 12.0 | 10.0 |
| α-Linolenic acid[c] | 7.8 | 0.6 | — | — | — | — |
| Ash | 10.0 | 8.0 | 12.1 | 7.5 | 10.0 | 9.0 |
| Water | 77.1 | 71.4 | 90.4 | 77.7 | — | — |

[a] Taken in part from Wolken (1967, p. 102).
[b] Heat-bleached at 37°C(33°–40°C).
[c] Erwin and Bloch (1963).

actosyl glycerides are the major chloroplast lipids. The galactosyl glycerides contain unusually high concentrations of polyunsaturated fatty acids and account for the major portion of these acids in chloroplasts.

Erwin and Bloch (1962, 1963) investigated the α-linolenic acid content from dark-grown and light-grown *Euglena gracilis*. They found that the α-linolenic acid increased from 0.6% in the dark-grown cells to 7.8% when the cells were grown in the light for 28 hours (Table 5.3). Upon analysis of the cells, 85% of the α-linolenic acid was found in the chloroplast fraction. Nonphotosynthetic cells were found to contain little α-linolenic acid. Therefore, the galactolipids as well as the unsaturated fatty acids are depleted when the organism is grown in the dark and are resynthesized by growth in the light (Carter *et al.*, 1961; Hulanicka *et al.*, 1964; Rosenberg, 1963). This suggested a correlation between the photosynthetic activity and the concentration of α-linolenic acid. From these analyses it was concluded that organisms which derive energy from photosynthesis and photophosphorylation require α-linolenic acid. On the other hand, γ-linolenic acid, arachidonic acid, and $C_{20}$ polynoic acids are required for organisms that depend on respiration and oxidative phosphorylation.

The galactosyl glycerides, the carriers of α-linolenate, are associated with the photosynthetic machinery in those organisms in which photosynthesis is of the advanced type, as in higher plants (Table 5.4). It appears, then, that in the evolution of photosynthetic systems biosynthesis of novel lipids as well

TABLE 5.4 THE COMPOSITION OF SPINACH CHLOROPLAST QUANTASOME[a]

| | | |
|---|---|---|
| Chlorophylls | | 230 |
| Chlorophyll a | 230 | 160 |
| Chlorophyll b | | 70 |
| Carotenoids | | 48 |
| β-Carotene | | 14 |
| Lipids | | 740 |
| Phospholipids | 164 | 116 |
| Sulfolipids | | 48 |
| Digalactosyl diglyceride | 460 | 114 |
| Monogalactosyl diglyceride | | 346 |
| Nitrogen atoms as protein | | 9380 |

[a] Taken in part from Park and Biggins (1964, p. 1010).

as a multiple of unsaturated acids occurs. The high degree of unsaturation of the fatty acids in these plant-specific lipids may be an important clue to the specific function of these molecules in photosynthetic electron transport (Bloch, 1965).

Another point to make about the lipids is that temperature variation alone, within the physiological range, can alter the physical state of fatty acids and phospholipids. Luzzati and Husson (1962) and Stoeckenius (1962) have studied the phase changes which occur in brain phospholipids *in vitro*. They found that lipids may exist in both an expanded lamellar form and a condensed hexagonal form, and that x-ray diffraction patterns indicate a condensation due to the breakdown of ordered structure as rising temperature increases kinetic activity and hence disorder. It is of interest then that the chloroplasts of *Euglena* are permanently removed at a temperature above 33°C within a generation. The result is a mutation, heat-bleached (HB) *Euglena* (Table 5.3). It seems reasonable to assume that temperature affects the synthesis of polyunsaturated acids and interferes with the chlorophyll synthesis and hence with the development of the chloroplast structure.

## The Chloroplast Structure

We now turn to the chloroplast to see how it is structured for function in photosynthesis. The chloroplasts of plant cells observed by light microscopy, particularly polarization and fluorescence microscopy, indicate considerable fine structure. Electron microscopy of chloroplasts in a variety of photosynthetic plants reveals that they consist of stacked discs or lamellae (Figures 5.5 and 5.6).

*Euglena* grown in the dark lose their ability to photosynthesize and the chloroplasts are disrupted. With continued growth in darkness the chloroplasts become no longer recognizable. Such dark-grown *Euglena* retain proplastid structures which are capable of forming new chloroplasts when they are restored to the light. *Euglena* taken from growth in the dark to the light begin, within hours, to synthesize chlorophyll, develop chloroplasts, and to photosynthesize (Table 5.5). *Euglena* permits us then to follow the structural development of the chloroplast along with chlorophyll synthesis.

A section through a single *Euglena* chloroplast is shown in Figure 5.5a and at a higher magnification in Figure 5.5b and c. The elongated chloroplast is seen to be made up of regularly spaced, electron-dense lamellae. The lamellae appear as a double membrane whose total thickness is of the order of 250 Å, and each membrane is from 50 to 100 Å in thickness. At high resolution these lamellae are not continuous membranes but consist of an array of particles (Figure 5.5c). The lamellae themselves are considered lipids, lipoproteins,

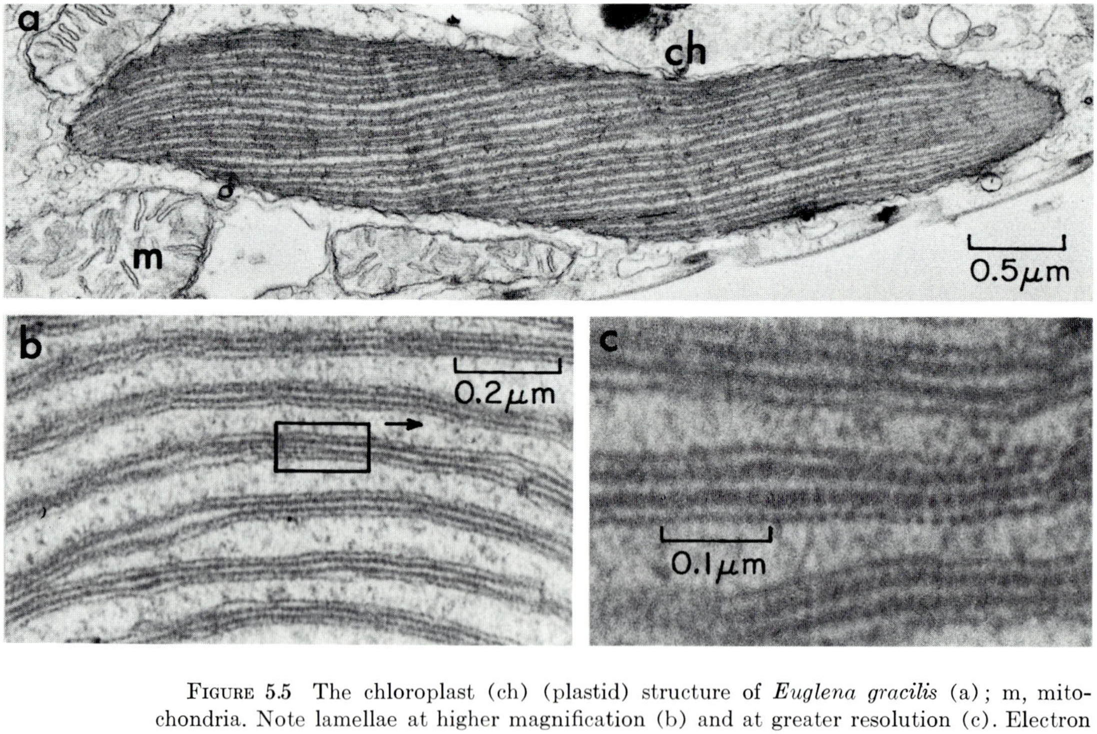


FIGURE 5.5 The chloroplast (ch) (plastid) structure of *Euglena gracilis* (a); m, mitochondria. Note lamellae at higher magnification (b) and at greater resolution (c). Electron micrographs.

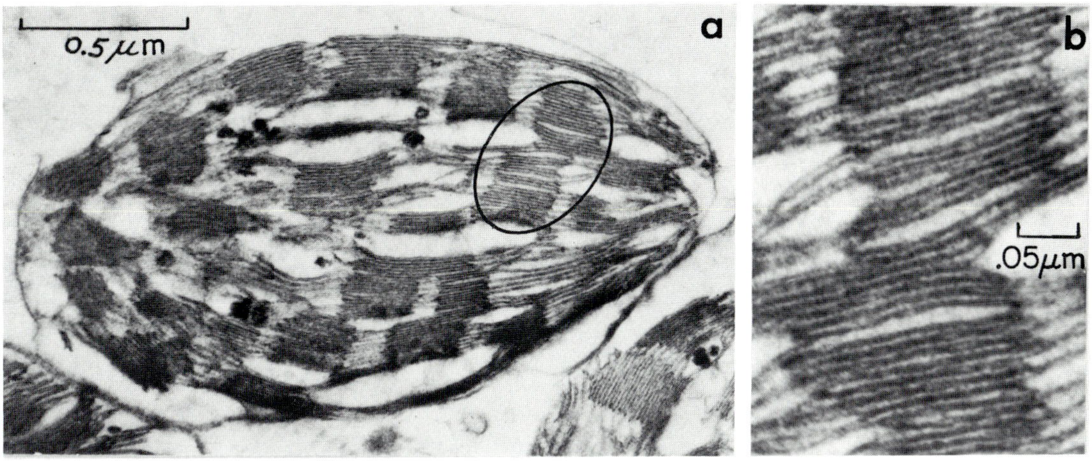


FIGURE 5.6 The chloroplast structure of *Elodea densa* (a). Note grana and enlargement in (b). Electron micrographs (courtesy of Dr. K. Mühlethaler, Eidgenossische Technische Hochschule, Zurich).

and proteins because of their affinity for the osmium tetroxide used in fixation for electron microscopy. The less dense spaces are thought to contain aqueous proteins, enzymes, and dissolved salts.

From numerous electron micrographs of *Euglena* chloroplasts, the diameter, length, thickness, and number of lamellae were measured and statistically evaluated (Table 5.6). The *Euglena* chloroplast was found to consist of an average of 21 lamellae. Each lamella is a double layer and therefore it was assumed

TABLE 5.5 PHOTOSYNTHETIC AND RESPIRATORY DATA FOR *Euglena gracilis*[a]

| *Experimental conditions* | *Chlorophyll conc.* (*mg/ml* × $10^4$) | $-CO_2/+O_2$ (*μl/hour*) | $P_Q/R_Q$[b] |
|---|---|---|---|
| Light-grown | 537.1 | 238/6.0 | 39.7 |
| Light-grown, HB mutant | <0.1 | 6.3/8.9 | 0.71 |
| Dark-grown | <0.1 | 4.5/49.2 | 0.09 |
| Dark-grown → light-grown | | | |
| 1 hour | 8.79 | 4.9/39.9 | 0.12 |
| 24 hours | 20.94 | 31.4/7.8 | 4.0 |
| 48 hours | 52.94 | 37.7/4.95 | 13.1 |
| 72 hours | 179.17 | 68.2/1.6 | 42.5 |

[a] From Wolken (1967, p. 111).

[b] $P_Q$ is the $CO_2$ uptake under photosynthetic conditions, and $R_Q$ is the $O_2$ uptake under respiratory conditions.

TABLE 5.6 DIMENSIONS IN THE CHLOROPLAST OF *Euglena gracilis*[a]

| | |
|---|---|
| Diameter | 1.23 μm (1.04–1.42)[b] |
| Length, $D$ | 6.50 μm (5.2–9.3) |
| Number of lamellae, $n$ | 21 (18–24) |
| Thickness of double layer of lamellae, $T$ | 242 Å (180–303) |
| Thickness of each lamella | 50–75 Å |
| Interlamellar space, $I$ | 374 Å (300–476) |
| Number of chlorophyll molecules/lamellar surface | $2.4 \times 10^7$ |
| Number of chlorophyll molecules | $1.02 \times 10^9$ $(0.88–1.36 \times 10^9)$[c] <br> $1.34 \times 10^9$[d] |

[a] Taken in part from Wolken and Schwertz (1953) and Wolken (1967, p. 61).

[b] Upper and lower extremes of measured values in parentheses.

[c] Calculated from chloroplast extract in solution.

[d] Calculated from a single chloroplast using the microspectrophotometer.

that there is $2n$ surfaces (42) available for the chlorophyll molecules. By dividing the chlorophyll concentration (Table 5.6) by twice the number of lamellae, there would be $2.5 \times 10^7$ molecules per lamellae. The number of chlorophyll molecules per unit area in the interfacial monolayers is the reciprocal of the area available for each molecule, that is, approximately $4 \times 10^{13}/\text{cm}^2$. Data for a variety of other plant chloroplasts indicate similar concentrations of chlorophyll in the chloroplast volume.

To establish that the chlorophyll molecules form a monomolecular layer on the surfaces of the lamellae calculations were made of the area available for the chlorophyllin (porphyrin) part of the chlorophyll molecule on the lamellar surfaces. The calculated cross-sectional area for the chlorophyllin part of the chlorophyll molecule was found to be 222 Å$^2$ for *Euglena gracilis* and 246 Å$^2$ for the chrysomonad *Poteriochromonas stipitata* chloroplasts (Wolken and Schwertz, 1953). For chloroplasts in a variety of plants, calculations of the chlorophyllin cross-sectional area in the chloroplast lamellae were found to be about 200 Å$^2$ (Elbers *et al.*, 1957). These calculations are then of the right order of magnitude for the cross-sectional area of a porphyrin molecule when spread on water–air interface.

## MOLECULAR MODEL OF THE CHLOROPLAST

On the basis of these analyses and calculations, a schematic molecular model of the chloroplast of *Euglena* was proposed (Wolken, 1959). In the

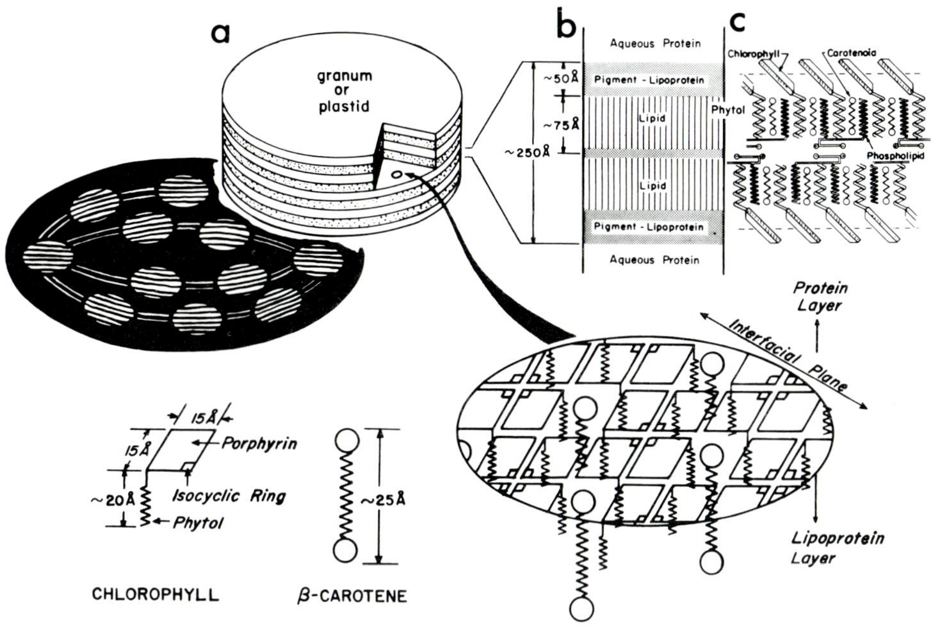


FIGURE 5.7 Schematic molecular model of the chloroplast.

model the chloroplast lamellae are lipid–protein membranes (Figure 5.7a) in which the chlorophyllin (porphyrin) is oriented as a monolayer on the protein (lipoprotein) surfaces and the phytol "tail" reaches into the lipid layer. The geometrical and analytical data, taken together, are consistent with such a model. The chloroplast lamellar network shows that four chlorophyll molecules are united to form tetrads and are oriented so that only one of the phytol tails is located at each intersection in the rectangular network (Figure 5.7). This arrangement has the advantage of leaving adequate space for at least one carotenoid molecule for every three chlorophyll molecules. Since the molecular weights of the carotenoid molecules are one-half to two-thirds of the molecular weight of the chlorophyll molecules, a weight ratio of chlorophyll to carotenoid of approximately 4:1 to 6:1 would be expected. On the other hand, the carotenoid molecules are slender linear molecules, about 5 Å in diameter, and therefore more than one molecule could conveniently fit into the 15 Å × 15 Å holes formed by the chlorophyll tetrads. From symmetry one might expect as many as four molecules per hole, but this would lead to very tight fitting which is energetically improbable. One can therefore put a lower limit on the number of chlorophyll to carotenoid molecules of roughly 1:1 and a weight ratio of 2:1.

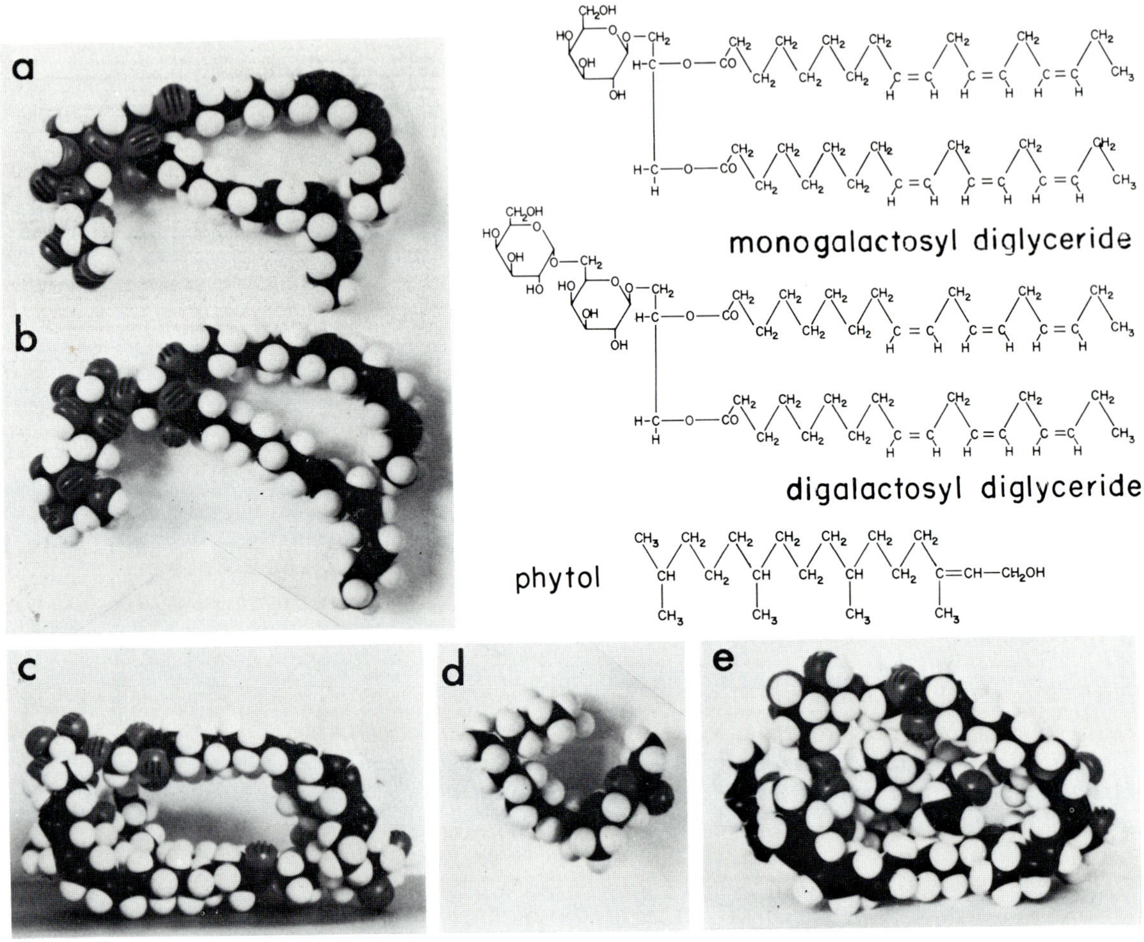


FIGURE 5.8 Molecular structures of mono- (a) and digalactosyl diglyceride (b) to form (c), and their association with the phytol of chlorophyll (d) to form (e) (structure a and b adapted from A. Rosenberg, 1967).

It will be noted in Table 5.4 that the mono- and digalactosyl diglycerides account for the major lipids in the chloroplast. The lamellae of the chloroplasts as observed with the electron microscope may be associated with the structure of the lipid molecules (Figures 5.5 and 5.6). Benson (1966) suggested that these lipids, because of their properties, can form a lipid or lipoprotein matrix for the chlorophyll monolayers (Figure 5.7). From spatial considerations the ratio of two galactosyl diglyceride molecules to one chlorophyll molecule could stabilize all the chlorophyll molecules in the monolayer. That is, there would be one phytol chain of chlorophyll for four *cis*-unsaturated acyl chains of galactosyl diglyceride (Rosenberg, 1967). Such a relationship is illustrated in Figure 5.8 and fits in with the proposed molecular model for the chlorophyll in the chloroplast lamellae (Figure 5.7).

This molecular model is consistent with the analytical data for the chloroplast. Also, such a highly ordered structure of the chloroplast, as depicted in Figure 5.7, greatly increases its surface area to provide not only for the chlorophyll and carotenoid molecules for energetic interaction, but also provides reactive sites for the enzymes.

Other molecular models for the chloroplast have been suggested (Calvin, 1959; Benson, 1966; Murakami and Packer, 1970), for there are several possible ways in which the chlorophyll molecules could be oriented in the chloroplast lamellae. If the porphyrin parts of the chlorophyll molecules lie at 0° as depicted in Figure 5.7, their greatest cross section would be available. If they are oriented at increasing angles to 45°, the cross-sectional area available would be decreased to about 100 $Å^2$. Since the chlorophyll molecules in the chloroplast are probably in a dynamic state, they would arrange themselves for maximum light absorption, that is, their greatest cross section would be made available for light capture.

From the model (Figure 5.7) and the analytical data (Table 5.6), the molecular weight of the chlorophyll–protein (chloroplastin) in the chloroplast can also be estimated. The assumption is that there is one chlorophyll molecule per protein molecule. An equation to calculate the molecular weight can be derived in a form containing only the experimental measurements:

$$M = \frac{\pi D^2 T s L n}{4P}$$

where $D$ is the length, $T$ is the thickness of the lamellae, $s$ is the density taken as 1.3 that of a protein (if a lipoprotein, the density would be closer to 1.1), $L$ is Avogadro's number, $n$ is the number of lamellae, and $P$ is the number of chlorophyll molecules. From this equation the molecular weight for the chlorophyll protein was calculated to be 21,000 (Wolken, 1958a). The molecular weight of an *in vivo* chloroplast in *Euglena* was estimated (using data obtained from interference microscopy and the chlorophyll concentration) to be 16,000 (Wolken, 1967). All estimated molecular weights for the chlorophyll–protein in a variety of plant chloroplasts indicate values from 20,000 to 60,000 (Frey-Wyssling, 1957; Thornber *et al.*, 1967b). Such differences in molecular weight may indicate species specificity for the protein (plastin) in the chloroplast lamellae.

## Chloroplastin

Ever since man recognized the photosynthetic process he has wanted to duplicate it outside the living cell. Even though this has not yet been accomplished there is no theoretical reason to believe that it cannot be achieved.

To accomplish this feat we must not only unravel the complicated photochemistry just described for photosynthesis (Figure 5.2), but also develop the structural molecular arrangement for the pigments, lipids, proteins, and enzymes as found in the chloroplast analysis according to our schematic molecular model for the chloroplast (Figure 5.7).

Attempts to isolate photoactive fractions from the chloroplast and to separate the fractions that contain photosystems I and II have been actively pursued for more than a decade. Chloroplasts can be isolated from photosynthetic cells and photoactive cells. Fragments can be obtained by sonication or other means. The chloroplast can also be dissociated by detergents [digitonin, sodium dodecyl sulfate (SDS), etc.] which solubilize the chloroplast. Of the nonionic detergents, digitonin, a digitalis glycoside, has a strong attraction for dye molecules (Figure 5.9). It has been useful in extracting the visual complex rhodopsin from the retinal rods of the eye, from which much of our knowledge of the biochemistry of vision has been obtained (Wald, 1953). It is thought that in the solubilization process the pigment–lipid–protein complex of the chloroplast lamellae is opened and that chlorophyll micelles are formed (Figure 5.9). These chlorophyll-containing micelles are referred to as *chloroplastin* (Wolken, 1961b).

The *Euglena* chloroplastin (in 1–2% recrystallized aqueous digitonin) absorption spectrum (Figure 5.10) is similar to that of the chloroplast (Figure 5.3).

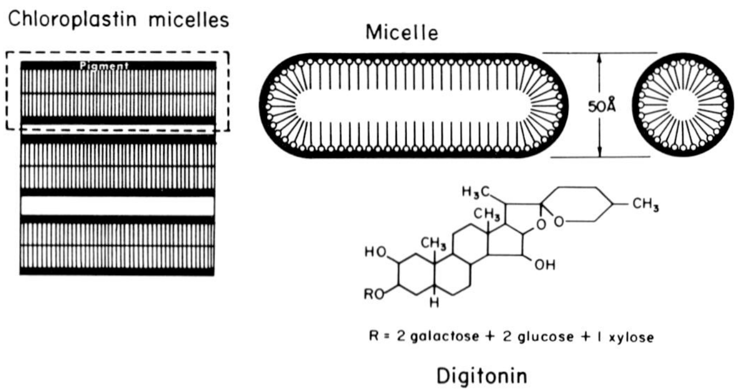


FIGURE 5.9 Digitonin micelle and structure of three such micelles as in chloroplastin. (Compare digitonin structural formula to cholesterol.)

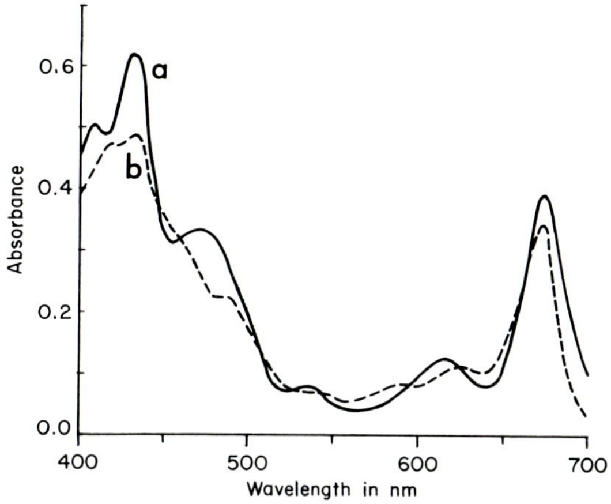


FIGURE 5.10 (a) Absorption spectrum of *Euglena gracilis* chloroplast (——) compared to (b) absorption spectrum of *Euglena* chloroplastin (------) in 1% digitonin.

Chloroplastin not fixed or stained and viewed in the electron microscope shows particles which range in diameter from 100 to 1000 Å (Figure 5.11). The average size compares well with the particles in the chloroplast lamellae

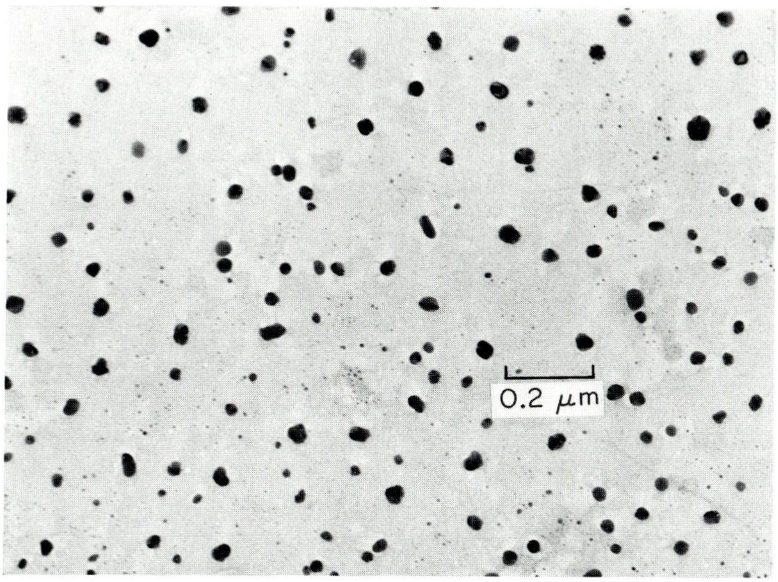


FIGURE 5.11 Chloroplastin particles (100–1000 Å), not fixed or stained, electron micrograph.

TABLE 5.7 ANALYSIS OF CHLOROPLASTIN[a,b] (*Euglena gracilis*)

| | | *1* | *2* | *3* |
|---|---|---|---|---|
| Dry weight, $w$ | mg/ml | 29.0 | 27.3 | 29.6 |
| Nitrogen, $w'$ | mg/ml | 0.37 | 0.36 | 0.35 |
| Chlorophyll, $p$ | moles/liter | $4.2 \times 10^{-5}$ | $6.3 \times 10^{-5}$ | $7.7 \times 10^{-5}$ |
| Chlorophyll, $p'$ | moles/liter | $10 \times 10^{-5}$ | $9.4 \times 10^{-5}$ | $10.2 \times 10^{-5}$ |
| Molecular weight, $M'$ (calculated from $w'$) | gm/mole | 59,000 | 38,000 | 30,000 |

[a] Taken in part from Wolken (1967, p. 124).
[b] $M'$, molecular weight of chloroplastin micelle, from ultracentrifuge data, 290,000; digitonin minimum weight, 155,000.

(Figure 5.5c). It is also possible to show that in the chloroplastin micelles, one chlorophyll molecule is associated with one protein molecule because the experimentally determined chlorophyll concentration $p$ is substantially the same as $p'$, calculated from the formula $p' = w/M'$. (See Table 5.7, chloroplastin analysis.)

Assuming that each chloroplastin micelle is 200 Å in diameter, they could each contain about 225 chlorophyll molecules, 55 carotenoid molecules, one cytochrome molecule, and one ferredoxin molecule together with protein and lipids. This compares to the spinach *quantasome* analysis in Table 5.4 (Park and Biggins, 1964). If three or more micelles aggregate to the order of a thousand chlorophyll molecules, their molecular weight would be a million. This approaches the estimate for the functional *photosynthetic unit* postulated by Emerson and Arnold (1932a,b) and the "chlorophyll holochrome" of Smith (1960). The micelle would then be equivalent to the chlorophyll-containing chromatophore of photosynthetic bacteria.

## PHOTOCHEMICAL ACTIVITY

Chloroplastin exhibits photochemical activity which shows some similarities to the photosynthetic process. For example, the photoreduction of the dye, 2,6-dichlorobenzenoneindophenol (DBIP), at 600 nm, exhibits a primary photochemical reaction (Figure 5.12). During photosynthesis, water-splitting reduction provides the chemical reducing power which is trapped by the dye, and consequently the dye is reduced to a colorless form. During dye reduction an absorption peak at 488 nm (one of the *Euglena* carotenoids) increases in the light and then decreases in the dark without further addition of dye (Eversole and Wolken, 1958).

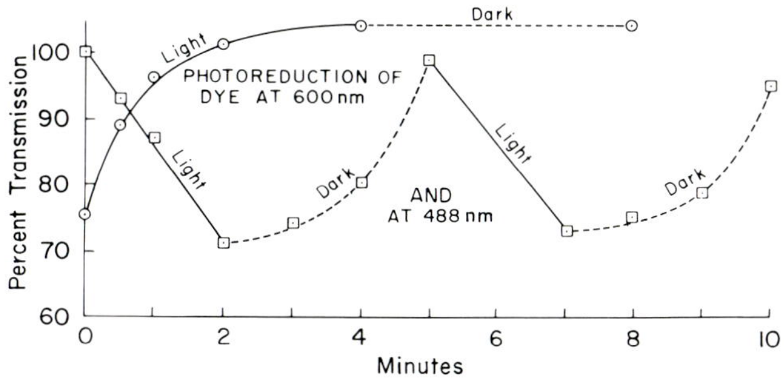


FIGURE 5.12 Chloroplastin absorption changes at two wavelengths (488 nm and 600 nm), with time in light and darkness, in the presence of the dye, 2,6 dichlorobenzenone-indophenol.

Chloroplastin which actively photoreduced the dye also caused photolysis, or evolution of oxygen (Figure 5.13). Photolysis was measured manometrically in completely anaerobic Warburg vessels made oxygen-free to permit a qualitative identification of oxygen with yields of 20 to 30 $\mu$l of $O_2$ in 2 minutes (Eversole and Wolken, 1958).

A light-catalyzed conversion of inorganic phosphate into labile phosphate was also observed over a 1-hour period in a similar anaerobic system (Figure 5.14) containing 6 cofactors and adenosine monophosphate (AMP). The reaction vessels contained 2 ml of chloroplastin with a chlorophyll concentration of about $10^{-5}$ $M$, 20 $\mu M$ of $Mg^{2+}$, 30 $\mu M$ of $\alpha$-ketoglutarate, 0.3 $\mu M$ of riboflavin 5-phosphate, 0.6 $\mu M$ of menadione (vitamin $K_3$), 2 $\mu M$ of ascorbate, 5 $\mu$g of cytochrome c, 55 $\mu M$ of adenosine monophosphate, and 4 $\mu$g of inorganic phosphate. When these experiments were immediately repeated with the addition of glucose and hexokinase, glucose 6-phosphate was formed. The inorganic

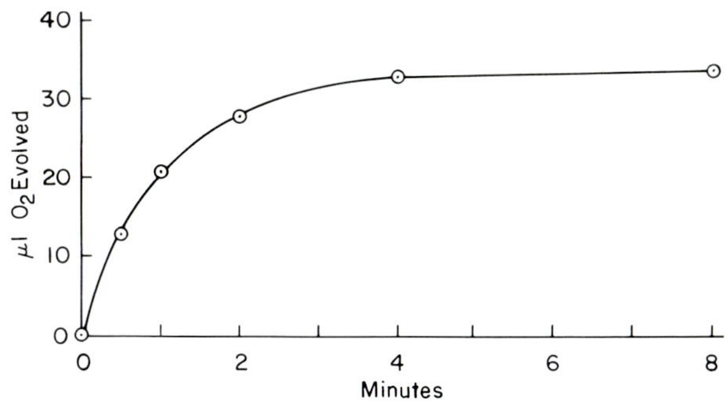


FIGURE 5.13 Rate of $O_2$ evolution of illuminated chloroplastin under anaerobic conditions. *Euglena gracilis.*

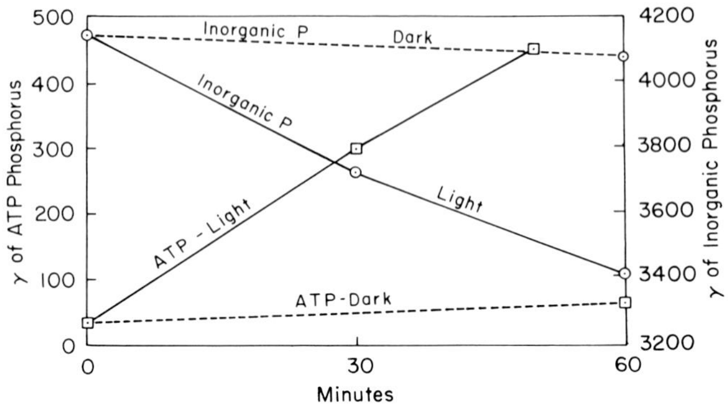


FIGURE 5.14 Phosphate conversion of illuminated chloroplastin under anaerobic conditions.

phosphate that disappeared (80–90%) was accounted for as labile phosphate (ATP). The phosphate conversion occurring in dark controls was only 3 to 4% of that found in light.

## CHLOROPLASTIN FRACTIONS

To isolate the two photochemical systems from chloroplastin, fractionation was attempted by sucrose density gradient high-speed ultracentrifugation and then by chromatography and electrophoresis. But the complete isolation of photochemical systems I and II has not yet been experimentally accomplished. However, Anderson and Boardman (1966) and Thornber *et al.* (1967a,b) have isolated two pigment fractions from spinach and other higher plant chloroplasts which they identified as complex I and II (Table 5.8). These complexes have properties which are related to the two photosystems (Figure 5.2).

*Euglena* chloroplastin when subjected to analytical ultracentrifugation shows a rapidly sedimenting component, 13.5 S, and a slower sedimenting component, 7.8 S. Similarly spinach chloroplastin shows two components, a fast component, 18.8 S, and a slow component 7.7 S (Figure 5.15). These can be considered to contain complex I and complex II.

The molecular weights estimated from the sedimentation ($s$) give about 33,000 (complex I) and 21,900 (complex II) for the spinach chlorophyll–protein complexes (Thornber *et al.*, 1967b). These values compare favorably to that calculated for *Euglena* chloroplastin (Table 5.8).

These data suggest that there are two pigment–protein complexes in the chloroplast lamellae, since two different sedimenting pigment components can be observed during analytical ultracentrifugation (Figure 5.15). The fast component with an 18 S has been observed during centrifugation of extract

TABLE 5.8 SEDIMENTATION CONSTANT IN SVEDBURG UNITS ($s_{20} \times 10^{-13}$)

| *Solution* | *S* | |
|---|---|---|
| Digitonin (1.0–1.8% in $H_2O$) | 6.35[a] | |
| | 6.30 | |
| | 6.24 | |
| | Slow component | Fast component |
| Chlorophyll a (in digitonin) | 7.37 | — |
| Spinach chloroplastin[b] | 7.70 (complex II) | 18.8 (complex I) |
| *Euglena* chloroplastin[c] | 7.85 | 13.5 |

[a] For calculated apparent molecular weights see Hubbard (1954).
[b] For calculated apparent molecular weights of spinach chloroplastin see Thornber *et al.* (1967a,b). 33,300 (complex I); 21,900 (complex II).
[c] Calculated apparent molecular weights of *Euglena* chloroplastin, 21,000–40,000 (Wolken and Schwertz, 1956).

from *Euglena,* indicated as a ribulose-1,5-diphosphate carboxylase (Anderson *et al.,* 1968). This sedimentation peak (18 S) appears similar to that of one of the extracts of green plants, designated as chloroplast Fraction I (Table 5.8). Chloroplastin subjected to polyacrylamide gel electrophoresis also shows

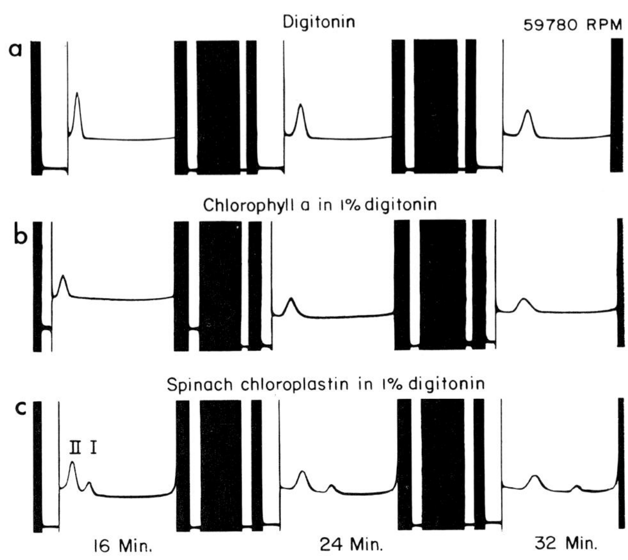


FIGURE 5.15 Analytical ultracentrifugation, Schlierin patterns. (a) Digitonin (% in water), (b) chlorophyll a in 1% digitonin, and (c) spinach chloroplastin in 1% digitonin. Data are summarized in Table 5.8.

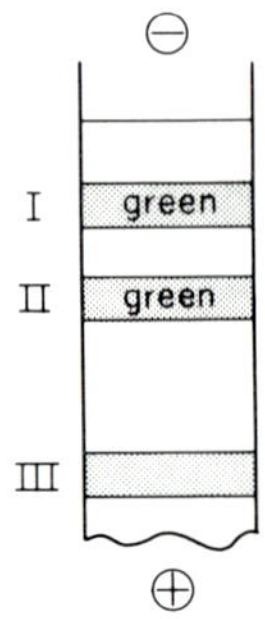


FIGURE 5.16 Spinach chloroplastin, 144,000 fraction, separated by electrophoresis on polyacrylamide gel into bands I and II containing chlorophyll and III containing protein.

two green chlorophyll-containing bands and in addition one protein band (Figure 5.16). Until these isolated fractions can be shown to have different chemical composition and photochemical properties, the question of function remains to be resolved.

It has not as yet been explained how the chlorophylls, carotenoids, lipids, enzymes, proteins, and other components of the chloroplast fit into the particular molecular arrangement for function in the two photosynthetic systems. However, by putting the bits of information together as indicated by these various approaches, answers may come to the question of how the photoexcitation and photochemistry takes place within the chloroplast to initiate the photosynthetic process.

It is obvious in this brief review of photosynthesis that many fascinating but perplexing problems remain to be solved—such as how the photosynthetic system evolved, how the physical and chemical reactions of these photosystems precisely function, and finally, when these are understood, how photosynthesis may be reproduced outside the living cell. These are challenging areas of research which need vigorous pursuit.

# VI
# Phototropism

> How a nerve comes to be sensitive to light, hardly concerns us more than how life itself originated; but I may remark that, as some of the lowest organisms, in which nerves cannot be detected, are capable of perceiving light, it does not seem impossible that certain sensitive elements in their sarcode should become aggregated and develop into nerves, endowed with this special sensibility.
>
> Charles Darwin, "The Origin of Species," 1859 (The Modern Library, New York, 1936, p. 133)

Instinctively all animals emerge to seek the sun and will move to orient themselves so that they maximize the sun's warmth. For example, the marine iguanas in the Galapagos Islands come out of the cold Pacific Ocean onto the rocks and continually change their position and thus their angle of exposure to the sun (Figure 6.1). One explanation is that they need to adjust their body temperature from 21°C when in the sea to 37°C on land. The sea gulls also align themselves on the beaches to gather up the sun's warmth. Go to any beach and observe the way all the human bodies are aligned toward the sun.

Cells, early in their history, evolved mechanisms for movement. This afforded them the means for protection and to search for food. Movement in response to the direction of light was also to find an optimum temperature for survival. But most important to photosynthetic cells, the ability to move enables them to search for light for more efficient use in photosynthesis.

Phototactic behavior can be classified into phototropism and photomotion. Phototropism is defined here simply as a positive orientation of one part of an organism toward light, as in the bending and twisting of the leaf of a plant to present its surfaces to the light. Photomotion, on the other hand, is applied only to freely moving organisms, as in the swimming of aquatic organisms upon stimulation by light.

FIGURE 6.1 Marine iguanas (*Amblyrhynchus cristatus*) oriented to maximize the sun's radiation. Punta Espinosa, Fernandina (Narborough) Island, Galapagos.

In plants there is evidence to suggest that phototropism was originally an adaptation of primitive photosynthetic cells. Consequently light does not merely elicit a tactic response, but in fact determines the cell's entire metabolism. As a result, the phototactic behavior of an organism may also influence the evolution of a species (Rockwell and Seiger, 1973). Therefore, to isolate the function of light in phototactic behavior becomes essential.

Pasteur (1857) said, "It would be of great interest to approach the problem of plant life by beginning with organisms in which simplicity of organization facilitates studies and renders conclusions more certain. In molds, the plant is, so to speak, reduced to the single cell state." And so, when undertaking the problem of phototropism, one should examine first the most simple cells that demonstrate the phenomena. These examples can be found in the fungi.

Throughout the fungi phototropism is widespread, for most will bend toward a light source. Fungi do not possess any chlorophyll and hence cannot photosynthesize; thus, we can isolate the organism's phototropism without the added complexity of photosynthesis. Their mode of nutrition is universally heterotrophic, being either saprophytic or parasitic. It is believed that fungi evolved from true photosynthetic plants; but they may have developed very early in the evolution of plants and played a vital role ever since in assisting saprophytically with the breakdown of dead organic matter. Thus, they assist in both the carbon and nitrogen cycles. Despite the probable early origin of fungi, their main evolution did not occur for a long time (Price, 1971).

A feature unique to the fungi is that whereas multicellular plants and animals are composed of tissues based on histologically distinct cells, fungal organs are all composed of undifferentiated tubular sporangiophores, in which numerous nuclei are found throughout the cytoplasm.

What kind of photoreceptor structure evolved for phototropism? What pigment is the photoreceptor molecule? And are these phototropic reactions the primitive counterparts of more highly evolved sensory systems? Let us review some experimental studies which we have pursued with the fungus *Phycomyces blakesleeanus* in search of information regarding phototropism and photosensory processes.

## A Model Cell: *Phycomyces*

The sporangiophore of the fungus *Phycomyces* is a large, single, aerial cell that matures in five distinct stages, I–IVb, and in the process may grow to be more than 10 cm in length and about 0.1 mm in diameter (Figure 6.2). The growth stages are related to the elongation of the sporangiophore and to the development of the sporangium. In the maturation processes, color changes are observed in both. The general appearance of the sporangiophore in stage IVb is that of a nearly transparent cylindrical filament supporting a spherical black sporangium. The growth time from stage I to stage IVb is approximately 16 hours; it is dependent on light, temperature, humidity, and growth media.

The development, growth, phototropism, and other sensory behavior of *Phycomyces* were reviewed by Shropshire (1963) and more recently by Bergman *et al.* (1969). The sporangiophore exhibits two light-sensitive phenomena, that of phototropism as well as a light-growth response. Light is not essential for growth, but growth is stimulated by light. So light functions as a signal to alter the growth in either space or time.

The sporangiophore exhibits phototropism when it is illuminated unilaterally with light; it is positively phototropic to wavelengths from 300 to 510 nm, and negatively phototropic to wavelengths shorter than 300 nm, with a peak of sensitivity at about 280 nm. By selection through mutation, strains can be produced that are (1) fully sensitive to light, (2) "night-blind," sensitive to only high light intensity as are the retinal cones of the vertebrate eye,

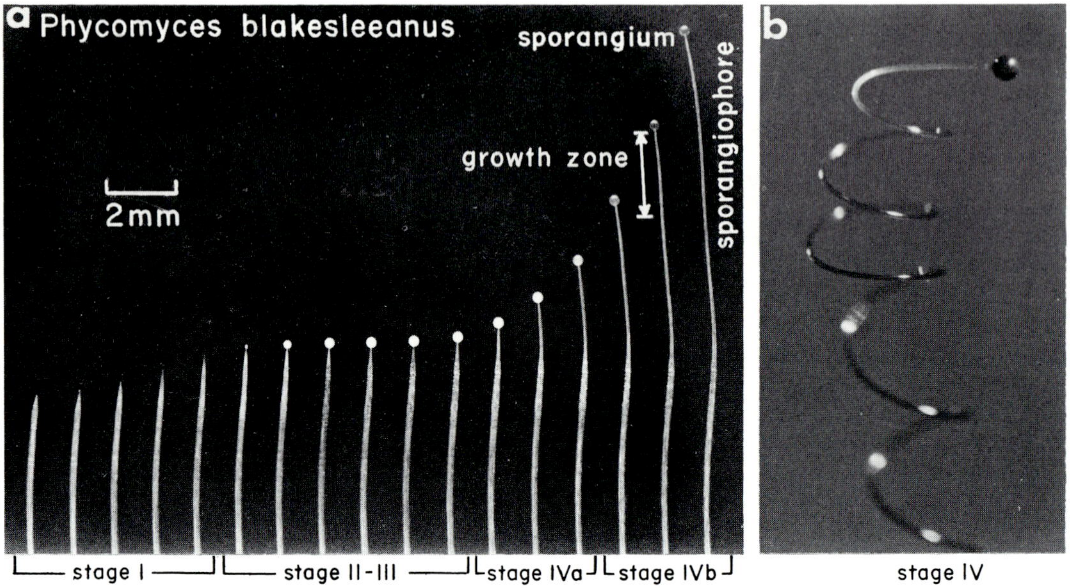


FIGURE 6.2 *Phycomyces blakesleeanus.* (a) Growth study; (b) demonstration of helical phototropism, stage IV (courtesy of Professor David Dennison, Dartmouth College, Hanover, New Hampshire).

or (3) insensitive to light and considered "blind." If one can establish here some genetic basis at the molecular level for the ability to "see" or for "blindness," then a tool is available to investigate pigment biosynthesis and photosensitivity by the methods of molecular genetics. From such investigation of these genetic mutants clues may be obtained about photosensitivity in general and about the visual pigments of animals in particular.

Another interesting example of *Phycomyces* photosensitivity is found when *Phycomyces* is cultured in darkness and completely isolated from light until it reaches stage I. Then a flash of light is given through a pinhole to the culture and it is restored to darkness until it reaches stage IVb. At this time the culture is examined and the sporangiophores are found to be bent in the direction of the light flash. Therefore in the absence of light, it continued to search for the light signal. In a sense it could be said that *Phycomyces* exhibits "memory"—for in stage I it was photosensitive, and its photoreceptor molecule was capable of "remembering" the light signal until it reached stage IVb.

We know that besides responding to light, the sporangiophore is also sensitive to gravity, to stretch, to touch, and to the presence of nearby objects. Why is all this sensory behavior built into such a primitive organism? Since it can so keenly sense its environment, what can be learned from it about the development of the sensory systems of more highly evolved plant and animal cells? As an organism it possesses no obvious structures that resemble an eye, a nervous system, or a brain. What kind of receptors and molecules does it have that permit it to detect environmental changes, to act as an integrated organism that exhibits behavior we normally associate with animals?

In search of clues to how photosensory systems evolved, it seemed that some of the answers could be found in the phototropic behavior of this primitive organism. Although *Phycomyces* has for a century been the object of research concerned with the sporangiophore's sensitivity to light, its photoreceptor molecule has not been identified.

## The Photoreceptor Molecule

More than a century ago, Julius von Sachs in 1864, observed that the bending of plants toward light is stimulated primarily by the blue end of the spectrum (von Sachs, 1868, 1882). Since light, to produce any effect, must first be absorbed, this meant that plant organs concerned with phototropic bending must contain blue-absorbing, or in other words, yellow pigments. One such pigment has been identified as a carotenoid (Figure 3.17). The evidence for this is that the action spectrum for phototropism resembles very closely the absorption spectrum of the carotenoid $\beta$-carotene extracted from plant tissues (Figure 3.15).

The action spectra for *Phycomyces* phototropism (Figure 6.3 b,c) and for its growth response show absorption peaks around 280, 365–385, 420–425, and 445–485 nm (Delbrück and Shropshire, 1960; Curry and Thimann, 1961). These action spectra indicate that the photoreceptor molecules for both phenomena should be similar.

The growth-zone is the sensory region of the sporangiophore. It extends from 0.1 mm to about 3 mm below the sporangium and occupies a surface area of about $6 \times 10^{-3}$ cm$^2$. The sporangiophore is most photosensitive in stages I and IVb. To determine the absorption spectrum of the growth zone the microspectrophotometer was used to scan up and down the sporangiophore. The absorption spectra obtained with the microspectrophotometer for stage IVb of the wild-type sporangiophore through the growth zone is shown in Figure 6.4. The spectrum of Figure 6.4a is typical of what is found from the bottom of the sporangium through 1 mm in the growth zone. In scanning down the sporangiophore from 1 mm to 2 mm below the sporangium, the absorption spectrum gradually shifts from the spectrum shown in Figure 6.4a to that of Figure 6.4b. Spectra taken beyond 2 mm below the sporangium are all typical of Figure 6.4b. The spectrum of Figure 6.4a is similar to that of the absorption peaks of $\beta$-carotene at 430, 460, and 480 nm (Figure 3.15). These peaks correspond to those in the visible part of the action spectrum of *Phy-*

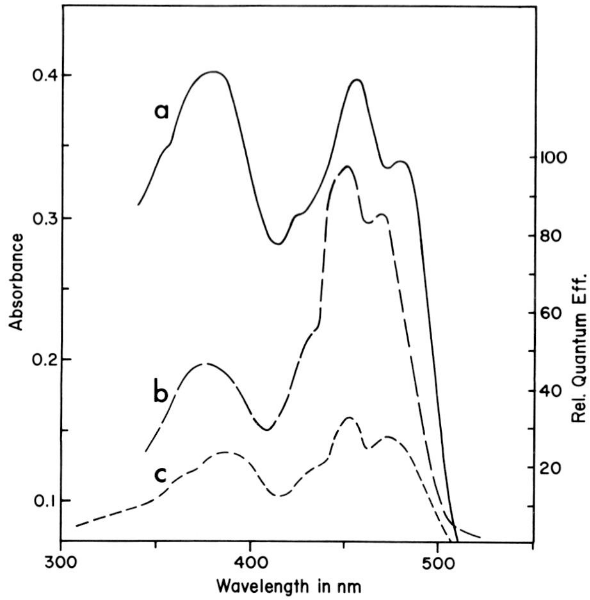


FIGURE 6.3 Absorption spectrum of an acetone fraction from *Phycomyces blakesleeanus* (a) compared to action spectra for *Phycomyces* phototropism obtained by (b) Curry and Thimann, 1961 and (c) Delbrück and Shropshire, 1960.

*comyces* phototropism. The absorption peaks in the ultraviolet near 280 and 370 nm and in the visible around 450 nm are similar to those of a flavin or flavoprotein molecule (Figure 3.14). It is interesting to note that the absorption peaks of Figure 6.4a,b together give absorption peaks at 280, 370, 435, 460, and 485 nm, which are similar to those of the action spectra for phototropism (Figure 6.3b,c).

*Phycomyces* albino mutants are deficient in β-carotene but are still phototropic, so their absorption spectra should be more informative. The absorption spectrum within the growth zone of the mutant has absorption peaks near 230, 267, and 370 nm (Figure 6.4c), which is a typical spectrum obtained in scanning throughout the growth zone of the sporangiophore. No absorption peaks typical of a carotenoid were found. It is important to reconcile these spectral findings to see more precisely the meaning of our absorption spectra.

Even though no detectable β-carotene spectra were observed by microspectrophotometry in the growth zone of the albino mutants, can β-carotene be ruled out as the photoreceptor pigment? If all the β-carotene found in the sporangiophores were concentrated in the growth zone, there would be about $10^{10}$ molecules. If such concentrations were found in the growth zone it would have been detected by our microspectrophotometer. However, β-carotene could

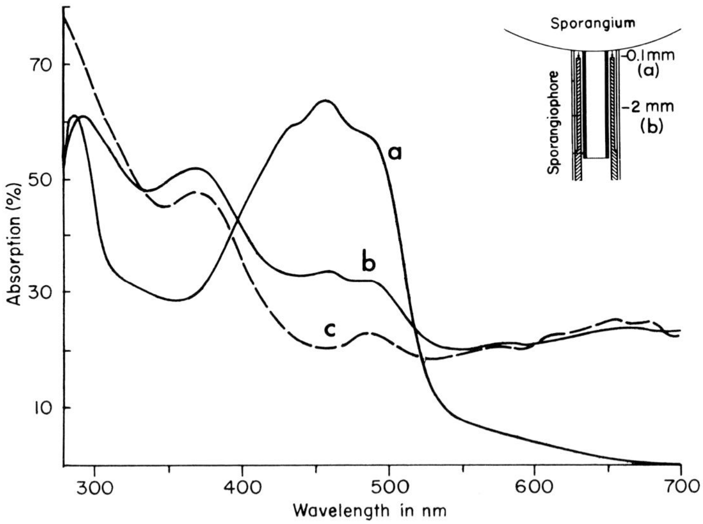


FIGURE 6.4 Microspectrophotometric scan showing absorption spectra in growth zone of *Phycomyces,* wild type, stage IVb, at (a) 0.1 mm to 2 mm and (b) from 2 mm and below the sporangiophore, compared to (c) albino mutant (C5) absorption spectrum in growth zone.

be present in very low concentrations and spread throughout the sporangiophore.

Neither the action spectra nor the absorption spectrum of the growth zone of *Phycomyces* have been sufficiently informative to clearly identify the photoreceptor molecule. Both are ambiguous enough to allow supporting evidence for interpretation that either a carotene or a flavin is the photoreceptor molecule. Therefore, it becomes necessary to extract the pigments, to identify them, and to see which pigment absorption spectrum matches the action spectrum for phototropism.

## Pigment Extraction

The evidence so far indicates that the photoreceptor molecule is probably present in very low concentration. To extract the pigment from the sporangiophore we first used a 2% saturated aqueous solution of digitonin, which is used to extract the chlorophyll complex from the chloroplast and rhodopsin from the retinal rods. Using a large quantity of freshly harvested stage IV wild-type sporangiophores, extraction with digitonin was carried out in the dark for 24 hours; then the solution was centrifuged. The supernatant showed only one absorption peak at 465 nm. To the pellet a series of graded acetones from 80 to 100% were used, and the process of extraction and centrifugation was repeated. When 100% acetone was reached, the resulting supernatant absorption spectrum showed peaks with maxima at 375, 465, and 480 nm (Figure 6.3a), which bears a striking resemblance to the action spectra of *Phycomyces* phototropism (Figure 6.3b,c).

It is difficult to identify which pigment or pigments were responsible for the absorption spectrum obtained. To isolate and identify the pigments, the extraction was again repeated with large quantities of sporangiophores, using various organic solvents as extractants. Phytofluene was extracted in petroleum ether and was identified by its absorption spectrum, for it showed maxima near 330, 345, and 370 nm, and the solution showed a strong grayish-green fluorescence. Phytofluene is a precursor in the synthesis of carotenoids (Figure 3.8). To extract the carotenoids a 1:1, v/v mixture of methanol and petroleum ether was used (extraction was carried out at 7°C for 20 hours). The solvent was then evaporated under vacuum at 7°C and the residue dissolved in hexane. The hexane solution was chromatographed on a column of alumina and the pigments were eluted from the column with petroleum ether and ether mixtures (Goodwin, 1952). At least eight different carotenoids were separated and identified. Of these carotenoids, β-carotene was found in highest concentration.

If β-carotene is not the photoreceptor molecule, what role does it perform in the photoreceptor process? At present the answer is not clear, for β-carotene can act as a filter to screen or shade in the wavelength range of 400–500 nm, to prevent photooxidation or destruction of the photoreceptor molecule at high light intensities. Under most conditions, the chief effect of a pigment that absorbs light of the same wavelengths as the photoreceptor is to reduce the amount of light that reaches the photoreceptor. β-Carotene may also participate as an accessory pigment molecule in the energy transfer process. An accessory pigment may also influence the action of the photoreceptor in a positive way if it is fluorescent, for example, a flavin.

There is experimental evidence to indicate that flavins or flavoproteins are involved in the photoreceptor process (Carlile, 1965; Berns and Vaughn, 1970; Wolken, 1969, 1972). Microspectrophotometry of stage IVb, wild-type sporangiophores showed what appeared to be a reduced flavin semiquinone spectrum (Figures 6.4b and 6.5). From these spectra it seemed very probable that *Phycomyces* sporangiophores contained flavins. It is difficult to distinguish between flavoproteins and carotenoids, since they have similar absorption spectra in the visible region, and even the fluorescence is inconclusive. Since no quantitative work had been done to establish identity of the flavins present in the sporangiophore the flavins were isolated and identified according to the method of Yagi (1962). The results indicated that the stage IVb sporangiophores of the wild type and the albino mutant contained a total of $13 \times 10^{12}$ and $5 \times 10^{12}$ flavin molecules, respectively. The flavins identified by chromatographic separation were riboflavin, lumiflavin, lumichrome, flavin adenine

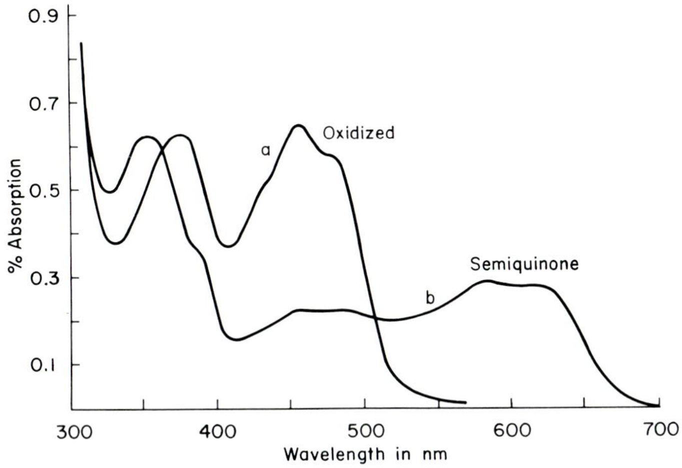


FIGURE 6.5 Absorption spectra of flavoprotein extracted from *Azotobacter;* (a) oxidized and (b) semiquinone.

dinucleotide (FAD), and flavin mononucleotide (FMN). How reasonable are these findings about the sporangiophore? Since the microspectrophotometry of the growth zone showed a shift in spectral characteristics from a typical carotenoid spectrum (Figure 6.4a) to a typical reduced flavin spectrum (Figure 6.4b), these chemical analyses are born out.

So far we have not specifically identified the photoreceptor molecule, but all the evidence would indicate that it is a flavoprotein which becomes photoreduced to a semiquinone.

## Location of the Photoreceptor

The location of the photoreceptor within the cell has been the object of numerous investigations—and the following results should have a bearing on that question.

Buder (1918) suggested that the sporangiophore of *Phycomyces,* since it is nearly transparent, acts as a cylinder lens. He found that sporangiophores which show positive phototropic response to visible light in air are negatively phototropic when immersed in mineral oil. In air, light is focused on the wall farthest from the source of illumination, and since light stimulates growth, the sporangiophore bends toward the light. In mineral oil, whose refractive index (1.47) is greater than that of the contents of the sporangiophore (1.38), light is no longer focused on the wall farthest from the light; consequently, the luminous flux on the wall closest to the light exceeds that on the farthest wall, and the sporangiophore bends away from the light. Buder's observations on *Phycomyces* have been confirmed and extended by Castle (1933), Banbury (1952), and Shropshire (1962).

The evidence for a lens effect has been interpreted as indicating that the photoreceptor is located in or near the cell wall. However, Castle (1933) has indicated that the effect would be the same if the photoreceptor were uniformly distributed within the cell. Cohen and Delbrück (1959) then showed that the primary effect of the light stimulus cannot be on the wall but must be on some structure that moved relative to the wall. This information necessitated a thorough microscopic investigation of the sporangiophore cell wall and a search for a cytoplasmic organelle that moved in the growth zone.

Electron micrographs of sections of the cell wall in the growth zone of *Phycomyces* show musclelike filaments (Figure 6.6). This suggested that a

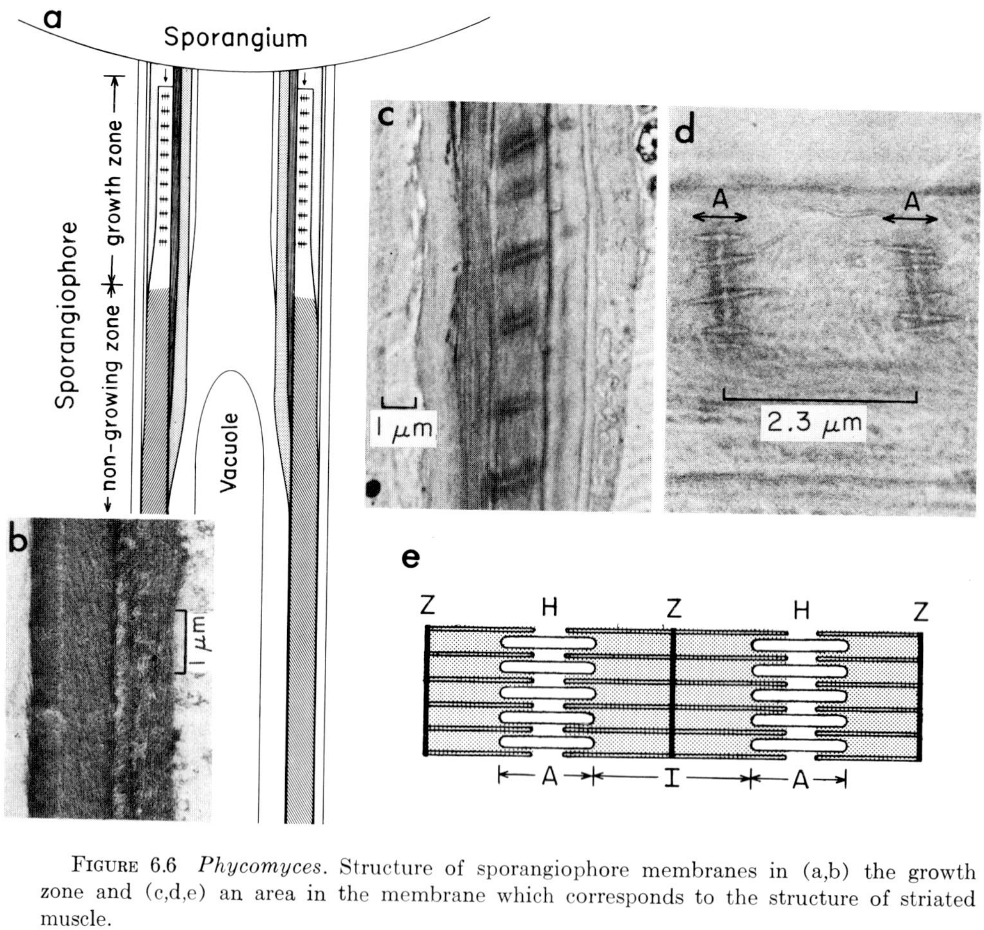


FIGURE 6.6 *Phycomyces*. Structure of sporangiophore membranes in (a,b) the growth zone and (c,d,e) an area in the membrane which corresponds to the structure of striated muscle.

contractile protein such as myosin should be found there. A myosinlike muscle protein has been claimed as a universal mechanochemical transducer. The existence of such a muscle system in *Phycomyces* is quite possible considering the elasticity and mobility of the sporangiophore. The mechanical behavior of the growth-zone changes as the sporangiophore matures to stage IV (Bergman *et al.*, 1969; Ahlquist and Gamow, 1973). The sporangiophore's mobility consists not only of bending toward or away from a light source, but also of the rotational growth of stage IV. During stage IVa, as the elongation of the sporangiophore resumes, it begins to rotate first in a counterclockwise direction (as seen from above), and then in a clockwise direction, reaching a steady state of one revolution every 30 minutes (Johnson and Gamow, 1972).

The mechanical behavior of the sporangiophore seems to be a musclelike response. The elastic growth zone of the sporangiophore responds to a mechanical deformation, but if the stimulus is sustained the response eventually ceases, as with vertebrate mechanoreceptors. Both light and mechanical stimulation elicit the twisting or bending response; these responses share the same effector system. This system could be the contractile protein myosin, located in the cell wall of the growth zone. The isolation of an ATP–ATPase system from the sporangiophore membranes indicates that a contractile protein like myosin is present and could be associated with sporangiophore movement. Muscle proteins have been isolated from the fungus, Myxomycetes, and from the plant, *Nitella flexilis* (Nagai and Rebhun, 1966; Berl *et al.*, 1973). These organisms contract and bend upon mechanical stimulation.

If such a contractile protein exists with the sensory receptor in the growth-zone, this would make the *Phycomyces* sporangiophore unique among photoreceptor systems. For it seems that the receptor and effector are built into the same structure and therefore the *Phycomyces* sporangiophore behaves as a photoneuromuscular system, responding not only to light but to a variety of stimuli.

## The Photoreceptor

Let us turn to the microscopy of the sporangiophore growth zone and examine in detail its cytoplasmic structures in order to determine which functions as the photoreceptor. If the absorption spectrum of any of these is similar to the action spectrum for phototropism then we can identify the photoreceptor.

### Crystals

One of the first observations we made with polarization microscopy revealed birefringent rod crystals, from 1 to 2 μm wide and from 5 to 10 μm in length, which appeared to be aligned near the vacuole in the growth zone (Figure 6.7a,b), the region where the photoreceptor is believed to lie (Wolken, 1969). When the sporangiophore was permitted to dry, numerous needle crystals appeared throughout the sporangiophore. With fluorescent microscopy both the rods and needle crystals were found to fluoresce (Figure 6.7c). Another

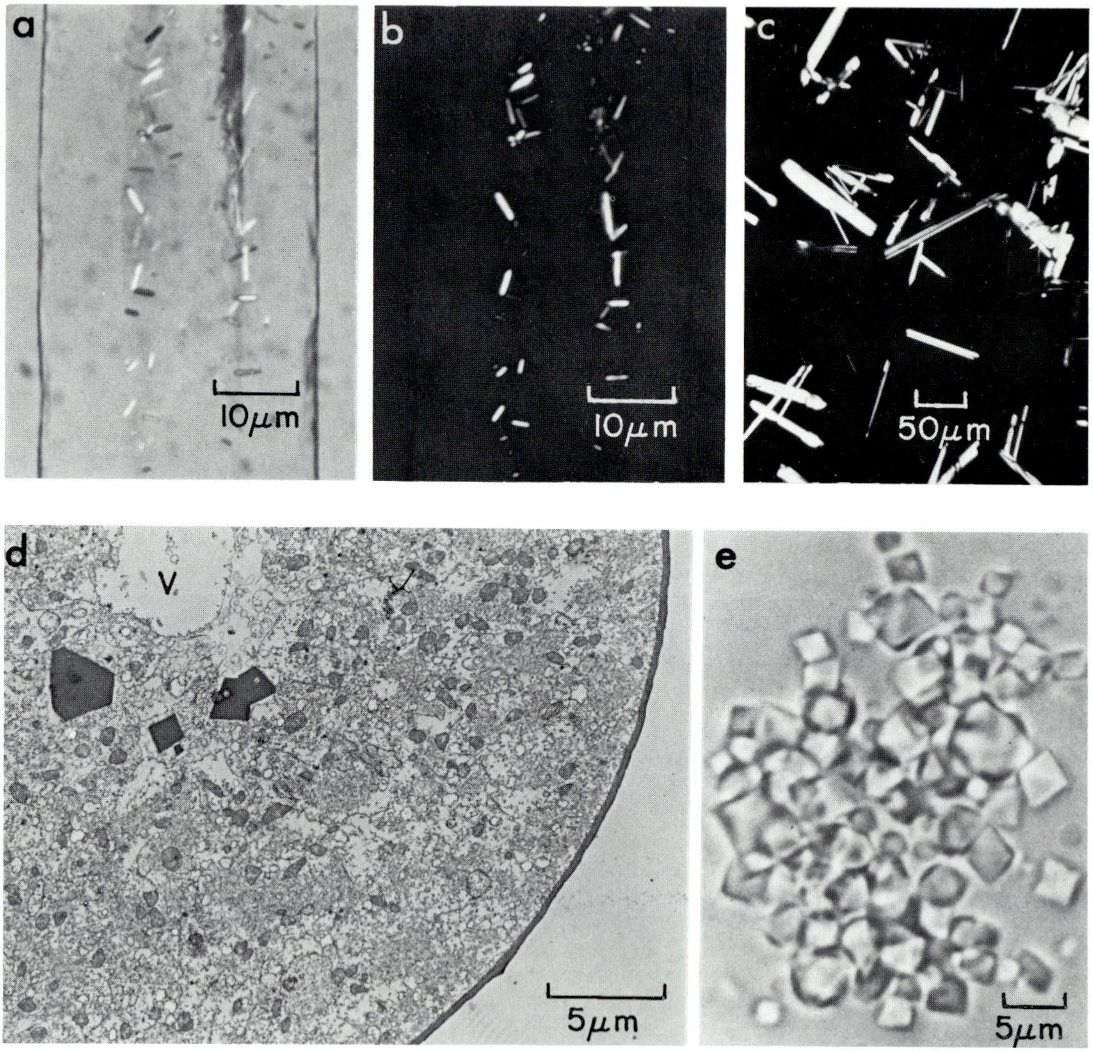


FIGURE 6.7 Various crystals found in the growth zone of *Phycomyces*. (a) Rods, (b) rods in polarizing light, (c) needles and rods in polarizing light, (d) electron micrograph of cross section through growth zone showing octahedral crystals, and (e) isolated octahedral crystals.

crystal was observed which was fluorescent but not birefringent. These crystals were octahedral shaped and appeared to tumble in the sporangiophore (Figure 6.7d,e). The finding that there are various-shaped crystals in the growth-zone of *Phycomyces* sporangiophores suggested to us that the rods or octahedral crystals could be the organism's photoreceptor.

Electron microscopy of fixed and stained sporangiophore sections taken from the growth zone shows numerous mitochondria, electron-dense granules, lipid globules, and ferritin particles (Figure 6.7d). None of these structures can be considered a photoreceptor. But surprisingly, in these sections the octahedral crystal position in relation to the cell wall, cytoplasm, and vacuole seemed more promising. This made it necessary to know under what environmental conditions the crystals are produced and how light affects them.

Microscopic observations of the sporangiophores during growth indicated that the octahedral crystals are found in all stages of sporangiophore development. The number of crystals per sporangiophore consistently increased with growth (stage II–III) and with increase of sporangiophore length until early stage IV, then gradually decreased. In cultures grown in darkness, the number of octahedral crystals per fresh weight of sporangiophores was the highest in stage I, and the lowest in stage IV (Figure 6.8). Also, the number of crystals in the sporangiophores was found to be greatly dependent upon the light intensity. For example, illumination with continuous white light reduced the crystal

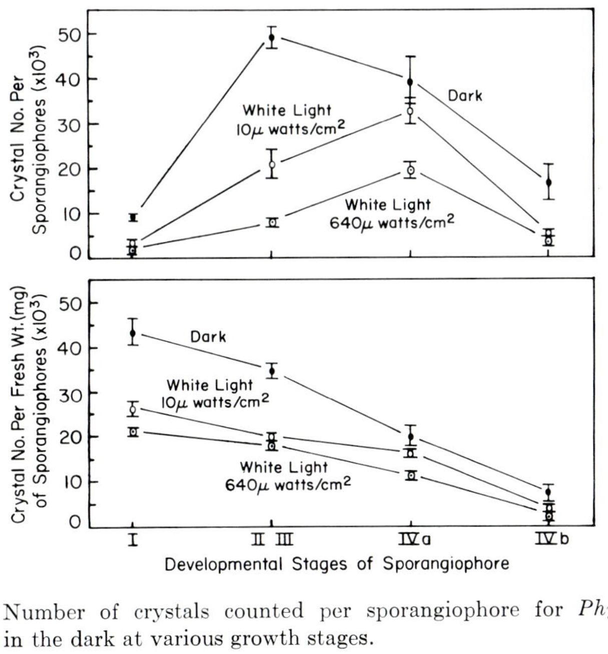


FIGURE 6.8 Number of crystals counted per sporangiophore for *Phycomyces* grown in the light and in the dark at various growth stages.

Table 6.1 Intracellular Location of Crystals, Stage I (5 mm Long) Sporangiophores[a]

| | Average crystal number/unit volume of sporangiophore ($5 \times 10^5$ $\mu m^3$) | | |
|---|---|---|---|
| *Regions of sporangiophore*[b] | *Wild type* | *Albino mutant (C5)* | *Albino mutant (C141)* |
| 0.0–0.1 | 0 | 0 | 0 |
| 0.1–0.2 | 10 | 3 | 1 |
| 1.0–1.1 | 15 | 7 | 3 |
| 2.0–2.1 | 13 | 5 | 2 |
| 3.0–3.1 | 13 | 8 | 2 |
| 4.0–4.1 | 13 | 8 | 5 |
| 4.9–5.0 | 0 | 0 | 0 |

[a] Sporangiophores were cultured under 160 $\mu W/cm^2$ of white light. Ootaki and Wolken (1973).

[b] Distance from sporangiophore tip in millimeters.

content in sporangiophores at all stages of growth, resulting in practically a twofold difference between dark- and light-grown organisms.

How is crystal formation related to the photoreceptor system and carotenoid synthesis? It is interesting to note here that there were more crystals in the wild type than in the β-carotene deficient mutants (Table 6.1), and the smallest number was found in the "night blind" mutant, which is only phototropic at high light intensity. Keeping in mind the possibility that β-carotene functions as a filter in the wavelength range of 400–500 nm, and that at high light intensity photodestruction of the crystals takes place, there should indeed be more crystals in the wild type which contains a larger amount of β-carotene, than in the mutants. But to implicate the crystals in the photoreceptor process required the isolation of the octahedral crystals in a relatively pure state and the determination of their photochemical properties.

## Crystals in Centrifuged Sporangiophore

It is possible to centrifuge at high speed a single sporangiophore from each of the growth stages, I–IVb. This is done by plucking the sporangiophore and tying it with surgical thread, then suspending the sporangiophore in a centrifuge tube of water with tip down. By this method the cytoplasmic particles become stratified according to their densities. At a density near 1.1 a double layer of crystals can be observed (Figure 6.9a,b). Some crystals are also found at the centrifugal tip of the sporangiophore (Figure 6.9c). In stage II–III spor-

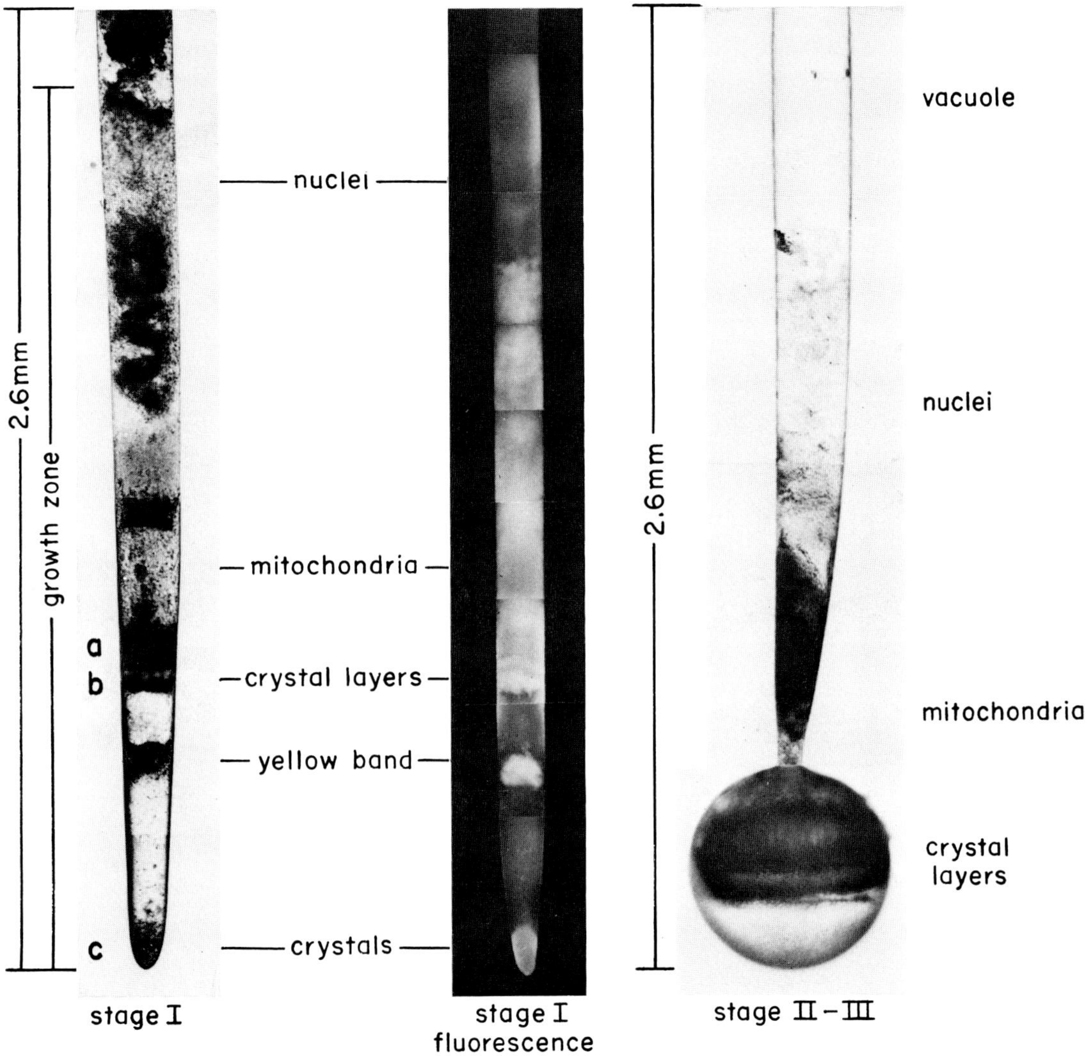


FIGURE 6.9 Stage I *Phycomyces,* (left) centrifuged at 100,000 *g* for 30 minutes; note crystal layers (a,b,c). Fluorescence microscopy of the same stage I sporangiophore (center). Stage II–III sporangiophore (right) centrifuged at 100,000 *g* for 30 minutes.

angiophores, the main crystal layers are found in the sporangium (Figure 6.9). The volume of the crystal layers is much greater in stage II–III than in stage I sporangiophores. Between layers b and c a pale yellow band is observed. With the fluorescence microscope the sporangiophores show an overall green fluorescence from stage I to stage IV. The crystal layers a and

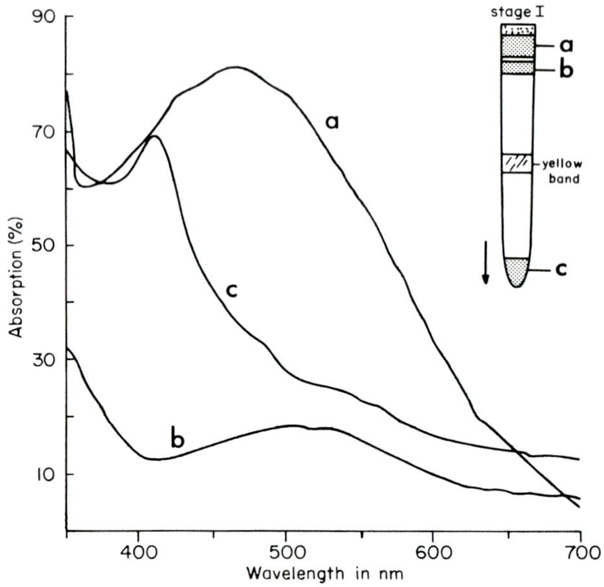


FIGURE 6.10 Absorption spectra of three crystal layers, a,b,c, in the centrifuged sporangiophore, stage I (Fig. 6.9), obtained with microspectrophotometer. Arrow indicates direction of centrifugal force.

b (Figure 6.9) emitted a moderate green fluorescence. The absorption spectra of the three crystal layers shown in Figure 6.10 were obtained with the microspectrophotometer. It is seen that crystal layer a (Figure 6.10) has an absorption maximum near 460 nm, layer b around 520 nm, and layer c at 412 nm, with a broad absorption band around 550 nm. The absorption spectrum of the c layer is similar to that of cytochrome c isolated from the sporangiophores of *Phycomyces* (Wolken, 1969).

The advantage of centrifugation is that these layers from the sporangiophore can be sectioned out and fixed for electron microscopy. We can also determine the effect of light on the structure and absorption spectra of the crystal layers. Finally, the crystals may be isolated, recrystallized, and their physical and chemical properties studied.

## THE ISOLATION OF CRYSTALS

### *Needle Crystals*

The needle crystals can easily be isolated from the sporangiophores because of their water solubility. The recrystallized needles are less than 1 μm wide and more than 20 μm long. These crystals were identified as gallic acid. The concentration of gallic acid was found to be 7.9 mg/ml in the cell fluid and 4.7 mg/gm of the fresh weight sporangiophores. The crystals are birefringent and fluorescent. The absorption spectrum of the recrystallized crystals dissolved in water shows only one absorption peak around 260 nm. The crystal solution is pH sensitive. At pH 7.0 it turns dark green, and its major absorption peak is shifted to 635 nm. This could be a polymer form of gallic acid accounting for the green color of the sporangiophore cell wall. The solution bleaches in light to pale yellow, and its absorbance in the near ultraviolet increases in the range of 300–350 nm. These color changes with pH and light indicate that gallic acid could function as a screening pigment for the photoreceptor. The action of gallic acid as a screening pigment is believed responsible for the reversal from positive to negative phototropism (Delbrück and Shropshire, 1960).

### *Rod Crystals*

The crystals of primary interest are the birefringent rods, for these were first observed in the growth zone of *Phycomyces* using polarization microscopy. *In situ* microspectrophotometry indicated their absorption spectrum to be similar to that of riboflavin (Wolken, 1969). To see if the rods do in fact contain riboflavin it was necessary to isolate the crystals and perform further analysis. The rod crystals can be isolated from water by precipitating them out of solution with 60% acetone. The isolated rod crystals (Figure 6.7c) were from 1 to 2 $\mu$m wide and from 5 to 10 $\mu$m in length. The absorption spectrum of a single crystal obtained with the microspectrophotometer showed two peaks, one near 300 nm and the other at 460 nm, and two shoulders at 260 and 360 nm (Figure 6.11). The rod crystals dissolved in water fluoresce. The fluorescence emission spectrum was broad and had a maximum at 490 nm and a shoulder at 525 nm; the excitation spectrum was more prominent, having peaks at 225, 265, 310, 365, 405, and 435 nm. The excitation spectrum strongly resembles the excitation spectrum of riboflavin. This confirmed our previous analysis that rod crystals do indeed contain riboflavin (Wolken, 1969).

### *Rhombohedral Crystals*

Another crystal not seen by microscopy, not even by electron microscopy, was obtained by extraction from the cell membrane with an acetone–water solution. These crystals are rhombohedral in shape, dark brown in color, birefringent, but not fluorescent. From 1 gm of fresh weight sporangiophores, 18.5 mg of the crystals were obtained. The absorption spectrum of a single crystal shows absorbance around 300 nm and a broad band from 500 to 550 nm with a maximum near 520 nm.

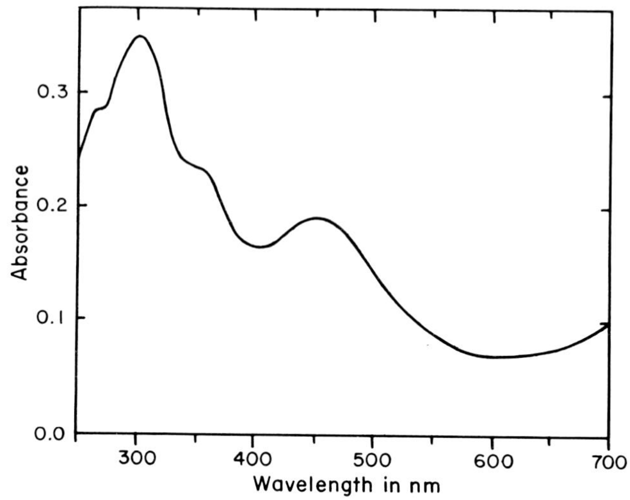


FIGURE 6.11 Absorption spectrum of rod crystal. (Obtained with microspectrophotometer.)

The cell membranes of *Phycomyces* sporangiophores consist mostly of chitin, a waxy or fatty material extractable with $CHCl_3$. Other substances are also extractable with alkaline solutions (Kreger, 1954). A yellow substance with an absorption maximum around 275 nm and two shoulders around 375 and 465 nm was extracted along with the rhombohedral crystals. This material was separated from the crystal by dissolving it in acetone. When the acetone solution was irradiated with white light the yellow color faded to pale yellow. A 375-nm maximum was found in the difference spectrum of the irradiated solution from the $NaBH_4$-reduced solution. These results indicate the presence of flavin or flavoprotein. The reduction with $NaBH_4$ gave the reduced form of flavin, and on photolysis, lumichrome was observed to be one of the products.

### *Octahedral Crystals*

This then brings us to the question of whether the octahedral crystals which are primarily found in the growth zone (Figure 6.7d,e) are the photoreceptors or participate in the photoprocess.

To isolate the octahedral crystals from the sporangiophore in a relatively pure state, a method of sucrose density fractionation was developed (Figure 6.12). In the isolation procedures the major crystal fraction was always found to occur between 75 and 80% sucrose layers. The density of these crystals was determined to be 1.4. From 1 gm of fresh weight sporangiophores 1 to

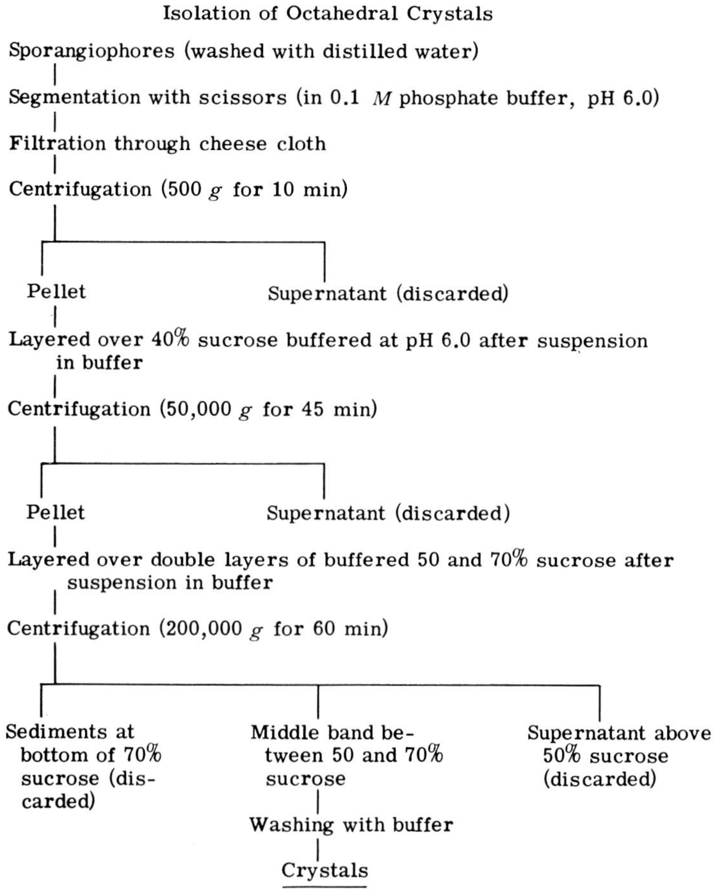


FIGURE 6.12 Method for the isolation of octahedral crystals from *Phycomyces* sporangiophores.

2 mg of crystals were isolated. The isolated crystals (Figures 6.7 and 6.13) were from 2 to 10 μm in a crystallographic axis. They can be placed in the isometric system, having properties of three equal crystallographic axes which cross at right angles to each other, forming a regular octahedron which has 25 symmetry elements (Figure 6.13).

The crystals can be solubilized only at pH's less than 2.5 and greater than 9.5. They are not soluble in 2% digitonin solution, but are soluble in 1% sodium dodecyl sulfate (SDS), pH 6.0 to 7.1.

Chemical analyses of the crystals indicate that they are about 95% protein, of which 1% by weight is tryptophan. In some analyses lipids as fatty acids, flavins, and carotenes were found.

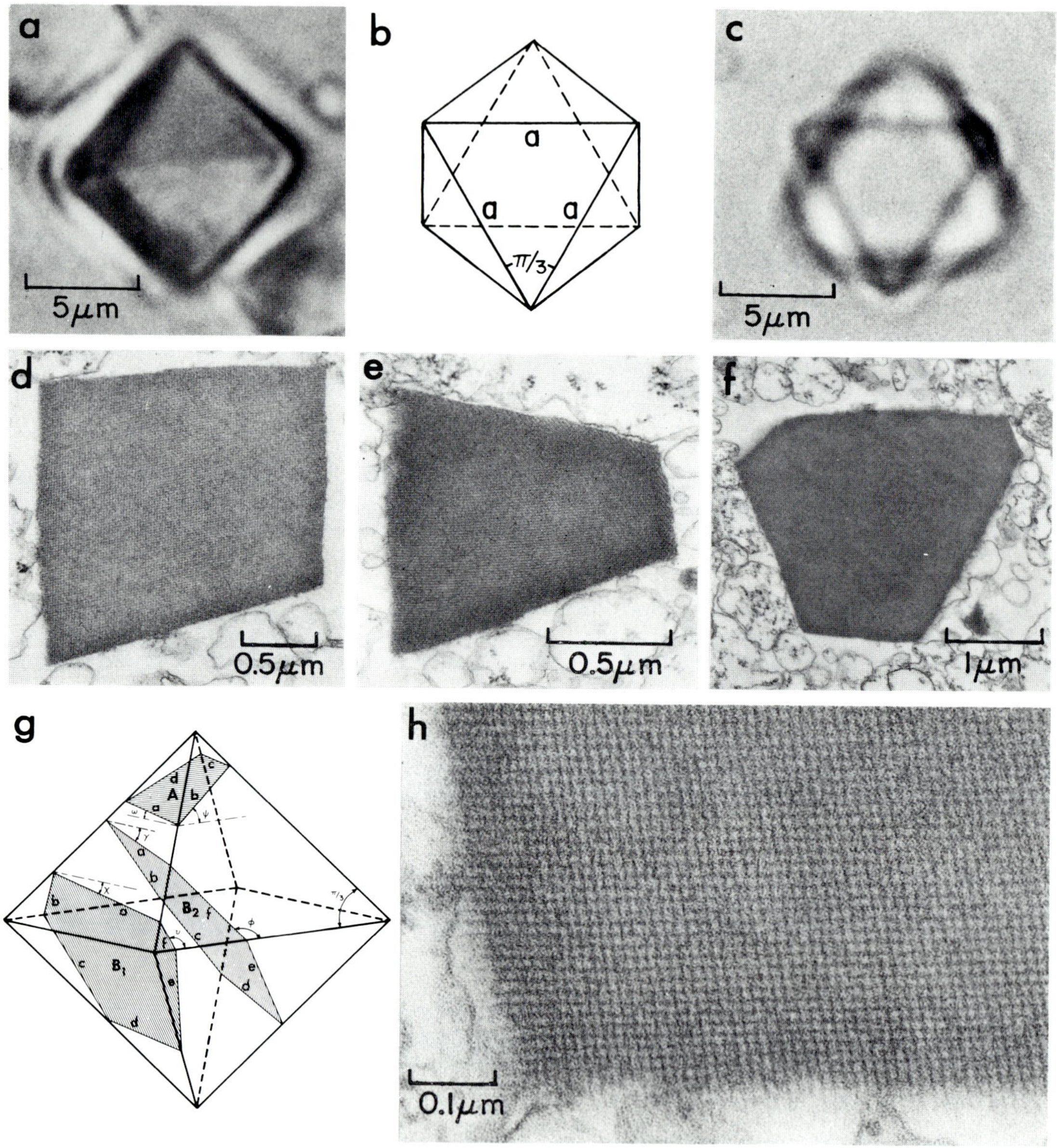


FIGURE 6.13 Octahedral crystal isolated from *Phycomyces*. Light microscopy (a) side view, (b) top view, showing equilateral triangular crystal faces, as in (c). Electron micrographs (d–f) showing the lattice structure obtained by various cuts through the octahedral crystal (g) and (h) enlargement of crystal structure.

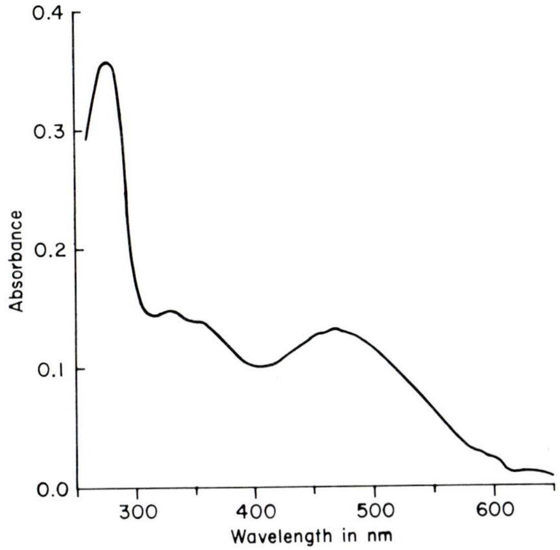


FIGURE 6.14 Absorption spectrum of octahedral crystal. (Obtained with microspectrophotometer.)

The absorption spectrum obtained of a single crystal (Figure 6.14) by microspectrophotometry showed absorption in the ultraviolet in the neighborhood of 280 nm, indicating a protein, near 350 nm, and in the visible with a peak near 460 nm (Ootaki and Wolken, 1973). The spectrum of the isolated crystals packed in a special absorption cell indicated an oxidized flavoprotein; when irradiated the spectrum changed to that of flavin semiquinone (Figure 6.15).

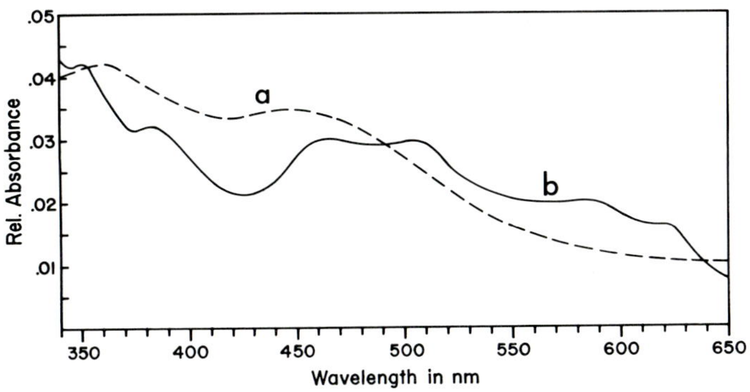


FIGURE 6.15 Absorption spectra of isolated octahedral crystals, stages I–III, packed in a special absorption cell; (a) dark (- - - - - -), (b) irradiated (——), compare to flavoprotein semiquinone (Fig. 6.5.)

TABLE 6.2 VARIOUS CRYSTAL SHAPES AND PROPERTIES

| *Crystal* | *Shape* | *Birefringent* | *Fluorescence* | *Identity* | $\lambda_{max}$ |
|---|---|---|---|---|---|
| Needles |  | + | + | gallic acid | 260 nm |
| Rods |  | + | + | flavin | 460 nm |
| Octahedrons |  | − | + | flavin | 350 and 460 nm |
| Rhomboids |  | + | − | unknown | 325 and 550 nm |

This similarity in spectra is seen in the flavoprotein isolated from *Azotobacter* (Figure 6.5).

x-Ray diffraction patterns of single crystals were not obtainable, but the electron micrographs could be optically diffracted. The results of the optical diffraction led us to a preliminary model for the crystal structure. The model proposes a chain of discs 120 Å in height which are in contact along the *c*-axis. From the volume of any one disc, which is approximately $6.8 \times 10^5$ Å$^3$, the molecular weight was estimated to be of the order of 450,000 (Wolken, 1972). Electrophoresis of the dissolved crystals indicates a subunit of 56,000, which would mean that the molecule consists of 8 subunits. Table 6.2 summarizes some characteristics of the various types of crystals found in *Phycomyces*.

We must still answer the question of whether the octahedral crystal really is the photoreceptor. There is some spectral evidence for this. The absorption spectrum of a crystal from stage IVb shows absorption in the ultraviolet, around 280 nm, and near 360 nm. However, in stages I and II–III, the crystal spectra show absorption near 275–285, 355, and around 465 nm in the visible (Figures 6.14 and 6.15). These are, in fact, the same absorption peaks found in the action spectrum and by microspectrophotometry of the growth zone of the sporangiophore (Figure 6.4). If either a flavoprotein or a carotenoid were adsorbed on the surfaces of these crystals $10^6$ to $10^7$ molecules could be accommodated and this is the right order of magnitude for a photoreceptor.

## Electrophysiology

If the sporangiophore of *Phycomyces* is a sensory cell analogous to receptor cells of animals, signals should be detected upon electrical stimulation, and as a photosensory cell, electrical signals should be detected upon light excitation. Therefore we set out to see if we could find analogies between the sporangiophore and a sensory cell.

The electrical phenomena accompanying the response to light of a photoreceptor may be divided into three types: (1) the early receptor potential (ERP) which is related to the photoreceptor pigment, (2) the receptor potential which is slow positive and/or negative going potential, and (3) the receptor potential related to nerve spikes (Tomita, 1970). These electrical signals have been studied extensively for animal photoreceptors and sensory cells. Do similar responses occur in plants? Early receptor potential type responses have been noted in most pigmented tissue including leaves (Arden *et al.*, 1966). Slow positive going potentials occurring after early receptor potential type responses have also been noted in plants (Ebrey, 1967) and negative going spikelike potentials have been recorded for *Mimosa* (Sibaoka, 1962).

The *Phycomyces* sporangiophore under continuous illumination decreases its sensitivity in response to a given flash of light. This loss of sensitivity as with most photoreceptors is some function of the incident intensity, for changes in level of adaptation to light are reversible, and they most likely represent changes in a photosensitive pigment which is bleached in the light and resynthesized in the dark.

In *Phycomyces* a sporadic early receptor potential type response is seen to occur upon illumination by a high intensity light source (60 W/sec). It has been found only for stage I sporangiophores and is related to the photoreceptor pigment. The most prominent response for all growth stages studied was a graded receptor potential, a positive wave whose amplitude is from 2 to 10 mV and 2 to 10 seconds in duration (Figure 6.16a,b,c). The amplitude correlated roughly with intensity. At equal intensities the amplitude showed a larger

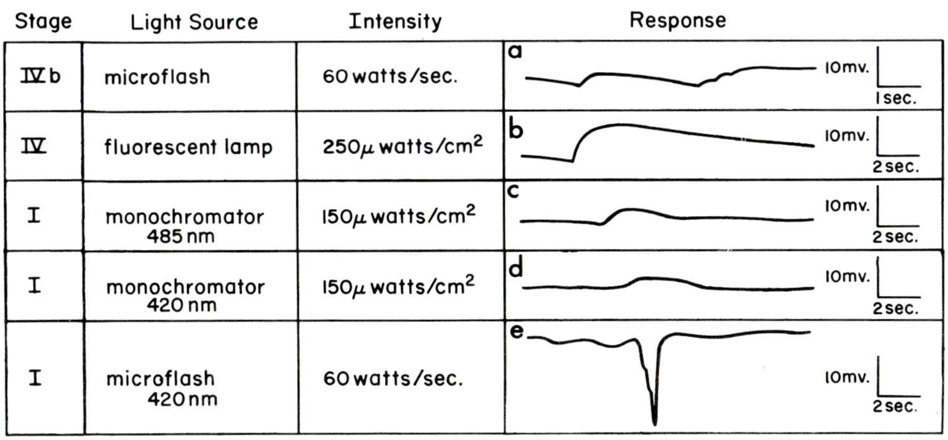


FIGURE 6.16 Electrical responses of *Phycomyces*.

response to a wavelength of 485 nm than to wavelengths of 420 and 385 nm (Figure 6.16c,d). A receptor potential was obtained in response to 500 nm in stage I, but not in stage IV.

One or more negative going spikes occurred on an average of 2.5 minutes after exposure to the high intensity light source. These spikes reached amplitudes of 8 to 20 mV and durations of 0.5 seconds (Figure 6.16e). Such spikes were also observed using a monochromatic light source (a lower intensity) for a wavelength of 420 nm. These spikes, as with the sporadic early receptor potential type responses, were observed to occur only in stage I.

To see if the tissue was electrically excitable, square wave pulses of 30 mV amplitude and 1 msec duration were applied to the sporangiophore. For stages I, IVa, and IVb, small amplitude, biphasic potentials were seen in response to the electrical stimulation. These rarely exceed 10 mV and occurred with an average latency of 1 minute.

As the sporangiophore matures through stage II to stage IVb the complexity and type of the observed electrical response appears to change. Although the receptor potential is seen for all stages studied, the sporadic early receptor potential type response and the spike are not. This bears a relationship to the pigment system of *Phycomyces,* since orientation of the photopigment molecule is related to the early receptor potential. The latency and time course of the receptor potential are slow when compared to that of other photoreceptors (Tomita, 1970). The amplitude of the response is related to both wavelength and intensity. Wavelengths of 385, 420, and 485 nm correspond to absorption maxima in the phototropic action spectra for *Phycomyces* (Figure 6.3b,c). Although 420 nm did not produce large amplitude receptor potentials, it was at this wavelength that spiking was seen in stage I sporangiophores. Spikes were observed for both monochromatic light and filters in combination with the high intensity light source. The average latency (2.5 minutes) suggests a relationship with the transient growth response, which occurs around 3 minutes after exposure to light (Bergman *et al.,* 1969).

The absence of the spike response in stage IV is also related to the pigment system of *Phycomyces.* It is possible that the photoreceptor-pigment system in stage IV may be in a different state from that in stage I. The absence of both spike and early receptor potential type responses as well as lack of response to illumination of 500 nm suggest this possibility. It is also possible that two different pigments may be responsible. Such an observation was made by Ebrey (1967) on the bean plant *Phaseolus vulgaris,* where a slow positive response was seen after detection of an early receptor potential type response. These results are suggestive of two different pigments or different mechanisms mediating these responses.

If the *Phycomyces* sporangiophore is to be likened a photo-neuromuscular-sensory system, the chemistry which mediates the response should be found.

We have already discussed that the membrane in the growth-zone have filaments that resemble that of a striated muscle (Figure 6.6). In addition we have found that a myosin-like protein can be isolated from the sporangiophore membrane that has ATPase activity which is associated with that of muscle. To establish that the behavior is that of a neuromuscular system, the neurotransmitter molecule acetylcholine should also be found (Figure 14.3). Experiments were carried out to determine the acetylcholinesterase activity and the order of $10^{-8}$ moles/l/gram membranes was found. It is interesting to note that brain tissue has activities of the order of $10^{-6}$ moles/l/gram. So we see that even in this relatively primitive organism the structure and chemistry is present that is found in much more highly evolved and specialized receptor systems of animals. Therefore the *Phycomyces* sporangiophore can be considered an analog of an animal receptor system (Mogus and Wolken, 1974).

# VII
## Phototaxis

The motility of microorganisms has fascinated microscopists since Antonie van Leeuwenhoek in 1674, using his simple microscope, described the swimming about of protozoa and algae:

> These animalcules had diverse colors, some being whitish and transparent; others with green and very glittering little scales; others again were green. . . . And the motion of most of these animalcules in the water was so swift and so various, upwards, downwards, and round about, that 'twas wonderful to see. . . .
>
> Leeuwenhoek, 1675 (in Dobell, 1958, pp. 110–111).

Such observations have continued to fascinate microscopists. For out of these observations have come some interesting relationships between the various stimuli, particularly light, and the patterns of motion which bear directly on our search for the mechanisms which underlie the function and evolution of photosensory cells.

The flagellated protozoa are the simplest organisms for quantitative studies of the relationship between various stimuli and their responses. *Euglena gracilis* is an ideal organism for such investigation. When *Euglena* are observed through a microscope various kinds of motion can be distinguished: pulsation, sidewise rotation, and forward swimming. The sidewise rotation and forward swimming are controlled by the whipping of the flagellum at the anterior end of the organism. Light acts as a stimulus and *Euglena* respond by swimming toward the light source; that is, it is phototactic.

I have classified phototactic behavior under the general term *photomotion,* and separated the responses for experimental reasons into *photokinesis* and *phototaxis.* These terms are commonly applied only to freely moving organisms to distinguish them from phototropism. Other terms used to describe phototactic behavior have been defined by Fraenkel and Gunn (1961).

Photokinesis as used here is the change in velocity or rate of swimming, upon illumination, without regard to orientation. Whereas phototaxis is the orientation of the entire organism with reference to light, and implies the ability to differentiate among light stimuli of differing intensities and wavelengths. The importance of phototaxis to animals is to provide a system to use light as a navigational signal in the search for food, the escape from enemies, and mating.

In search of evolutionary clues to how the photosensory cell developed, it is worth examining a relatively primitive photoreceptor system found in the protozoa. *Euglena* possess a photoreceptor and an effector that can alter the nature of its movements. We have already described *Euglena* structure and noted that it possesses an eyespot region, the *stigma,* and close to the eyespot and protruding from its gullet, a flagellum. Englemann (1882) speculated from *Euglena* reaction to light that the eyespot (stigma) is the primary region for light reception. Later, Mast (1911) suggested that a photosensitive pigment, which functioned as a photoreceptor analogous to a retinal cell of the eye, lay inside the eyespot region.

Let us then examine more closely *Euglena* photosensory structures and photobehavior before we consider the photoreceptors of more highly evolved animals.

## The Photoreceptor System of *Euglena*

### Eyespot Region

The eyespot is an orange–red body at the anterior end of the organism. It is an agglomeration of numerous granules, each of which is about 0.1 to 0.3 $\mu$m in diameter. The granules of the eyespot are located just below the membrane of the reservoir, a chamber with smooth walls that follows the ridged gullet from where the flagellum originates (Figure 7.1a). Between the granules and the membranes of the reservoir one sees a system of regularly spaced fibrillae (Figure 7.1b,c,d). A dense, homogeneous body is seen attached to the flagellum and facing the eyespot; it is known as the *paraflagellar body* and is believed to be the *Euglena* photoreceptor (Figure 7.1b,c,e). The para-

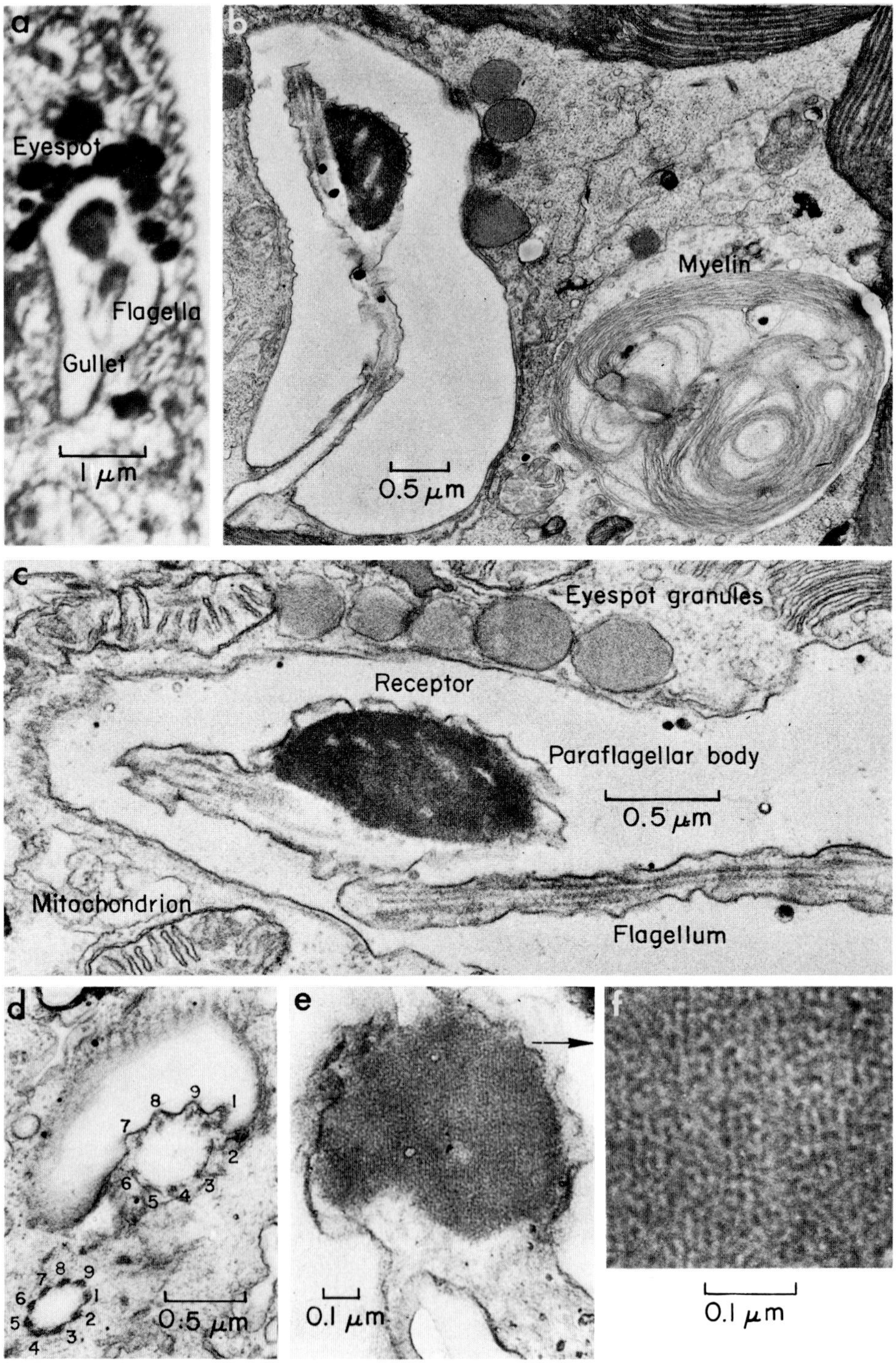
a
Eyespot
Flagella
Gullet
1 μm
b
Myelin
0.5 μm
c
Eyespot granules
Receptor
Paraflagellar body
0.5 μm
Mitochondrion
Flagellum
d
8 9 1
7
2
6
5 4 3
7 8 9
6 1
5 2
4 3
0.5 μm
e
0.1 μm
f
0.1 μm

TABLE 7.1 NEUROHORMONES IN PROTOZOA[a]

| | *Acetylcholine* | *Acetylcholinesterase* | *Catechola mines* |
|---|---|---|---|
| *Euglena gracilis* | + | | |
| *Paramecium* | + | + | + |
| *Trypanosoma rhodesiense* | + | | |
| *Tetrahymena geleii* | | + | |
| *Tetrahymena pyriformis* | | | + |
| *Crithidia fasciculata* | | | + |
| *Noctiluca miliarus* | | | + |

[a] Data taken in part from Lentz (1968, pp. 112–113).

flagellar body as observed with the electron microscope is a crystalline structure and appears to have a repeat unit of about 100 Å (Figure 7.1f). This observation is important, for it supports our hypothesis that all photoreceptors are highly ordered crystalline structures.

In addition, in the eyespot region a myelin structure is observed resembling an axon bundle (Figure 7.1b). The whole eyespot region appears to be structured like a neurosensory cell.

If indeed the *Euglena* cell has a mechanism of excitation analogous to that of a neuromuscular system, it should possess the chemistry of a nerve cell. Neurohormones are known to be present in other primitive nervous systems (Table 7.1). Acetylcholine, epinephrine, norepinephrine, and 5-hydroxytryptamine are neurotransmitter molecules associated with the chemistry of nerve excitation and are found in protozoa and invertebrate systems (Lentz, 1968). Experimental analysis shows that there are $3.85 \times 10^2$ acetylcholine molecules per *Euglena* cell. It seems possible that because these substances are generally found in nervous systems, including those of the protozoa, the basic secretory capacity of these prenervous cells was modified, during evolution, for the coupling of the excitatory and conductile properties to allow transmission to occur after stimulation.

## THE FLAGELLUM

The photoreceptor may have evolved as the result of the differentiation of ciliary or flagellar processes of a cell with which the photosensitive pigment became associated. In *Euglena* the elongated flagellum is of the order of 30 μm or more in length and arises from the gullet or cytopharynx. Attached to

FIGURE 7.1 *Euglena gracilis.* Electron micrographs of eyespot–flagellum area. (a) Eyespot–gullet area, (b) paraflagellar body, flagellum, and myelin structure, (c) enlargement of paraflagellar body and flagellum, (d) base of the flagella, (e) enlargement of paraflagellar body, and (f) further enlargement of paraflagellar body; note crystalline structure.

the flagellum are small lashlike fibrils called *mastigonemata* (Figure 4.2j). The flagellum appears to have numerous junctions along its entire length (Figure 4.2k). In cross section the flagellum is from 0.25 to 0.40 $\mu$m in diameter and consists of a number of elementary filaments, *axonemata*, which are embedded in a matrix and covered by a membrane (Figure 7.1c,d). Their internal structure, shown by the electron microscope, consists of an axoneme containing the "9 + 2" pattern of filaments running the length of the flagellum. The two central filaments are not always seen but appear to be typical microtubules, while the nine peripheral filaments appear to be microtubular doublets (Figure 7.1d). The base of the flagellum is situated in a vacuole very close to the paraflagellar body.

Recent studies of the chemistry of the algal flagella indicate that they are composed of myosin, the contractile protein of muscle (Lewin, 1962). These flagellar proteins are called tubulins (Witman *et al.*, 1972a,b). When large quantities of *Euglena* flagella are isolated and placed in $10^{-3}$ *M* solutions of ATP, they display vigorous beating (Wolken, 1967). Therefore the flagella contain a mechanochemical mechanism that converts the chemical energy of ATP into movement. But it is not known how the excitation is carried from the light-absorbing pigment in its photoreceptor pigment molecule to the flagellum.

## Photokinesis and the Photoreceptor Molecule

Numerous investigators, particularly Jennings (1906) and Mast (1911), have studied the rate of movement of *Euglena* and other protozoa. It was found that the swimming rate of *Euglena* was dependent upon the absolute amount of light energy absorbed, that a constant rate of motion was observed at a constant light intensity, and that the rate of motion varied as the intensity was varied. Action spectra for *Euglena* photokinesis should reveal the absorption characteristics of the molecule or molecules responsible for such photobehavior.

The rate of swimming of *Euglena* was measured by suspending the organism in culture media which had a viscosity of 0.987 centipoises at 25°C. The measurements were made under a microscope in a dark room at 25°C by timing the distance an individual organism moved in a calibrated Levy counting chamber. The counting chamber was illuminated uniformly and the intensity of light was controlled. The velocity of free forward swimming without any sidewise rotation was measured by counting the number of squares on the chamber through which the organism swam in a given length of time. A typical swimming pattern is shown in Figure 7.2. The average distance which *Euglena* travelled was found by plotting on graph paper the actual patterns of numerous organisms.

It was observed that swimming velocity did not immediately change with

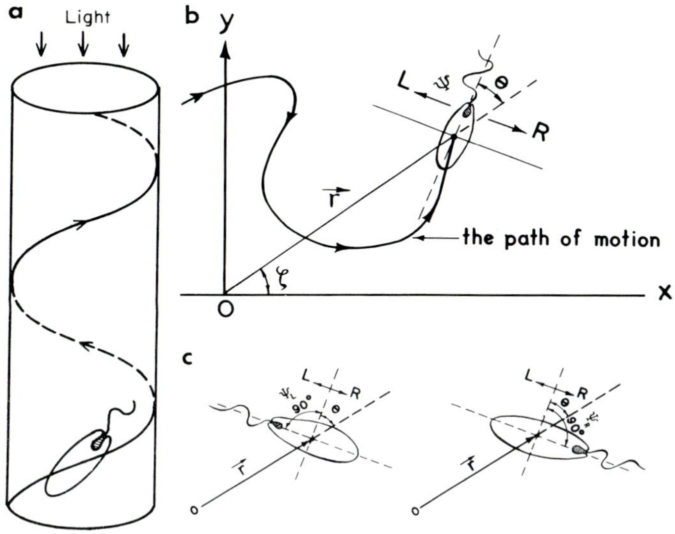


FIGURE 7.2 Swimming patterns of *Euglena gracilis* in response to light.

light intensity, but that it took 10 to 15 minutes for the organism to adapt to the new conditions before a regular pattern of motion could be observed. A lag period of 10 to 15 minutes for a change in velocity following a change in illumination was previously observed for *Euglena*, *Chlamydomonas*, *Volvox*, and many other microorganisms (Holmes, 1903; Mast, 1911; Loeb, 1918). A possible interpretation of this reaction is that it depends upon a mechanism that is only indirectly affected by illumination.

The relationship between velocity and light intensity is shown in Figure 7.3a. It will be noted that the mean velocity or rate of swimming rises sharply from 0.11 mm/sec at 3 $\mu W/cm^2$ until it reaches a maximum rate of 0.16 mm/sec at 60 $\mu W/cm^2$, the saturation intensity; it then starts to decrease slowly as the light intensity is raised above 60 $\mu W/cm^2$ to 225 $\mu W/cm^2$. However, when the light is passed through a polarizer, the velocity steadily increases from 0.10 to 0.18 mm/sec at 20 $\mu W/cm^2$ and then remains constant to 45 $\mu W/cm^2$ as shown in Figure 7.3b.

The swimming rate in white light reaches a maximum average velocity of 0.16 mm/sec only when the intensity of light is raised to 60 $\mu W/cm^2$. The significance of this value is that the number of light quanta at this particular intensity is sufficient to cover all the molecules which have a thermal energy equal to or greater than the minimal thermal energy, which is defined by the illumination and the absolute temperature. Beyond this intensity, the absorption rate will remain constant without being disturbed by the extra number of quanta falling on the molecules, since the maximum absorption capacity has already been reached. The same effect was observed for various wave-

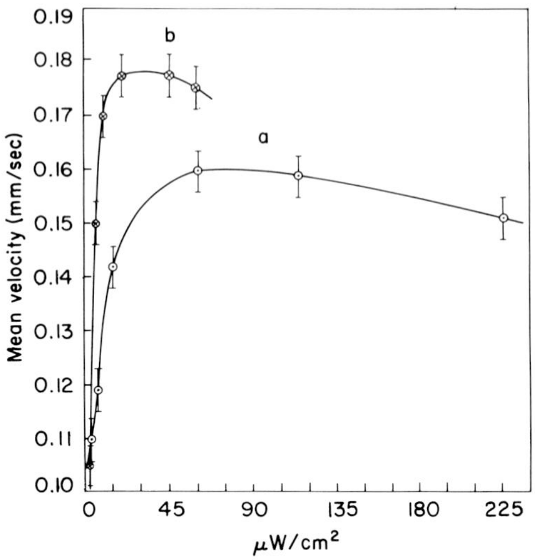


FIGURE 7.3 *Euglena gracilis.* The effect of different light intensities on the mean velocity or rate of swimming; (a) in nonpolarized white light and (b) in polarized light.

lengths, although the saturation intensity was different in each case. The greater mean velocity of 0.18 mm/sec, observed at 20 $\mu W/cm^2$ of polarized light, would indicate that polarized light is more efficient than is white light.

In evaluating the relative effectiveness of various wavelengths on the rate of swimming, it was found that the mean velocity (mm/sec) plotted against light intensity gave straight lines for light intensities of less than 22 $\mu W/cm^2$. The photokinesis action spectrum plotted for the rate of swimming (mean velocity in mm/sec against wavelength, at 6 $\mu W/cm^2$, intensity) is illustrated in Figure 7.4a. It will be observed that there is a major peak at 465 nm and another peak near 630 nm. This action spectrum is indicative of the absorption spectrum of the pigment involved in regulating the rate of swimming. The peak at 465 nm is suggestive of a carotenoid ($\beta$-carotene), whereas the peak at 630 nm is suggestive of chlorophyll, or its precursor protochlorophyll (Figure 3.17), or another molecule. The peak at 630 nm may actually be the result of energetic interference between two pigments, or the presence of other light-absorbing molecules within the organism.

In the dinoflagellate *Gyrodinium dorsum* prior irradiation with red or far-red light was found to affect the organism's stop response. This stop response can be used as a criterion for measuring the effect of light on the photoreceptor. Another observation was that brief irradiation at 470 nm elicited a stop response in cells anytime during 6-minute intervals; around 620 nm the cells are reactivated, but at 700 nm there is a reversal of reactivation (For-

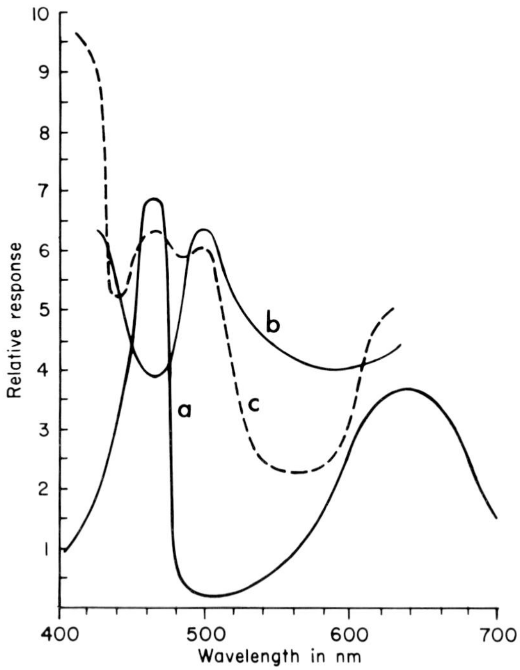


FIGURE 7.4 *Euglena gracilis*. Action spectra for (a) photokinesis, (b) phototaxis, (c) phototaxis in polarized light.

ward and Davenport, 1968). The blue-absorbing pigment at 470 nm was indicated to be a flavin, but their observations also suggest that a phytochrome is associated with the blue-absorbing pigment and is in fact the active pigment molecule. The spectral data associated with the photobehavior of *Gyrodinium* imply that a flavin and a phytochrome participate in the process. However a phytochrome pigment has so far not been isolated from *Euglena*.

## Phototaxis and the Photoreceptor Molecule

The degree to which *Euglena* respond to different wavelengths was studied experimentally by obtaining the spectral sensitivity, or phototaxis action spectrum (Mast, 1911; Loeb, 1918; Manten, 1948; Clayton, 1953; Bünning and Schneiderhöhn, 1956, Batra and Tollin, 1964; Halldal, 1964; Diehn and Tollin, 1966a,b).

To obtain more information about the phototactic response of a population of *Euglena,* the organisms were dark-adapted and a known population was placed into a long piece of glass tubing. Various filter combinations were inserted along the length of the tube and all were adjusted for equal light intensities. The tube was shaken and the organisms permitted to swim freely to a colored area of their choice. Their accumulation in terms of density of organisms in front of each colored filter was determined at various time intervals. The results after 15 minutes showed that a greater density of organisms accumulated in a narrow region near 465 nm, and in the region from 570 to 600 nm. The major peak at 465 nm agreed well with that found in the photokinesis action spectrum (Figure 7.4a).

For a quantitative study of phototaxis, a spectrophotometer was constructed, as illustrated in Figure 7.5. The apparatus was a completely enclosed black box; the only light entering the test cell was from the light source $B_2$. The light source $B_1$ was collimated by the convex lens $L_1$ to the filter $\lambda_1$. The glass plates, G, were used to adjust the intensity of the filtered light passing through the experimental cell ($W_1$ and $W_2$). In these experiments a suspension of *Euglena* was placed in the cell compartment $W_1$, and the culture medium in $W_2$ as a reference. The light transmitted through both $W_1$ and $W_2$ passed through the slit $S_1$, the lens $L_2$, and was then reflected 90° at the half-reflecting prism P to the eyepiece, E. $W_2$ counterbalanced the deflection of the light pass-

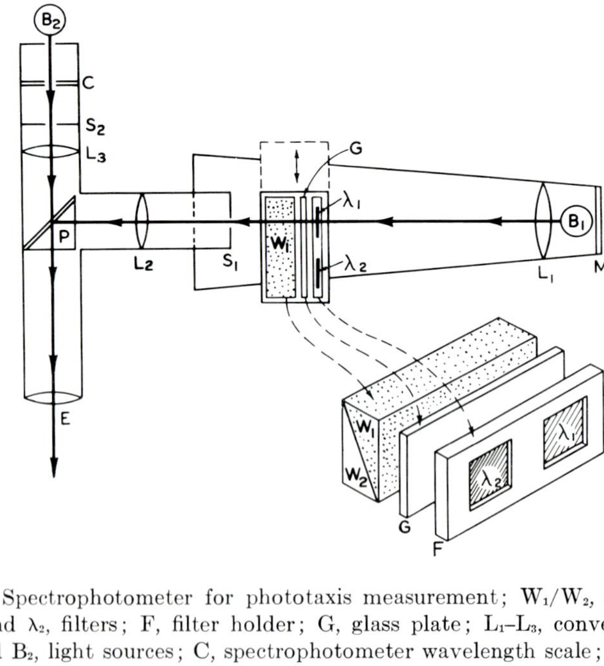


FIGURE 7.5 Spectrophotometer for phototaxis measurement; $W_1/W_2$, experimental cell chamber; $\lambda_1$ and $\lambda_2$, filters; F, filter holder; G, glass plate; $L_1$–$L_3$, convex lenses; $S_1$ and $S_2$, slits; $B_1$ and $B_2$, light sources; C, spectrophotometer wavelength scale; P, half-reflecting prism; M, mirror.

ing through the prism formed by the suspension of organisms inside $W_1$. The light from source $B_2$ passed through the scale plate C, the slit $S_2$, and the lens $L_3$, which projected the scale onto the eyepiece where it could be read.

The principle upon which this measurement is based is given by the equation:

$$I_0/I = Ae^{-kx}$$

where $I_0$ is the light intensity before entering $W_1$, $I$ is the light intensity after passing a distance $x$ through the suspension of organisms inside $W_1$, and $k$ and $A$ are constants characteristic of the organism and the medium, respectively.

Two filter combinations corresponding to two different dominant wavelengths, $\lambda_1$ and $\lambda_2$, can be placed simultaneously in the filter holder F, so that one half of $W_1$ is illuminated with $\lambda_1$ and the other half with $\lambda_2$. If at any instant, the concentration of organisms in $W_1$ on the side of $\lambda_1$ is greater than that on the side of $\lambda_2$, the slit image will shorten in vertical length more on $\lambda_1$ than on $\lambda_2$. Thus the relative spectral sensitivity for the wavelength $\lambda_1$ is defined as greater than that for $\lambda_2$. Similarly, observation can be made for another filter combination corresponding to dominant wavelengths $\lambda_1$ and $\lambda_2$ (Wolken and Shin, 1958).

The advantage of using the wedge-shaped cell is that the light intensity transmitted is a negative exponential function of the concentration times the distance of transmission, and the nonuniform vertical distribution of the concentration of the organisms is counterbalanced by making the angle of the wedge-shaped compartment $W_1$ sufficiently large. This makes it possible to obtain a sharper boundary at the top of the slit image.

By this method the relative spectral sensitivities, or the action spectrum, for phototaxis were obtained for dark-grown, dark-adapted, and light-grown *Euglena*. The action spectrum for light-grown *Euglena* shows a major peak at 490 nm, with some sensitivity near 420 nm and a rise beyond 600 nm (Figure 7.4b). For dark-adapted and dark-grown *Euglena*, similar spectral peaks were obtained. This indicates that similar pigments are probably responsible for the phototactic response. In polarized light there are maxima at 468 and 508 nm and a rise near 600 nm (Figure 7.4c). The maxima at 468 nm and the one approaching 600 nm almost correspond to the peaks of the photokinesis action spectrum (Figure 7.4a). The polarized light effects may indicate that there are two light-absorbing pigments and these would be $\beta$-carotene and a flavin (Diehn, 1973). If there are two pigments, the resulting polarized light effect could be due to a mutual energetic interference between them.

To relate the action spectra obtained for photokinesis and for phototaxis to the absorption spectrum of the photoreceptor pigment molecule, it becomes necessary to isolate the pigment from the eyespot region.

## The Eyespot Pigment and the Photoreceptor Molecule

It was generally believed that the pigment of the eyespot is a carotenoid mainly because the action spectrum for phototaxis resembles the absorption spectrum of a carotenoid.

From light-grown *Euglena gracilis* three main carotenoids can be extracted. These are β-carotene, lutein, and neoaxanthin. When these pigments are dissolved in petroleum ether, the absorption spectrum lies within the spectral range of 415 to 475 nm; in carbon disulfide, a nonpolar solvent, the absorption spectrum range is from 450 to 510 nm. These or similar carotenoids could then account for the absorption peaks found in the action spectra (Figure 7.4). In the dark-grown *Euglena,* lutein and β-carotene are found in highest concentration. Chromatographic analysis of the *Euglena* carotenoids showed that of these carotenoids, 80% are lutein or antheraxanthin, 11% are β-carotene, and 7% are neoaxanthin; also there are small amounts of γ-carotene, crytoaxanthin, echinenone, and two ketocarotenoids, euglenanone and hydroxyechinenone (Krinsky and Goldsmith, 1960). In addition, Krinsky (1964) studied the carotenoid pigments following illumination under aerobic and anaerobic conditions and found that under anaerobic conditions a photochemical conversion occurred from epoxide to nonepoxide carotenoids:

$$\text{Antheraxanthin} \underset{\text{dark}}{\overset{\text{light, } N_2}{\longleftrightarrow}} \text{Zeaxanthin}$$

This reaction was reversed in the dark and could be inhibited by aerobiosis (see for example Figure 3.8). Both the carotenes and neoaxanthin remained essentially unchanged under these conditions (Bamji and Krinsky, 1965).

When Tischer (1936, 1941) isolated from the red *Euglena sanguinea* astaxanthin, $C_{40}H_{44}O_4$ (3,3′-dihydroxy-4,4′-diketo-β-carotene), it was thought to be the photoreceptor pigment molecule (Figure 3.6). Astaxanthin was until then found only in animal tissue, particularly in the integuments of crustacea (familiar as the red pigment of boiled lobster), and as a screening pigment in the retina of birds. The absorption spectrum of crustacean astaxanthin, unlike those of the common plant carotenoids, is a single broad band, maximal in the blue–green at about 500 nm (Goodwin, 1952). Tischer (1936, 1941) also isolated the red pigment haematochrome from *Euglena heliorubescens* and from *Haematococcus pluvialis,* and showed that the principal component of hematochrome is euglenarhodon ($C_{40}H_{48}O_4$), a ketonic xanthophyll. Euglenarhodon is also believed to be the red pigment of *Euglena rubra* (Johnson and Jahn, 1942). Euglenarhodon structurally resembles and could be identical to astacin which has been isolated from crustacea. Astacin is a ketonic carotenoid (3,4 tetraketo-4,4′-β-carotene), found together with astaxanthin from which it is formed by oxidation. The absorption spectrum of astacin in pyridine has its maximum around 500 nm, similar to that of astaxanthin. If the *Euglena* eyespot pigment is astaxanthin it would nicely fit the peculiar position in the biology of the photosynthetic flagellates on the plant → animal continuum, for they would have both the plant pigment chlorophyll and the animal pigment astaxanthin. But astaxanthin has not been identified in photosynthetic or nonphotosynthetic *Euglena gracilis* mutants (Goodwin and Jamikorn, 1954). Furthermore, there is no evidence to suggest that astaxanthin participates in the photoreceptor processes of animals, either in the visual pigment chemistry of the retina or in the metabolism of vitamin A.

To circumvent the difficulties of isolating and extracting the eyespot with solvents, the absorption spectra of the eyespot can be obtained *in situ* with the use of a microspectrophotometer. Absorption spectra from *Euglena* eyespots show a broad absorption band from 440 to 520 nm (Gössel, 1957; Strother and Wolken, 1961; Wolken, 1967). More recent studies (Wolken, 1971) of absorption spectra of the eyespot region show that the peaks lie near 430, 465, and 495 nm, and near 350 nm in the ultraviolet (Figure 7.6a). In the heat-bleached (HB) mutant that lacks chloroplasts, and hence chlorophyll, the spectrum of the eyespot area shows peaks near 430, 465, and 510 nm and at 340 nm (Figure 7.6b).

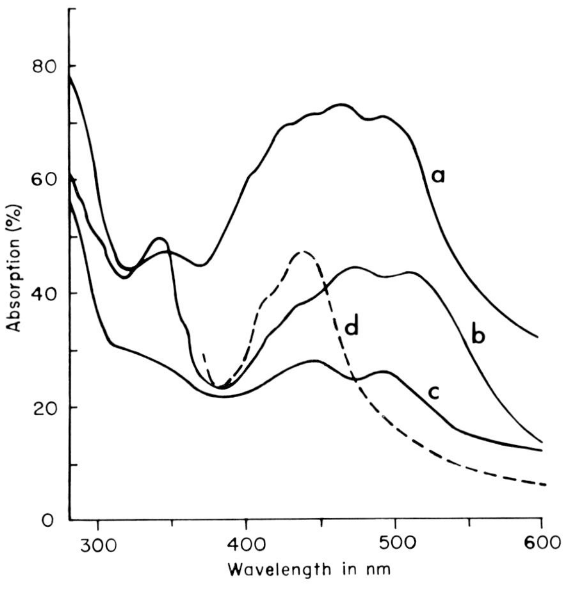


FIGURE 7.6 *Euglena gracilis.* Absorption spectra for (a) eyespot area, light-grown, (b) eyespot area of heat-bleached mutant, (c) eyespot area after dark adaptation for 1 hour, and (d) eyespot area after 5 minutes white light. Compare to Figure 7.4. (Obtained by microspectrophotometry.)

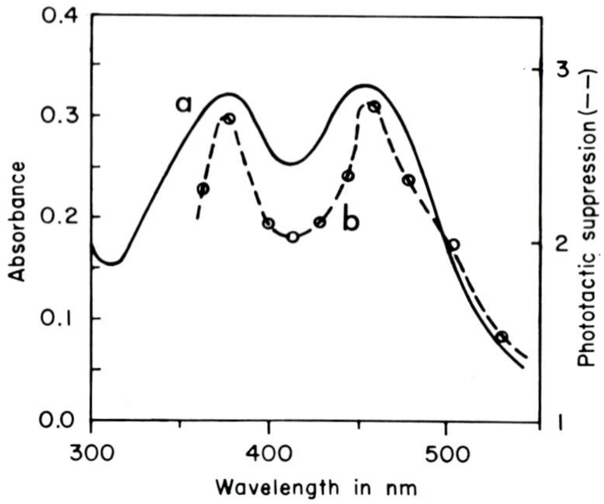


FIGURE 7.7 (a) Absorption spectrum of riboflavin compared to (b) phototactic suppression action spectrum of *Euglena* (from Tollin and Robinson, 1969).

Spectra closer to the base of the flagellum and near the paraflagellar body show absorption peaks at about 440 and 490 nm and two smaller peaks near 558 and 590 nm (Figure 7.6c). When these light-grown *Euglena* are dark-adapted for 1 hour and mounted on the cold stage (5°C) of the microspectrophotometer, and the same area is illuminated with strong white light for 1 to 5 minutes, the absorption peak around 490 nm bleaches, accompanied by an increase in the absorption peak at 440 nm (Figure 7.6d).

Interpretation of these spectra to establish the identity of the photoreceptor pigment is, however, extremely difficult. Phototactic action spectra by Tollin and Robinson (1969) indicate that the eyespot pigment acts as a shading device or filter for a photosensitive pigment, located in or near the paraflagellar body. Their action spectrum resembles that of a flavin with absorption peaks at about 460 and 370 nm (Figure 7.7) and suggests a flavin or flavoprotein to be the photoreceptor pigment. This is not too surprising, for flavins have also been implicated as the pigments responsible for chloroplast movement (Haupt, 1973). From *Euglena* phototactic studies Diehn (1973) also concluded that there are probably two photoreceptor molecules, a carotenoid which functions in the visible range and a flavin in the ultraviolet part of the spectrum. Such a pigment system would be similar to that which we have described for *Phycomyces* phototropism.

If a flavin is the photoreceptor molecule, then an analysis of *Euglena* flavins would give us some idea of the photoreceptor concentration. The total flavins extracted from both light- and dark-grown *Euglena* are of the order of $10^8$ molecules per cell, which falls within the range found for the number of photoreceptor molecules.

All the evidence at present then would indicate that there are two pigment molecules in *Euglena*: a carotenoid in the eyespot granules which acts as a light filter or accessory pigment, and at the base of the flagellum, near or in the paraflagellar body, a flavoprotein which becomes photoreduced to flavin semiquinone. We conclude then that a flavoprotein functions as the primary photoreceptor molecule. So we see that the photoreceptor pigments for photomotion of *Euglena* are similar to the pigments for *Phycomyces* phototropism.

## The Photoreceptor System and Excitation

Let us now consider *Euglena* as a primitive photosensory cell and examine more closely the eyespot region (stigma and flagellum) as the organism's device for nervous control. As already mentioned, in *Euglena* the paraflagellar body is associated with the flagellum which serves as its photoreceptor–effector system. By means of the intensity and wavelength of light we can "communicate" with the organism to the extent that its speed and direction of motion are controlled. This in turn suggests a sensory cell or its analog, a photocell. *Euglena's* eyespot–flagellum system may therefore be regarded as a servo or feedback mechanism which endeavors to maintain an optimal level of illumination on the organism. The eyespot and flagellum are somehow linked so that light falling on the "eye" produces motion. This translation of an internal effect into a surface action produces problems similar to those involved in the origin of nervous impulses in animal photoreceptor cells. If we extend this point of view, we can look upon the paraflagellar body and flagellum region as a primitive retinal photoreceptor cell (see Figures 7.1 and 7.8).

Knowing the area of the photoreceptor, the effective wavelength, and the light intensity, the energy necessary to produce a response can be roughly cal-

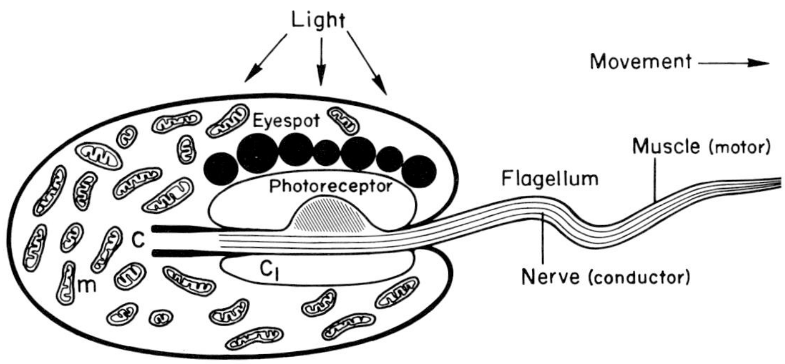


FIGURE 7.8 Photosensory cell, based on the eyespot–flagellum system of *Euglena;* m, mitochondria; C, microtubule in axoneme of cilium; $C_1$, centriole.

culated. Castle (1935), studying phototropism in *Phycomyces*, estimated that the energy is $1.9 \times 10^{-7}$ ergs/cm$^2$/sec, assuming that 10% of the radiation was absorbed. This estimation shows that the energy necessary to produce a perceptible mechanical response in the cell is small. The reaction time of *Phycomyces* under most favorable conditions is at least 2 minutes and may be as much as 10 minutes (Delbrück and Reichardt, 1956). Our estimation for *Euglena* is $1.7 \times 10^{-11}$ ergs/cm$^2$/sec, or a quantum efficiency of 14%. The number of photons which can excite the "eye" at this frequency is seven. It is interesting to note in comparison that the human eye can detect a minimum of four photons at around 500 nm.

The absorption spectrum of the eyespot implies that the velocity of swimming is proportional to the number of light quanta absorbed at the eyespot and that the swimming motion is energetically controlled by light absorption at the eyespot. The shape of the intensity-dependence curve in Figure 7.3, showing a gradual rise with increase in intensity and the appearance of plateaus at higher intensity values, is very similar to the current-intensity curve of a photoconductive cell. It has been suggested that the creation of nerve impulses in visual processes may be considered an event in which some electrical charge is produced by means of reactions derived from photoactivation of rhodopsin (Rosenberg, 1958). Whether there really exists a characteristic threshold potential, and whether the energy transfer is done electrically, may be experimentally determined by measuring the potential drop between the eyespot and the flagellum; such measurements, however, are difficult to make.

In the photoreceptors of higher animals, the photoexcitation triggering the optic nerve, and for the most part the energy contained in a pulse, are derived from chemical energy. Thus, the number of electronic charges involved in forming one such pulse is much larger than the minimum number of light quanta required to trigger the optic nerve. In the case of *Euglena*, however, such an amplification mechanism is not necessary. The minimum number of quanta required to excite the eyespot is comparable to the power involved in the swimming motion. This means that one light quantum which is effectively absorbed at the eyespot can be associated with approximately one electronic charge formed at the base of the flagellum. At the saturation intensity of about 60 $\mu$W/cm$^2$, the swimming velocity is about 0.18 mm/sec in a medium of viscosity of 0.987 centipoises. Using the cross section of the eyespot, the intensity of 60 $\mu$W/cm$^2$ at wavelength 465 nm (which is equivalent to about $2 \times 10^4$ quanta/cm$^2$/sec), and the average radius of the *Euglena* cell, we have estimated the threshold potential to be of the order of 0.01 to 0.1 mV. This is small when compared with the values found for nerves of higher animals but within the order of magnitude found for certain insect eyes (Naka, 1960), and is comparable to the early receptor potential (ERP) found in most excitatory tissue.

## The Flagellum and Excitation

I would like to add my own speculations as to how the flagellum functions. The flagellum performs a whipping motion very similar to that of a cowboy whipping his horse. The whipping reaction is probably adjusted so that the organism can move forward or backward and faster or slower. The light, being absorbed by the pigment within the eyespot area, acts as a continuous energy source for the motion. The energy flow is assumed to be noncontinuous but supplied to the flagellum in unit pulses. The fact that the motion of the flagellum is very smooth indicates that the total length of the flagellum, as a whole, forms a certain regular pattern of motion. One whipping is not caused by one single pulse but is caused by many small pulses being fired successively along a certain length of the flagellum.

Although no one has seen how the whipping causes the organism to perform with such a streamlined motion, it is suggested that the nine outside fibrils are perhaps wound helically around the two continuous central ones, although the electron micrographs do not indicate this. First, the flagellum was thought of as a simple propeller. In Figure 7.9 the propeller is rotating clockwise, and the direction of the flow of the medium is shown by arrows. Thus, the medium flows in the direction opposite to that of the motion of the beat of the propeller, either at the front or the tail of the organism. The surface of the propeller is so inclined that it pushes the medium to the back when it rotates clockwise and forward when it rotates counterclockwise. If the nine fibrils were wound

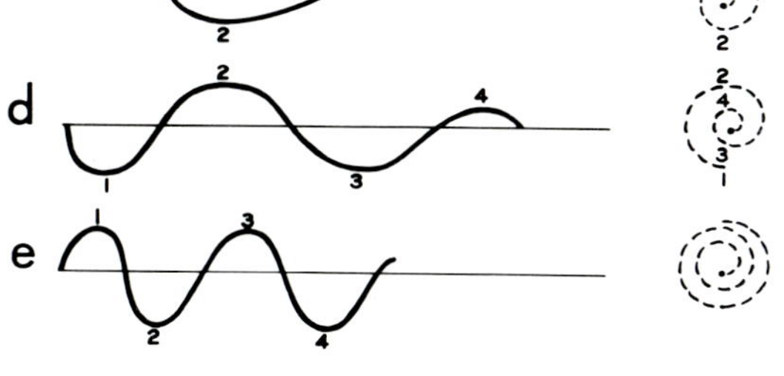


FIGURE 7.9 Whipping patterns of the flagellum, a,b,c,d,e, position and movement.

helically with a certain constant advancing angle and the whipping were made by twisting the flagellum clockwise, the medium in which the organism swims would be pushed backward. If the flagellum were twisted counterclockwise, the medium would be pushed forward. In the former case the organisms would move forward and in the latter, backward. The velocity would depend on the strength of the thrust, the number of whippings per unit time, and the distribution of impulse firings along the length of the flagellum. To keep the whipping action going at a constant rate, it is necessary that the thrusts occur at certain intervals of length along the flagellum at a critical instant. Also, if a thrust is given at a point to the right, the next one should be given to the left to keep the motion at a constant whipping radius. Therefore, it is conceivable that there is a certain feedback mechanism at work which lets the pulse firings occur at the side where the external mechanical stimulus is greatest.

Using the same model, the two central fibrils are thought to be energy-pulse carriers as well. One of them is designated the main pulse carrier and the other the feedback-pulse carrier. The nine outside fibrils fire the pulses at various junctions along the flagellum, while the central fibrils bring the pulses to the junction. The direction of the pulse transfer of the main pulse carrier is from the eyespot down to the flagellum, and that of the feedback-pulse carrier from the flagellum to the photoreceptor. The central fibrils and the nine outside fibrils are responsible for locating the right junctions at which to fire pulses. Thus, the feedback unit at a junction forms a feedback mechanism with the other outside fibrils. If a sudden shock is given to a point on the flagellum, it will cause a sudden reflective impulse at that point. Therefore, a sliding filament mechanism analogous to a muscle system appears to provide the most satisfactory basis for a simple feedback mechanism (Brokaw, 1972).

That the energy flow from the receptor to the effector is not continuous but is in unit pulses could be due to a certain threshold phenomenon associated with the photo- and electrochemistry at the base of the flagellum, close to or at the paraflagellar body. These are observed as successive trial and error movements in the course of hunting for the place of optimum illumination. The sidewise oscillations or the gyrating motion of *Euglena* may be regarded as part of the hunting process, similar to the rotating antenna of an antiaircraft gun.

If we consider that the information received at the eyespot in terms of light quanta of various frequencies and intensities is converted into some other form of communication, and that these new information codes are transmitted to the effector (flagellum) in such a way that the resulting responses of action are consistent with the initial information received at the eyespot, we can conceive of the following simple type of mechanism of transmission of information from the eyespot (photoreceptor) to the flagellum (effector) in *Euglena*. The

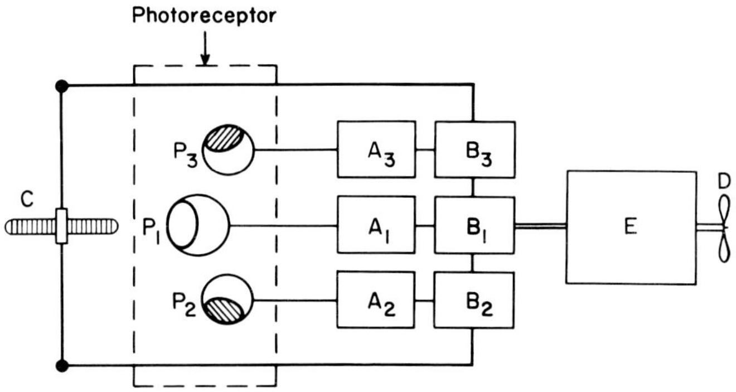


FIGURE 7.10 An analog "machine" of the photoreceptor system that is able to perform the same kind of motion and light searching as *Euglena*.

information is transmitted by means of pulses of a constant height, where the time interval between two pulses defines the information being transmitted; each pulse carries a certain constant amount of energy so that the net amount of energy transmitted down to the flagellum per unit time is directly proportional to the number of pulses carried per unit time; a change in frequency of the light incident on the eyespot brings a change in the time interval between two pulses (the time interval decreases as the spectral effectiveness increases); a change in the intensity of light brings a change in the time interval between two pulses; and the information for selecting the direction of movement, or the "steering mechanism," is good enough only for selecting between two directions and no more, i.e., between left (L) and right (R), as shown in Figure 7.2b,c.

The rate and the direction of swimming of *Euglena* can be practically controlled by external light conditions. This then permits us to conceive of a machine which may be able to perform exactly the same kind of motion as *Euglena* on being illuminated by light. One possible machine of this type is a photocell which is schematically drawn in Figure 7.10. In the diagram, $P_1$, $P_2$, and $P_3$ are photocells, of which $P_2$ and $P_3$ are the phototactic receptors effective in determining the directional preferences, while $P_1$ controls the speed of motion resulting from the engine E and the propeller D. $A_1$, $A_2$, and $A_3$ are the proper amplifiers which will amplify the signals conveyed from $P_1$, $P_2$, and $P_3$ to $B_1$, $B_2$, and $B_3$, respectively. $B_1$, $B_2$, and $B_3$ are the devices through which the signals from $P_1$, $P_2$, and $P_3$, respectively, are observed as mechanical motion at D and C. It is supposed that the photosensitive pigment in $P_1$ has a characteristic spectrum like that of photokinesis (Figure 7.4a), while $P_2$ and $P_3$ contain a photosensitive pigment which has a characteristic effectiveness spectrum like that of phototaxis (Figure 7.4b).

We conclude from the photobehavior of *Euglena* and its structures for photoreception, that the eyespot, the flagellum, and the paraflagellar body function together as its photoreceptor system. We can then look upon *Euglena* as a primitive "retinal cell." Whether suitable analogies can be drawn between *Euglena* photoreceptor structures and the retinal photoreceptors of imaging eyes may become apparent as we proceed in our comparative structural analysis from the flagellated protozoa phylogenetically to the vertebrate eye.

# VIII
# Invertebrate Eyes

In the country of the blind the one-eyed man is King.

Greek Proverb

Darwin, wondering how the eye evolved from the phototactic protozoa through the arthropods and molluscs to the vertebrate eye, stated:

> When we reflect on these facts, here given much too briefly, with respect to the wide, diversified and graduated range of structure in the *eyes of lower animals;* and when we bear in mind how small the number of living forms must be in comparison with those which have become extinct, the difficulty ceases to be very great in believing that *natural selection* may have converted the simple apparatus of an optic nerve, coated with pigment and invested by transparent membrane, into an optical instrument. . . .
>
> Charles Darwin, "The Origin of Species," 1859

The invertebrates are the most numerous and diverse among all animal species, and their photoreceptors and visual organs are equally diverse. From an evolutionary standpoint their history goes back to Cambrian times, about 500 million years ago. Therefore it is of considerable interest to investigate the photoreceptors and eyes of these animals. Their structure and chemistry may have evolutionary significance and may link the development of simple receptors to that of invertebrate eyes and from these to the vertebrate eye.

Let us see if we can trace the evolutionary development of the eye by examining the eye phylogenetically from primitive photoreceptor systems to simple imaging eyes, compound eyes, and the refracting type eye of vertebrates.

The first eyes were merely regions of the outer skin which were sensitive to light. The phototactic responses, phototropism and phototaxis, could have been more sensitive to heat than to light, and there is some evidence to indicate that the touch receptors preceded the visual receptors. Gregory (1967) proposed that the first retinal images received by the original nervous system were for pattern touch, and that the compound eye could be a multiplication element, sensing down a single channel to explore space in a manner similar to a touch probe.

Haldane (1966), in discussing the evolution of the eye, wrote:

> There are only four possible types of eye if we define an eye as an organ in which light from one direction stimulates one nerve fiber. There is a bundle of tubes pointing in different directions and three types analogous to three well-known instruments, the pin-hole camera, the ordinary camera with a lens, and the reflecting telescope. A straightforward series of small steps leads through the pin-hole type to that with a lens, and it is quite easy to understand how this could have been evolved several times.

The flatworms evolved the pinhole eye, which is the simplest of imaging eyes for it uses the principle by which a small hole in the wall of an opaque chamber allows the passage of a very narrow beam of light from each point on an object, forming an inverted image on the opposite wall of the chamber (Figure 8.1a). This type of eye is not very efficient as an image-forming device. The pinhole eye allows only a small fraction of the light from an object to reach the photoreceptor surface. If the hole is made larger to increase the amount of light, image definition is lost, and if it is made smaller to improve the definition, diffraction effects become a problem. However, this eye has the advantage of simplicity because no focusing is required for near or distant objects, and the size of the image is inversely proportional to the distance of the object.

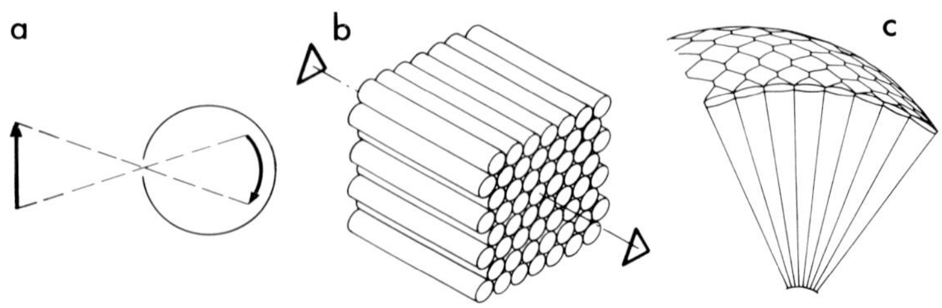


FIGURE 8.1 Imaging systems. (a) pinhole, (b) parallel tubes, and (c) tubes arranged as in the compound eye.

The common planarian, *Platyhelminthes,* has two such pinhole type eyes, each containing pigment cells and photosensory cells. The pigment cells function to shade the sensory cells from light in all but one direction. This enables the animal to respond differentially to the direction of light, that is, to move away from the light. The animal's tendency to avoid light seems to be controlled by the balance of nervous impulses from the photosensory cells of the eye. Such sensory cells were described for *Dugesia lugubris* and *Dendrocoelum lacteum* (Röhlich and Török, 1961; Röhlich, 1966) and for the marine planarian *Convoluta roscoffensis* (Keeble, 1910; Wolken, 1971). A pinhole eye is also found in *Nautilus,* a cephalopod mollusc. These animals all have photosensory cells of a specialized structure formed of lamellae for light capture.

For example, in the earthworm, an annelid, there are light-sensitive cells embedded in the body wall, each containing a lens surrounded by a neurofibrillar network. In many leeches such cells are gathered into cups shielded by pigment cells. These cup-shaped clusters of light-sensitive cells may lie under a lens, which is a specialized area of cuticle tissue. The complete structure of a lens and photosensory cell clusters is referred to as an *ocellus* or simple eye (Figure 8.2).

Insects possess from one to three ocelli. The median or dorsal ocellus of many insects consists of a layer of photoreceptor cells, a synaptic zone in which the axons of the photoreceptor cells come into contact with dendrites of the ocellar nerve fibers, and an ocellar nerve that leads from the eye to the brain (Chappell and Dowling, 1972).

A second kind of rudimentary optical system that forms an image utilizes a bundle of tubes. Only light falling on a particular tube in the direction of its axis can proceed to the end of the tube where it forms an upright image of the object (Figure 8.1b). The image formed is the same size as the object, regardless of distance. Therefore only those objects equal to or smaller than the device can be imaged completely, and no impression of distance by perspec-

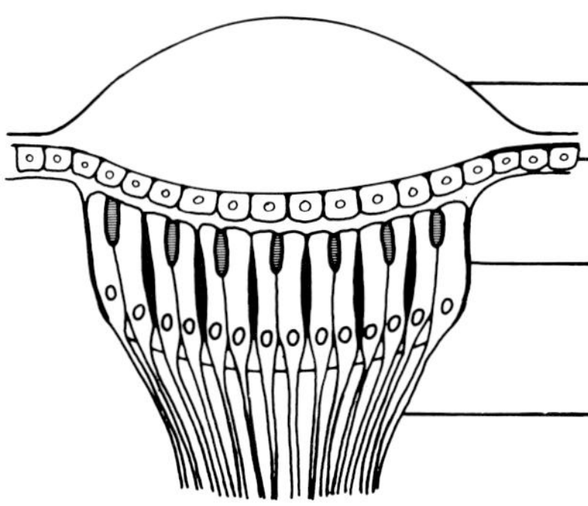


FIGURE 8.2 Simple eye, an ocellus.

tive is conveyed since the image size remains unchanged with object distance. However, if the tubes are disposed around the surface of a sphere or sphere segment with their axes pointing toward the center, as developed in the structure of image-forming compound eyes, then many of these disadvantages are avoided (Figure 8.1c).

## Arthropods

### The Compound Eye

The arthropods include insects, arachnids, and crustacea. Not only do they possess eyespots and ocelli, but almost all arthropods have well-developed, image-forming compound eyes. The image-forming compound eye evolved in

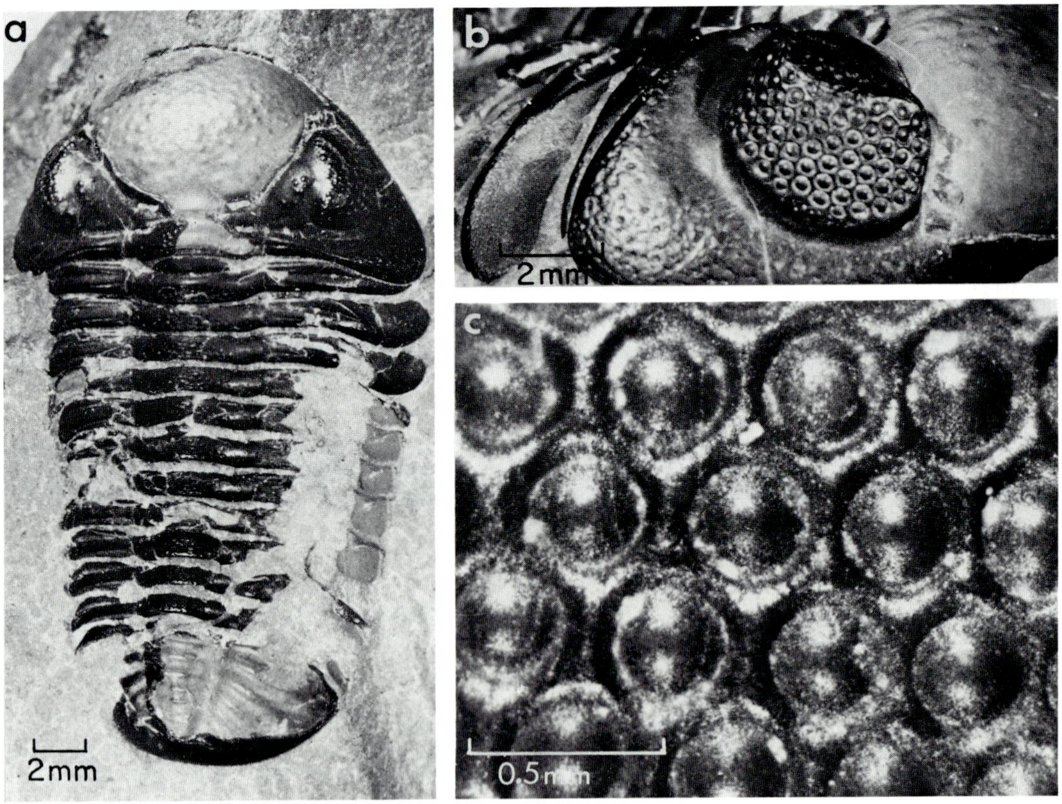


Figure 8.3 *Phacops rana*. (a) Trilobite fossil, Devonian silica shale, from Ohio; compound eye structures (b,c). (Trilobite courtesy of Carnegie Museum, Pittsburgh, Pennsylvania.)

the Cambrian era some 500 million years ago, as found in the eyes of a trilobite fossil, *Phacops* (Figures 8.3 and 8.4).

The arthropod eye is very efficient for short-range vision and for detection of movement in the total visual field. Many arthropods orient to the plane of vibration of polarized light, which would indicate a polarized light analyzer within the eye. Some can discriminate colors and therefore should have visual pigments with different absorption maxima.

Let us now take a look at the compound eyes of a few selected insects to see how they are structured for vision. In the generalized structure of the compound eye, the eye facets are called *ommatidia*. The number of ommatidia varies from only a few in certain species of ants, to more than 2000 in the dragonfly. Each ommatidium is a complete eye in the sense that it has a cor-

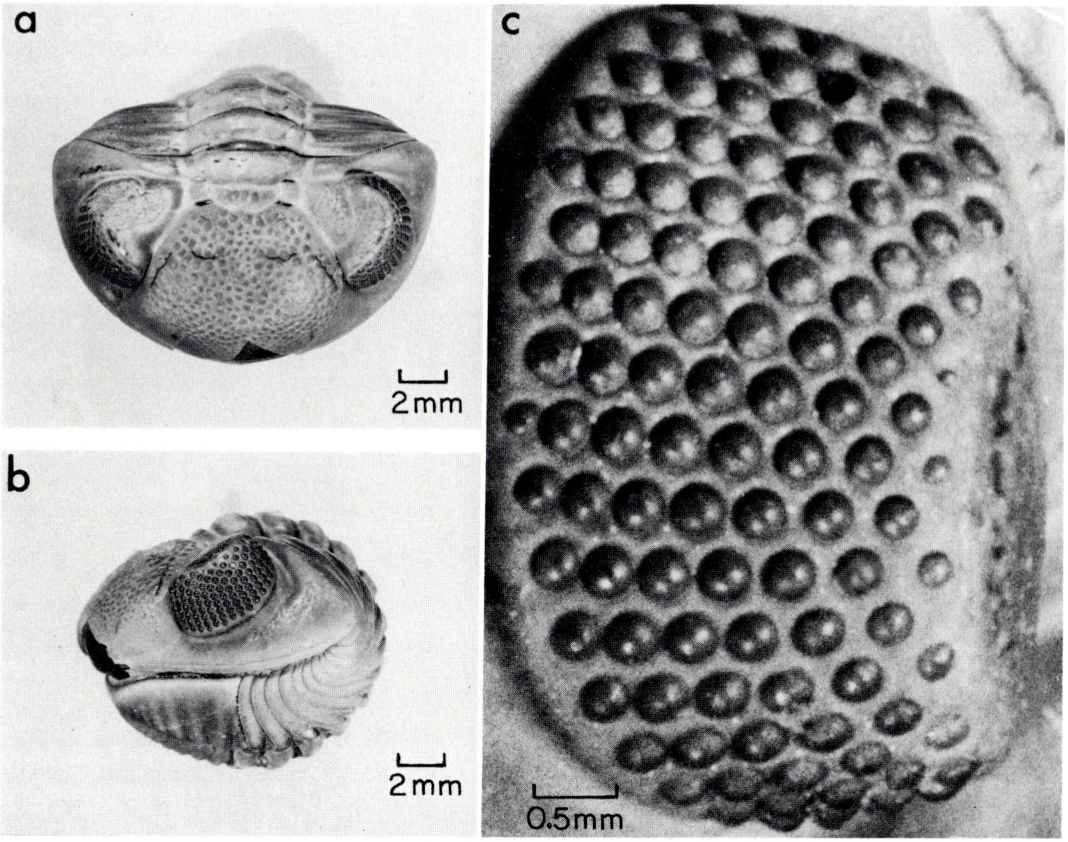


FIGURE 8.4 *Phacops rana,* trilobite fossil, Devonian, from New York; compound eye structures (a,b,c). (Trilobite courtesy of Carnegie Museum, Pittsburgh, Pennsylvania.)

neal lens, a crystalline cone, and retinula cells (Figure 8.5). Depending upon the insect, there are from three to 11 retinula cells per ommatidium. The retinula cell contains a differentiated photoreceptor structure, the *rhabdomere*, which is analogous in function to the retinal rod outer segment of the vertebrate eye. Collectively, the rhabdomeres form the light-sensitive *rhabdom*, which serves as the photoreceptor area within each ommatidium.

Exner (1891) described two anatomically distinct types of compound eye structures, the *apposition* and *superposition* eyes. In Exner's model, apposition eyes are those in which the rhabdomeres forming the rhabdom lie directly beneath or against the crystalline cone, and in which an inverted image is formed at the level of the rhabdom (Figure 8.5a). Each ommatidium is entirely sheathed by a double layer of pigment cells. Therefore, only light striking the lens within 10° of the perpendicular reaches the rhabdomeres. Light striking the lens at a more oblique angle is either reflected by the lens or absorbed by the pigment sheath. Light falling upon the lens of an ommatidium reaches only the rhabdomeres of that ommatidium. This type of compound eye structure was believed to be characteristic of diurnal insects.

Exner's concept of a superposition eye was one in which the receptor cells reside a certain distance away from the crystalline cone (Figure 8.5b). The term "superposition eye" has been loosely applied to any compound eye in which, during the dark-adapted state, the tips of the crystalline cones are separated from the rhabdom layer by an apparently clear space. The superposition eye was believed to be characteristic of nocturnal species, because its mechanism was thought to be important for the requisite increase in light-gathering power. However, superposition eyes have been found in both diurnal and nocturnal species.

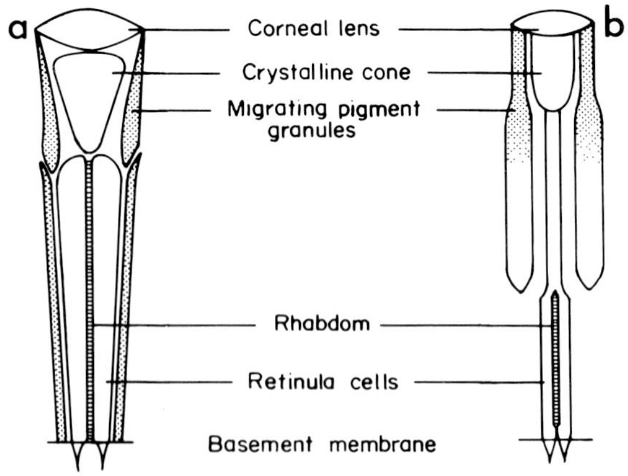


FIGURE 8.5 Ommatidium of a compound eye showing the (a) apposition type eye and (b) superposition type eye.

## Insects

### *Drosophila melanogaster*

The fruitfly, *Drosophila melanogaster,* is a diurnal insect. Its compound eye is composed of over 700 ommatidia. The ommatidium is approximately 17 μm in diameter and from 70 to 125 μm in length. Each ommatidium consists of a corneal lens, a crystalline cone, retinula cells, and a sheath of pigment cells that extends over its entire length (Figure 8.6a). In cross section there are seven retinula cells radially arranged that form the rhabdom (Figure 8.6b). The retinula cell has a medial portion extending toward the center of the ommatidium and terminating in a dense circular rhabdomere. Each rhabdomere is distinct with respect to its retinula cell, and appears to be attached to it with a fine membrane.

The rhabdom consists of seven individual rhabdomeres, $R_1$ to $R_7$, situated in a relatively clear fluid cavity, one of which is asymmetrical (Figure 8.6c). The rhabdomeres average 1.2 μm in diameter and are more than 60 μm in length. A definite fine structure of lamellae, separated by less dense interspaces, is found within each of the rhabdomeres. These dense laminations originate at the line of attachment and terminate in a scalloped border on the medial side of the rhabdomere. This lamellar structure is observed in all cross sections of the rhabdomeres, while the tubular structure is seen in some oblique and longitudinal sections (Figure 8.6d,e). A single structure of microtubules (Figure 8.6e) can produce both of these geometric structures in thin section, depending upon the orientation of the individual rhabdomere with respect to the plane of cutting (Wolken *et al.,* 1957a). Therefore the rhabdomere is structured of microtubules, each being about 500 Å in diameter.

### *The Cockroach*

The cockroach, a nocturnal insect, is believed to be one of the more primitive of the unspecialized insects. The structure of the compound eye of two large species of cockroaches, *Periplaneta americana* and *Blaberus giganteus,* was investigated by electron microscopy (Wolken and Gupta, 1961). In each of these compound eyes there are approximately 2000 ommatidia which are separated from one another by a pigment sheath. A cross section through the *Periplaneta* eye shows many ommatidia surrounded by pigment granules (Figure 8.7a). Each ommatidium appears also to be made up of seven retinula cells.

Electrophysiological studies have indicated that the cockroach eye contains two receptor systems which differ spectrally (Walther and Dodt, 1957, 1959). The question arises whether there are any structural differences between these

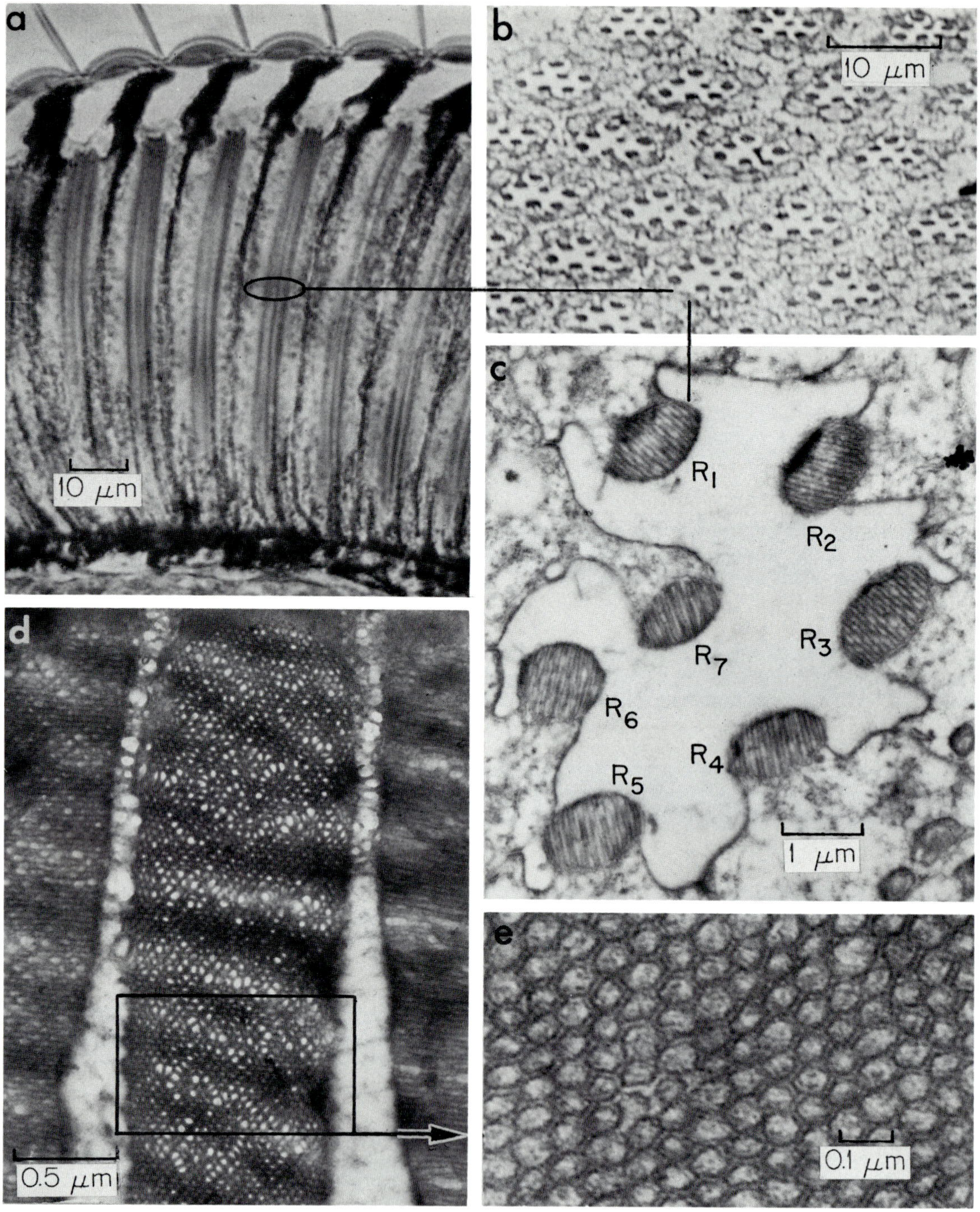


FIGURE 8.6 *Drosophila melanogaster,* compound eye. (a) Longitudinal section through several ommatidia showing the corneal lens, crystalline cone, rhabdom, and pigment sheath; (b) cross section through ommatidia; (c) cross section through the rhabdom to

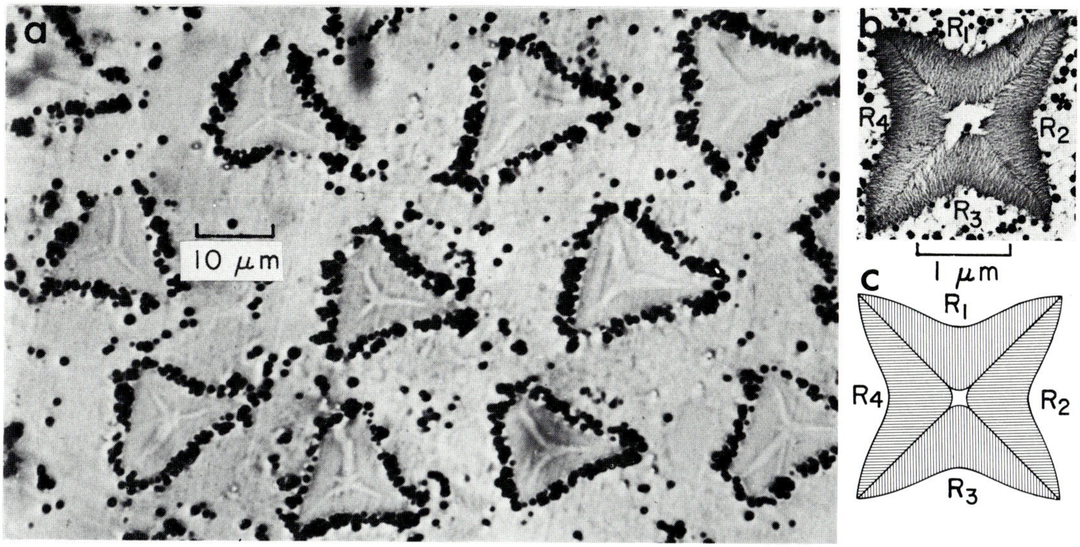


FIGURE 8.7 *Periplaneta americana,* the cockroach. (a) Cross section through many ommatidia, (b) rhabdom formed of four rhabdomeres ($R_1$–$R_4$), and (c) diagram of fused rhabdom.

two systems. The number of retinula cells in the upper and lower halves of the cockroach compound eye does not differ, but there are minor variations in the relative size of the cells (Walther, 1958). Each retinula cell is from 7 to 9 μm in diameter with a large elliptical nucleus. Although only seven retinula cells were observed in *Periplaneta,* eight retinula cells were found in the cockroaches *Blatta* (*Stylopyga*) *orientalis* and *Blatella germanica* (Jörschke, 1914; Nowikoff, 1932). The eighth cell is probably a rudimentary structure located close to the basal membrane (Dietrich, 1909). It does not extend the entire length of the rhabdom, and no rhabdomere is differentiated from it. Aggregates of intracellular pigment granules, which do not seem to be affected by dark-adaptation, surround the rhabdoms and extend the whole length of the retinula cells (Figure 8.7a). Numerous mitochondria are dispersed through the retinula cell. Depending upon the location and angle of cut of the eye section, the rhabdom appears either rhomboidal or triangular (with sides measuring from 5 to 12 μm in length). Each rhabdom is made up of four rhabdomeres ($R_1$–$R_4$) which lie in close proximity and form a regular pattern of organization (Figure 8.7b). This is schematized in Figure 8.7c. A rhabdomere averages 2 μm in diameter and is about 100 μm in length. Each cockroach rhabdomere is a single geometrical structure of packed microtubules. Each tubule is about 500 Å in diameter with walls of the order of 100 Å in thickness.

illustrate the orientation of the rhabdomeres ($R_1$–$R_7$); (d) longitudinal section at the distal end of the ommatidium showing three adjacent rhabdomeres; (e) enlarged area of the rhabdomere showing the fine structure of microtubules.

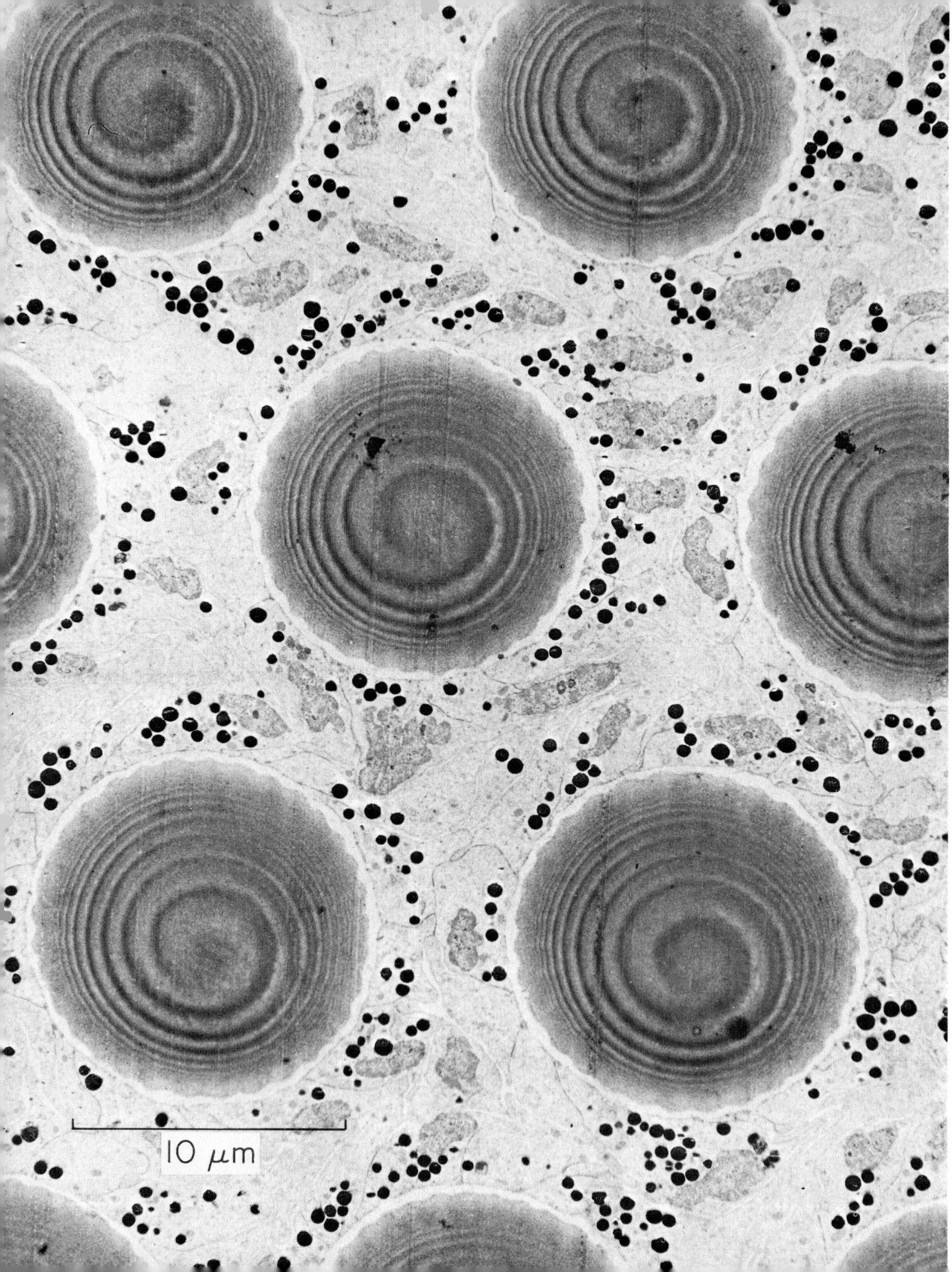
10 μm

There are approximately 400 tubules in one square micrometer of surface, or about 80,000 tubules in a single rhabdomere (Wolken and Gupta, 1961).

The closed-type rhabdom observed here in the cockroach (Figure 8.7b) is most likely an efficiency mechanism used for light capture by nocturnal insects. Evidence supporting this view is found in the fact that the cross-sectional area of the cockroach rhabdom is about five times that of the *Drosophila* rhabdom.

### *The Firefly*

The firefly has attracted man's curiosity for centuries. They are found on every continent except Antarctica. More than two thousand species are known and of these about one hundred inhabit the United States. The two most common species are *Photuris pennsylvanica** and the North American *Photuris pyralis*.

Their attraction is due to the fact that they luminesce. This ability to emit light by organisms, called bioluminescence, is discussed in Chapter XV.

Of interest to us here is the structure of their compound eyes, for the compound eye of the firefly provides an example of an ommatidium which is structurally unique among insect eyes. The firefly was the model used by Exner (1891) to derive his theory of the superposition type compound eye (Figure 8.5b). However, the microanatomy of the firefly eye was never fully described, and only recently was the stucture of the compound eye of the firefly *Photuris* studied (Horridge, 1968, 1969; Wolken, 1971, 1974).

All compound eyes have two functional systems: the dioptric or optical system, as well as the photoreceptor or rhabdom system. In its optical system the firefly eye differs from other insect eyes in that the corneal lens extends into the region normally occupied by the crystalline cone (Figures 8.8 and 8.9a). Eyes exhibiting this type of corneal lens structure have been called an exocone or pseudocone type of eye.

The compound eye of *Photuris pennsylvanica* consists of several hundred ommatidia, and the corneal lenses in cross section are seen to form spirals (Figure 8.8). In a longitudinal section these appear as laminations (Figure 8.9b). The corneal lens is surrounded by many distal pigment cells. Below the lens are situated the nuclei of four cone cells which extend to the basement membrane as fine filaments similar to the cone cell extensions found in other insects (Waddington and Perry, 1960; Goldsmith, 1962; Horridge, 1966; Perralet and Baumann, 1969). These filaments or cone cell threads (Horridge,

* *Photuris pennsylvanica* is the symbolic insect of Pennsylvania.

FIGURE 8.8 The firefly, *Photuris pennsylvanica* section through lenses. Note structure.

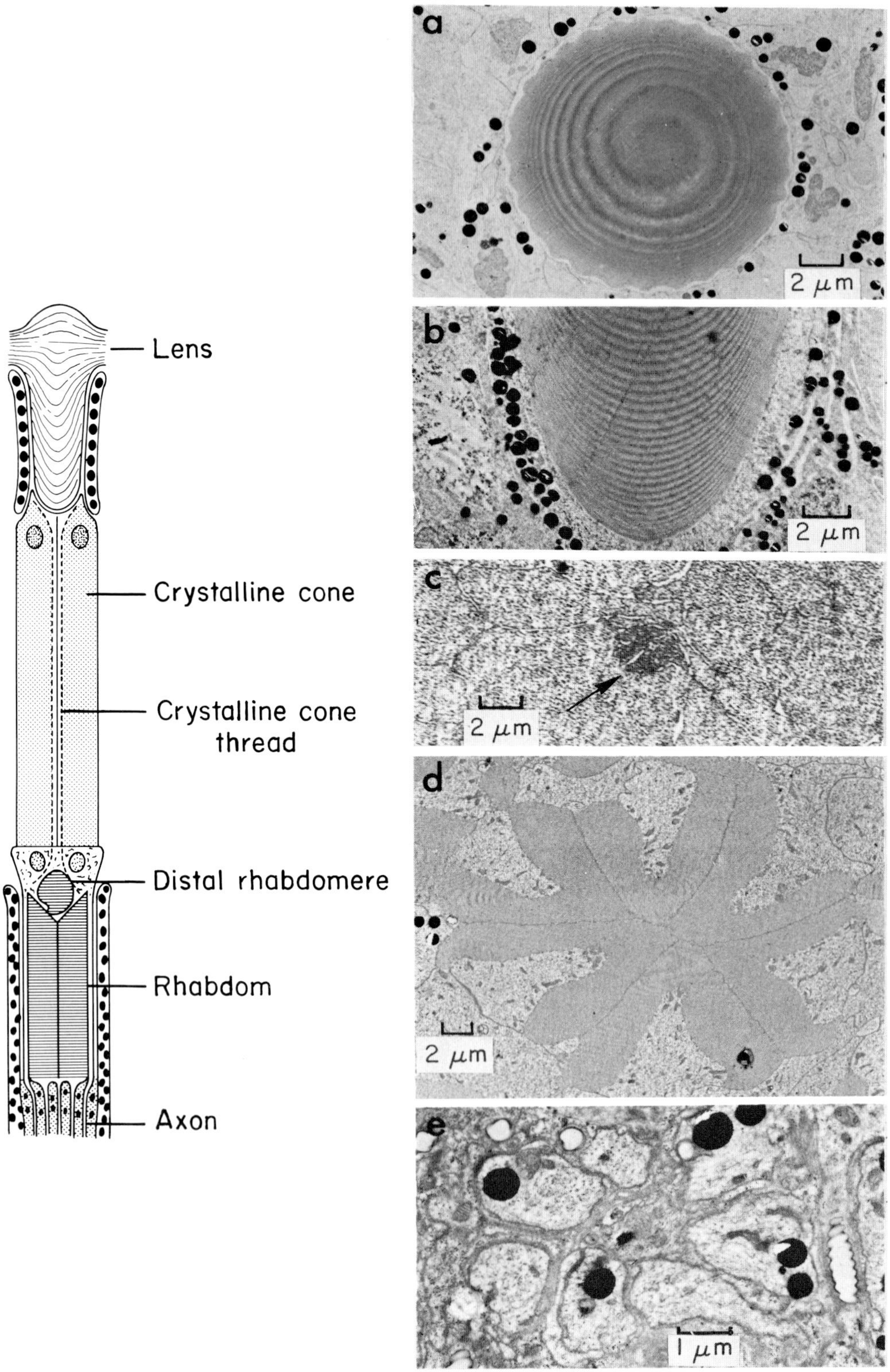
Lens
Crystalline cone
Crystalline cone thread
Distal rhabdomere
Rhabdom
Axon
a
2 μm
b
2 μm
c
2 μm
d
2 μm
e
1 μm

1968) are believed to be fiber optic elements functioning as wave guides directing light to the photoreceptor (Døving and Miller, 1969). They are seen as structures of increased electron density in an amorphous cellular medium where it is difficult to distinguish the boundaries of individual ommatidia (Figure 8.9c). In the eyes of *Phacops*, a trilobite fossil. Stuermer (1970) found filamentary structures leading from the corneal lens to the photoreceptor, the rhabdom, and has shown that these are in fact light guides. This would indicate that the adaptation of fiber optics for a visual detector evolved in the Cambrian era.

The photoreceptor system of retinula cells that form the rhabdom in the firefly occupies only one-fourth of the ommatidial length. The system has two nuclear layers, a distal layer of retinula cell nuclei and a basal layer of nuclei. For each ommatidium there appears to be a single basal cell which surrounds the axonal terminals of the retinula cells (Figure 8.9e). These basal cells could serve to insulate the retinula cell axons in the manner of a Schwann cell. The ommatidium contains eight retinula cells which give rise to six rhabdomeres that form the rhabdom (Figure 8.9d). The small distal rhabdomere is formed by two rhabdomeres from two retinula cells. The major proximal portion of the rhabdom is formed from six retinula cells and has a six-arm configuration (Figure 8.10). Each arm has a V-shaped rhabdomere formed from the rhabdomeres of adjacent retinula cells. The fine structure of the rhabdomere also consists of microtubules.

Firefly eyes are well adapted for light-gathering. The major portion of the rhabdom occupies nearly the entire cross-sectional area of the ommatidium and is tightly packed against the rhabdoms of neighboring ommatidia (Figure 8.10). In general, nocturnal insects such as moths and cockroaches (Fernández-Morán, 1958; Wolken and Gupta, 1961) have rhabdoms that occupy a large portion of the ommatidium, while daylight active insects have relatively small rhabdoms (Wolken, 1968b; Goldsmith, 1962). The almost solid layer of rhabdom found in fireflies, however, is unique, as are the exocone and crystalline thread dioptric system (Horridge, 1968, 1969).

### *Other Insects*

We can now compare what has been described and illustrated for the compound eye structure of the fruitfly, cockroach, and firefly to the rhabdom structure of some other insects.

The housefly, *Musca domestica*, compound eye ommatidium has an open-type rhabdom of seven rhabdomeres (Fernández-Morán, 1956; Eichenbaum and Goldsmith, 1968) similar to *Drosophila* (Figure 8.6c).

FIGURE 8.9 The firefly, *Photuris pennsylvanica*, ommatidium. Electron micrographs of various sections through the ommatidium; (a) exocone lens, cross section; (b) exocone lens, longitudinal section; (c) crystalline cone thread; (d) rhabdom; and (e) axon.

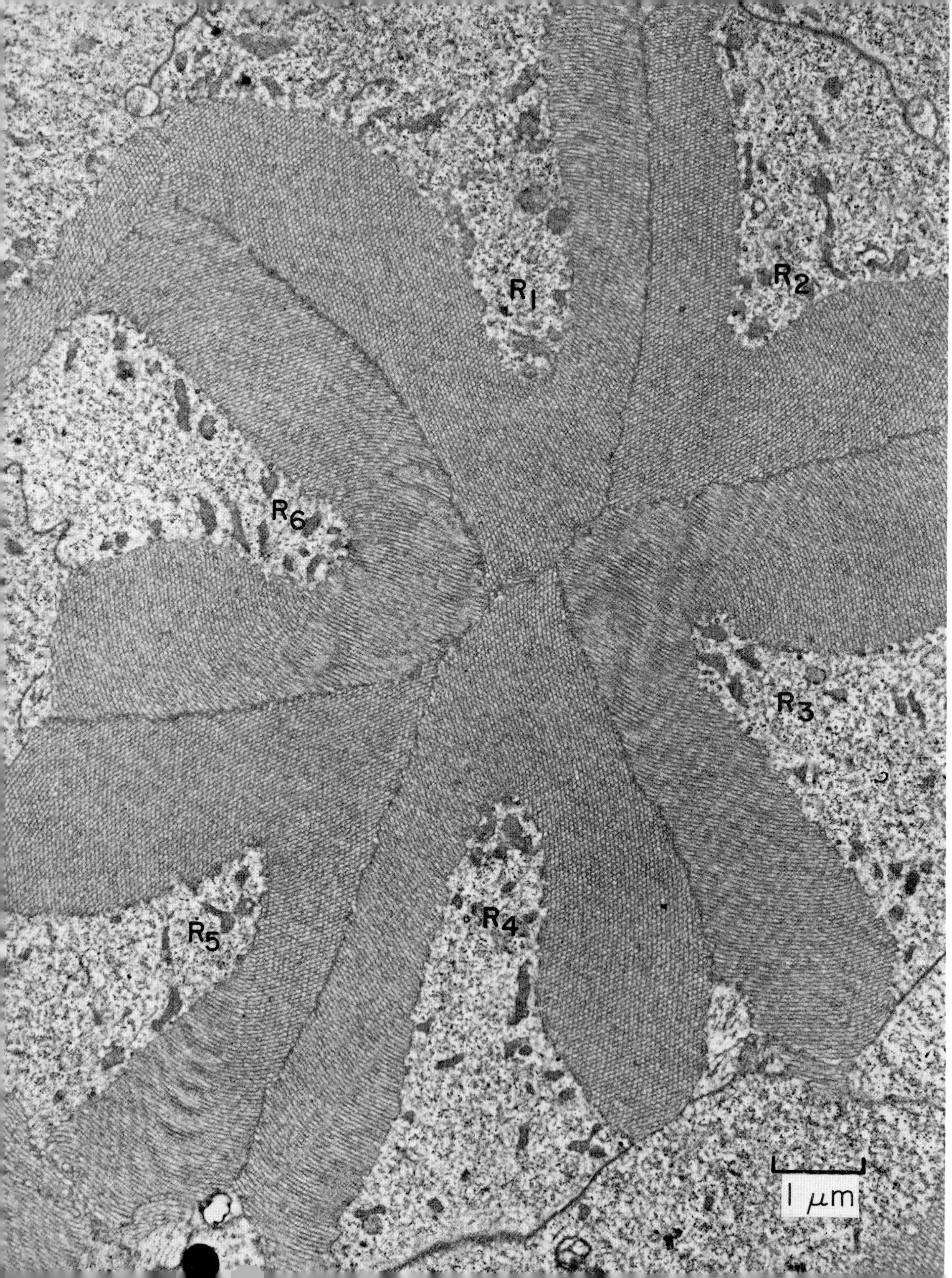
R1
R2
R6
R3
R5
R4
1 μm

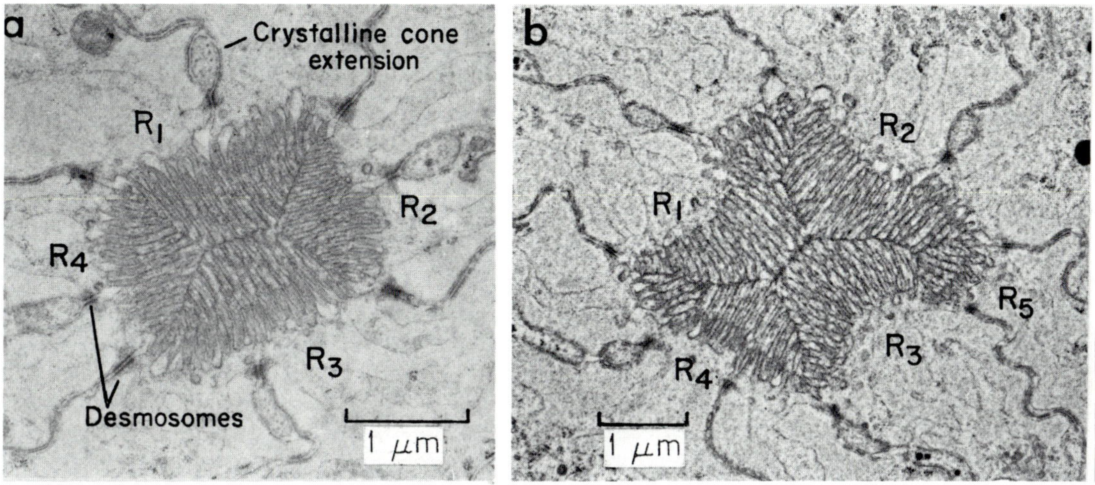


FIGURE 8.11 The hornet, *Vespa maculata,* rhabdom. (a,b) Cross sections showing the fused rhabdoms, crystalline cone extensions, and desmosomes.

The baldface hornet, *Vespa maculata,* compound eye ommatidium contains eight retinula cells. In a few ommatidia there is a ninth retinula cell located more distal than the other eight. The rhabdom is composed of four pairs of rhabdomeres (Figure 8.11). The crystalline cone is built of four cellular segments which taper into long tubular cone cell extensions (this is also found for the clothes moth ommatidium), ending just distal to the basement membrane. There are no connections between these cone cell extensions and the nerve cells. On the other hand, the cone cell extensions have a definite structural relationship to the pigment cells which appear to make synaptic connections with them.

The honeybee, *Apis mellifera,* compound eye structure has several thousand ommatidia. The rhabdom of this worker bee consists of four rhabdomeres, although it is formed from eight retinula cells (Goldsmith, 1962). The rhabdom structure is similar to that of the hornet (Figure 8.11) and each rhabdomere is a closely packed parallel array of microtubules with their axes perpendicular to the axis of the rhabdom. The microtubules in adjacent rhabdomeres of the rhabdom are mutually perpendicular.

In the carpenter ant, *Camponotus herculenus pennsylvanicus,* the compound eye structure appears to represent a transition from the apposition-type eye (Figure 8.5a) to that of the pseudocone-type eye with a crystalline thread, as seen in the firefly (Figure 8.9) and the hornet (Figure 8.11).

The retinula cells of the carpenter ant, like those of the hornet, show two zones, a clear zone near the rhabdom and a peripheral cytoplasmic zone. The

---

FIGURE 8.10 The firefly, *Photuris pennsylvanica.* The rhabdom enlarged to show geometric arrangement of its rhabdomeres ($R_1$–$R_6$).

rhabdom of the carpenter ant (Figure 8.12a,b) is circular and occupies a much larger portion of the ommatidial cross section than in the hornet (Figure 8.11). Despite its size, the rhabdom is formed from only six retinula cells as compared with seven to nine in other insects (Vowles, 1954).

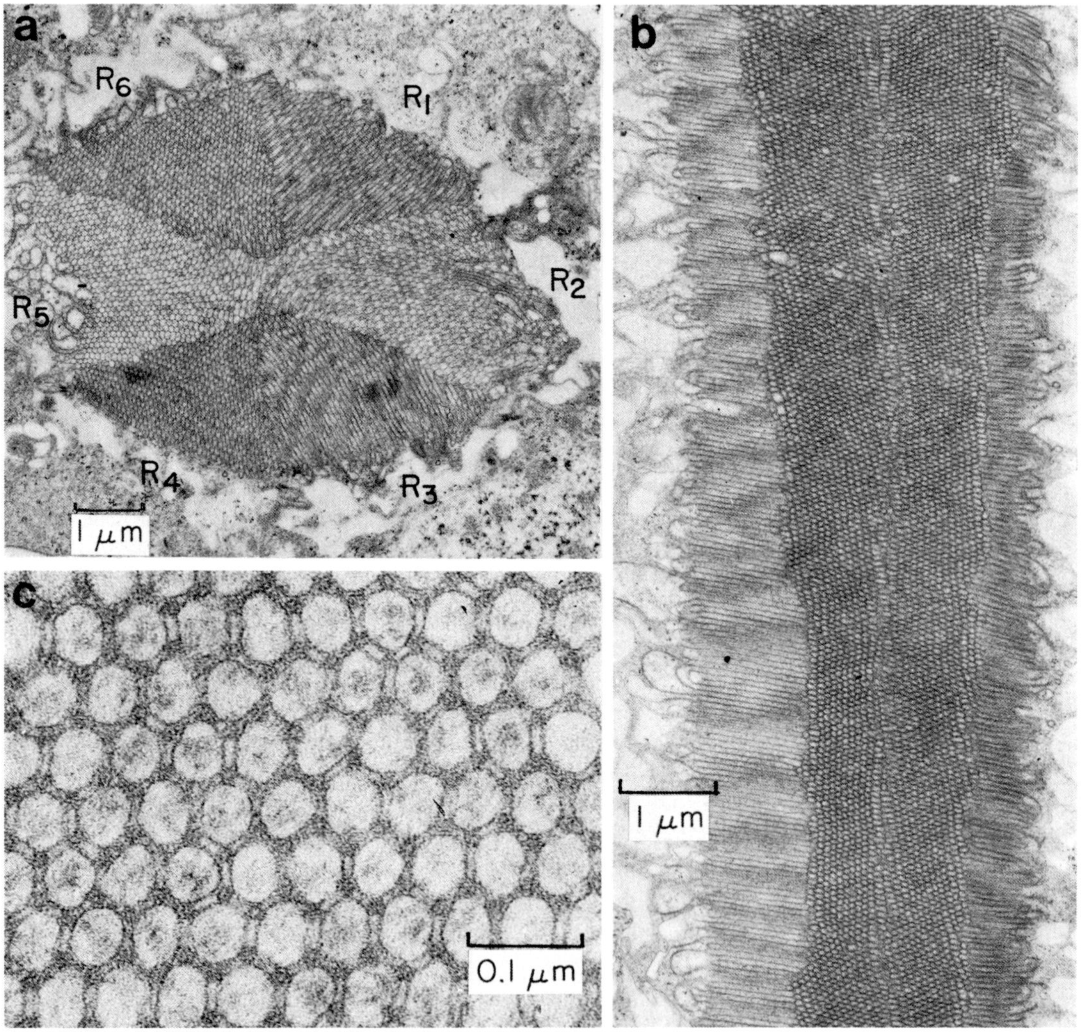


FIGURE 8.12 The carpenter ant, *Camponotus herculenus pennsylvanicus.* (a) Cross section through the rhabdom showing rhabdomeres ($R_1$–$R_6$), (b) longitudinal section of the rhabdom, and (c) higher resolution of a cross section through many rhabdomere microtubules, electron micrographs.

The microtubules of the rhabdomere have an elaborate fine structure. There appears to be intratubular material at the center of which is a star-shaped, electron-dense body (Figure 8.12c). At the angles of adjacent microtubules, electron-dense intervillous bridges are found which bind adjacent microtubules.

The nerve bundle from an ommatidium contains seven axons, but we have only been able to identify six retinula cells. It is not clear whether the additional axon is from a vestigial retinula cell or whether this is a second order neuron.

Cajal (1918), on tracing out the nervous systems of vertebrates, thought that the general plan for all visual systems would be found in the insect eye, but after studying the insect eye, he wrote: "The complexity of the nerve structure for vision is even in the insect something incredibly stupendous."

Only a few examples of insect compound eyes were selected here to illustrate their photoreceptor structures. Although we do not completely understand how the compound eye functions in insect vision, our picture of the anatomical structure of the optical and photoreceptor systems is finally coming into focus.

## Crustacea

Among the crustacea are found a variety of animals, from the lobsters and crabs, which are the giants of the class, to the tiny water flea, which is only a few millimeters long. Like the insects, some species of crustacea possess simple eyes and some possess compound eyes. As in our survey of the compound eyes of insects, we will describe several species of freshwater and marine crustacea that show some interesting features in their structure.

### *Daphnia*

The freshwater *Daphnia,* a water flea, possesses two eyes, a compound eye and a simple nauplius eye (Wolken, 1971). The compound eye of *Daphnia pulex* and *Daphnia magna* consists of about 22 ommatidia enclosed in a capsule. The eye is in continuous oscillatory movement. The corneal lens, as seen in the insect ommatidium, and the distal pigment cells are not usually found. The ommatidium has a crystalline cone and elongated retinula cells which are surrounded by pigment granules. There are eight retinula cells that give rise to seven rhabdomeres (Figure 8.13a) to form the closed-type rhabdom (Röhlich and Törö, 1965).

The rhabdomeres that form the rhabdom consist of microtubules about 500 Å in diameter which show lamellae in longitudinal sections (Figure 8.13b), not unlike those observed for the insect rhabdomeres. The microtubules within each rhabdomere are precisely arranged with their longitudinal axes regularly aligned in a given direction for one set of rhabdomeres, and in the perpendicu-

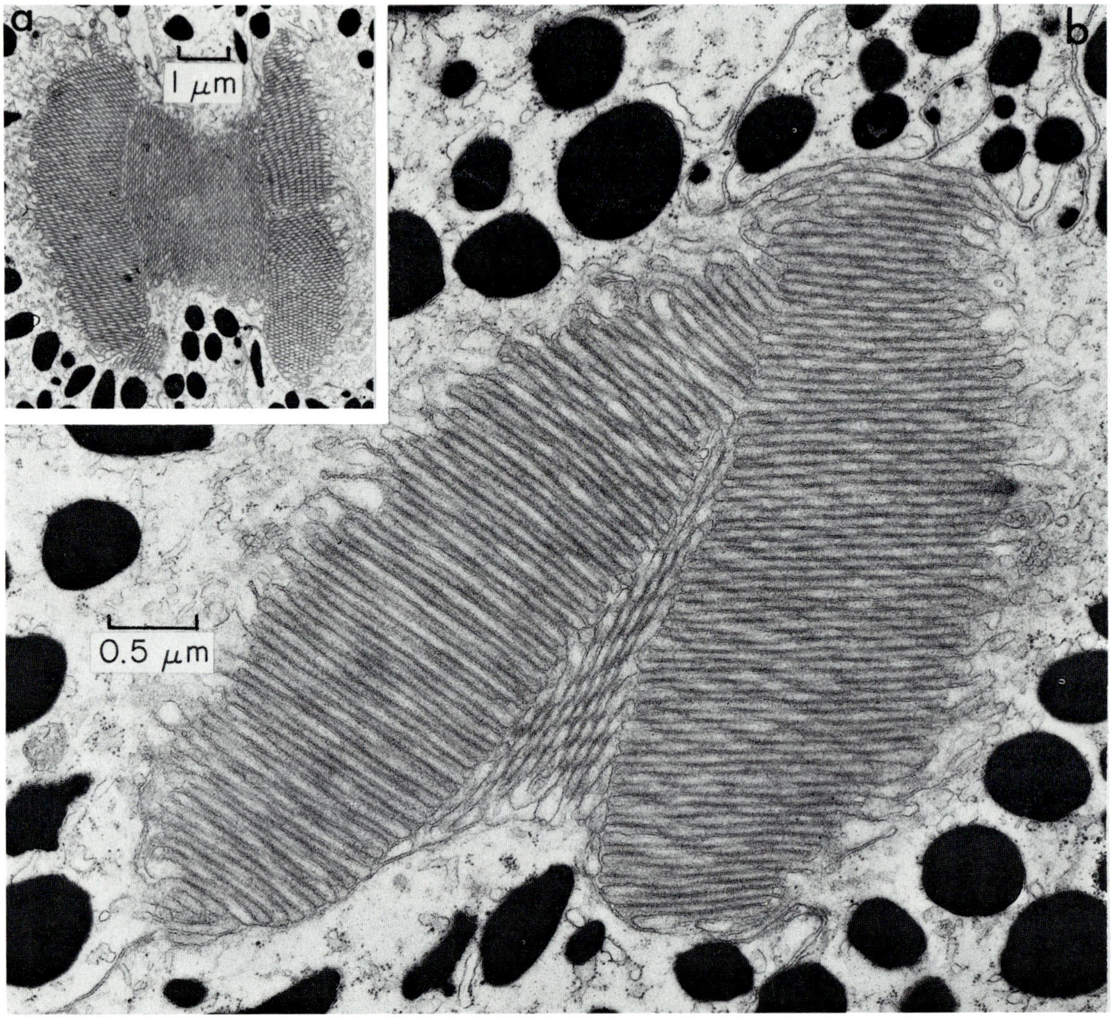


FIGURE 8.13 The waterflea, *Daphnia pulex*. Rhabdom (a), and (b) enlarged rhabdom to show microvilli and their relation to the cytoplasm.

lar direction for alternate layers. This type of structure was observed for other crustacea, for example, in the land crab, *Cardisoma,* and the swimming crab, *Callinectes* (Eguchi and Waterman, 1966). Such an arrangement of the rhabdomeres with their microtubules at right angles could be a structural basis for the ability of *Daphnia* and other crustacea to analyze the direction of polarized light (Waterman *et al.*, 1969).

*Leptodora*

Another freshwater crustacean is the carnivorous water flea, *Leptodora kindtii,* which is relatively large (18 mm in length) in comparison with *Daphnia.* The animal is almost completely transparent with one median spherical compound eye (Figure 8.14a). *Leptodora* is believed to have better vision than most crustacea. The basis for this assertion is the rapidity with which it can capture copepods as large and fast as *Cyclops* for food.

The entire eye of *Leptodora kindtii* is contained within the transparent exoskeleton at the anterior end of the organism, and the eye is free to move and can rotate 10° in either direction. A small area behind the eye with connecting neurons to the brain accommodates the optic processes (Gerschler, 1911, 1912; Scharrer, 1964).

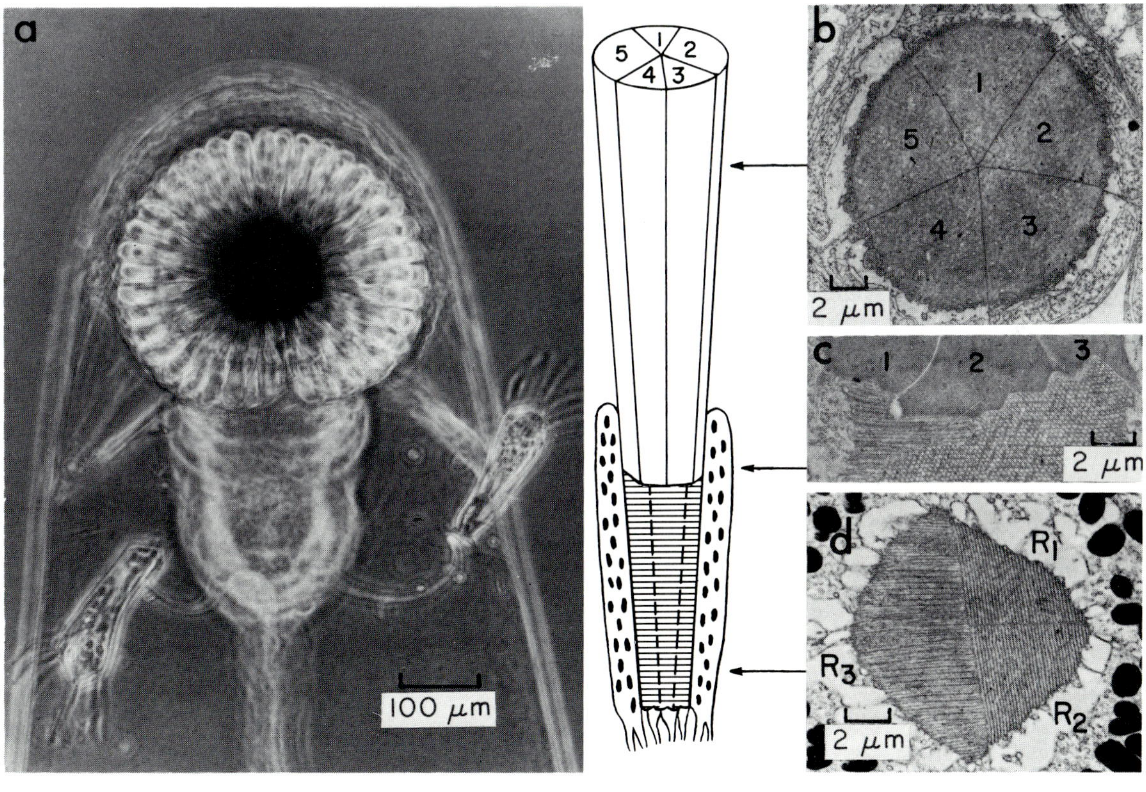


FIGURE 8.14 The waterflea, *Leptodora kindtii.* (a) The eye, (b) cross section of crystalline cone, composed of five wedge-shaped segments (1–5). (c) longitudinal section showing the connection of the crystalline cone with the rhabdom, and (d) cross section of the rhabdom indicating rhabdomeres $R_1$–$R_3$.

The *Leptodora* compound eye is composed of approximately 500 ommatidia radially arranged (Figure 8.14a). The ommatidia are large conical structures; an ommatidium measures about 180 $\mu$m in length and narrows from 30 $\mu$m in diameter at the outer portion to about 2 $\mu$m at the base. A schematic ommatidium is illustrated in Figure 8.14, and cross sections through the crystalline cone and rhabdom are shown in Figure 8.14b–d. The crystalline cone constitutes about two-thirds of the ommatidial length. Although it is rounded at the outer end, there is no evidence of a distinguishable lens cap. There is, however, a completely transparent interstitial space between the surface of the eye sphere and the external chitinous wall which functions as a common lens for all the ommatidia. The crystalline cone is composed of five equal wedge-shaped segments, formed from five crystalline cone cells which serve to concentrate the light into a narrow beam (Figure 8.14b,c).

Whether or not this segmentation promotes a system of total internal reflection is not known. Although structural data show that the crystalline cone continues proximal to the surface of the rhabdom as in the apposition-type eye, observations of pigment migration indicate that under certain conditions of dark-adaptation, "crossing" among adjacent crystalline cones could result in the formation of a superposition image. As the crystalline cones continue inward, the space between them increases and is filled with pigment cells.

The rhabdoms are affixed directly to the ends of the crystalline cones (Figure 8.14c). The four radially arranged retinula cells that form the rhabdom show only three rhabdomeres (Figures 8.14d and 8.15a). Cross sections through numerous ommatidia in the rhabdom area reveal the rhabdom structure. One of the rhabdomeres ($R_3$) is large in comparison with the other two ($R_1$ and $R_2$) and appears to be two rhabdomeres that have fused (Wolken and Gallik, 1965). Another instance in which four retinula cells yield only three closed-type rhabdomeres for the rhabdom is observed in the dragonfly (Goldsmith and Philpott, 1957; Naka, 1960).

The rhabdomere fine structure of *Leptodora* is that of tightly packed microtubules which average about 500 Å in outside diameter and whose double-membraned wall is about 100 Å in thickness. The microtubules of the small rhabdomeres $R_1$ and $R_2$ are arranged perpendicular to those of the large rhabdomere $R_3$. The ends of the microtubules appear continuous with the cytoplasm (Figure 8.15b), and this structural connection with the cytoplasm may be of some importance to visual excitation (Lassansky, 1967).

### *Copilia*

One of the most interesting crustaceans is the relatively rare marine copepod, *Copilia*. Its eye has been of considerable curiosity since it was first found in the Bay of Naples and described by Grenacher (1879). Exner (1891), working

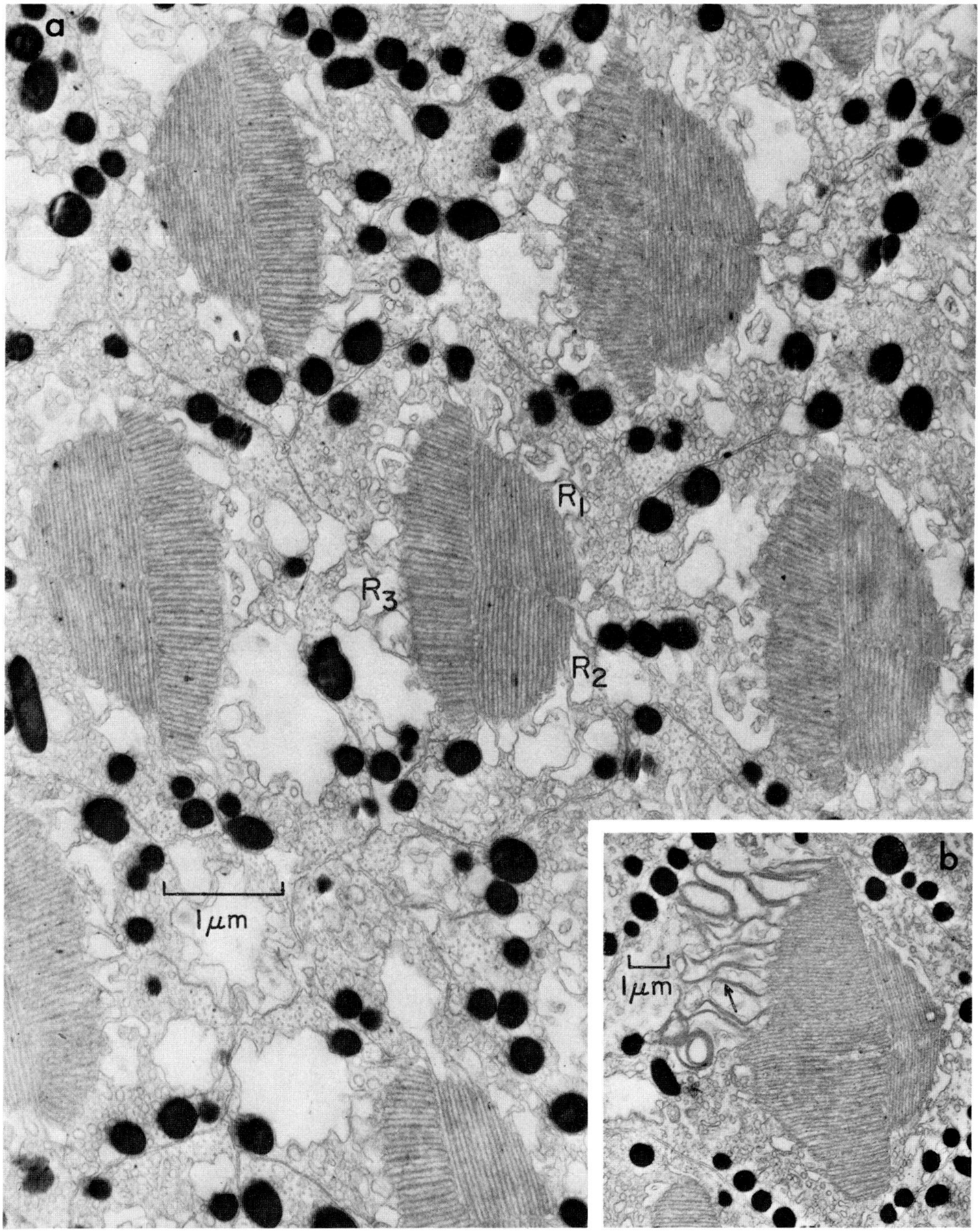


FIGURE 8.15 The waterflea, *Leptodora kindtii*. (a) Cross section through many ommatidia and their rhabdoms, and (b) rhabdom, cross section at higher magnification (arrow) showing the relationship of the rhabdomere microvilli with the cytoplasm.

in Naples, rediscovered *Copilia* and was fascinated by its strange pair of eyes which he likened to a telescope with two lenses. The eyepiece lens is deep in the animal and is in continual motion. It seems important to learn more about the optics and imaging properties of such scanning eyes, and to see how their photoreceptors are structured in comparison to other arthropod eyes.

This copepod (which is about 1 mm wide and 3 mm long) is found in both the Mediterranean, *Copilia quadrata,* and the Caribbean, *Copilia mirabilis.* It is only the female of the species which possesses these scanning eyes, for the male is blind. The *Copilia* eye structure has more recently been described by Vaissière (1961), Gregory (1966, 1967), and Wolken and Florida (1969).

The *Copilia* eye resembles an ommatidium of the compound eye with a corneal lens (biconvex anterior lens, $L_1$) and at some distance away, the crystalline cone (posterior lens, $L_2$). Attached to the crystalline cone are the retinula cells which give rise to the rhabdomeres that form the rhabdom. The rhabdom lies in an L-shaped, orange-colored stem which is the only pigmented part of the body (Figure 8.16). It is this stem that oscillates back and forth in a saw-toothed pattern, varying from about 1 scan per 2 seconds to 5 scans per second (Gregory *et al.*, 1964; Moray, 1972). The stems from both eyes move synchronously and rapidly toward each other, then separate slowly. Gregory (1966) has likened such scanning to a television camera: "It seems that the pattern of dark and light of the image is not given simultaneously by many receptors, as in other eyes, but in a time-series down the optic nerve, as in the single channel of a television camera."

It was expected that *Copilia,* since it lives at a depth (200 to 300 meters) where the light level is near that of moonlight, would have a closed-type rhabdom with the significantly greater effective cross section necessary for better light gathering efficiency. But the *Copilia* rhabdom is the open type which is common to most diptera that navigate at high light levels.

In *Copilia quadrata,* the retinula cells lie directly behind the crystalline cone and are followed by the rhabdomeres that comprise the rhabdom (Figure 8.17a–c). The rhabdom measures 11 $\mu$m $\times$ 17 $\mu$m and is completely surrounded by pigment granules. It extends 60 $\mu$m in length from the retinula cells to the bend of the stem. Only five rhabdomeres ($R_1$–$R_5$) can be identified in the rhabdom. One of them ($R_1$) is an asymmetric rhabdomere, located at a nodule on the side of the stem facing the brain and lying at the base of the crystalline cone (Figure 8.17b). The asymmetric rhabdomere appears to lie at nearly 45° with respect to the stem. Rhabdomeres $R_2$–$R_5$ are ellipsoids measuring 1 $\times$ 2 $\mu$m in diameter and about 58 $\mu$m in length. These rhabdomeres lie with their longest dimension parallel to the stem. Mitochondria and structures resembling nerve vesicles are also found in this region. Rhabdomeres $R_1$ to $R_3$ are separated by screening pigment granules, whereas rhabdomeres $R_4$ and $R_5$ are not. The rhabdomere fine structure is that of packed microtubules about 500 Å

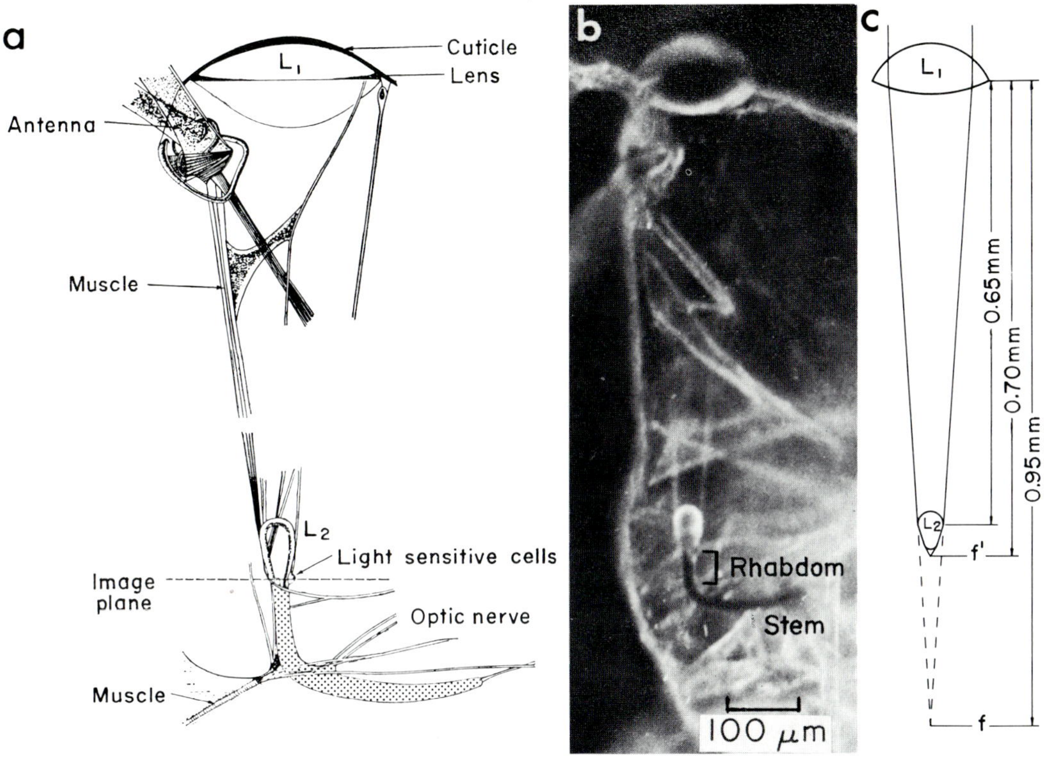


FIGURE 8.16 (a) The structure of the *Copilia* eye, as drawn by Grenacher (1879), (b) *Copilia quadrata,* dark-field photomicrograph of the eye, (c) the optical system showing the positions of the corneal anterior lens ($L_1$) and the crystalline cone, posterior lens ($L_2$); f, focal point of the corneal (anterior) lens; f′, focal point of the total optical system.

in diameter (Figure 8.17c), which is similar to that already described for the arthropod rhabdomeres.

The *Copilia* eye can be considered analogous to the superposition-type ommatidium in which the crystalline cone lies at some distance from the corneal lens (Figure 8.5b). In addition, the crystalline cone forms a convex interface with a fluid of lower refractive index. The concentration of this material varies across the diameter, the greatest concentration being in the center (Figure 8.17a). Therefore, the crystalline cone would have the properties of a lens. The *Copilia* eye with its corneal lens, $L_1$, and its crystalline cone, $L_2$, may then be considered a two-lens optical system in which the posterior lens is positioned a short distance in front of the rhabdom (Figure 8.17). The optical system is so arranged that the image is focused on the rhabdom (Wolken and Florida, 1969).

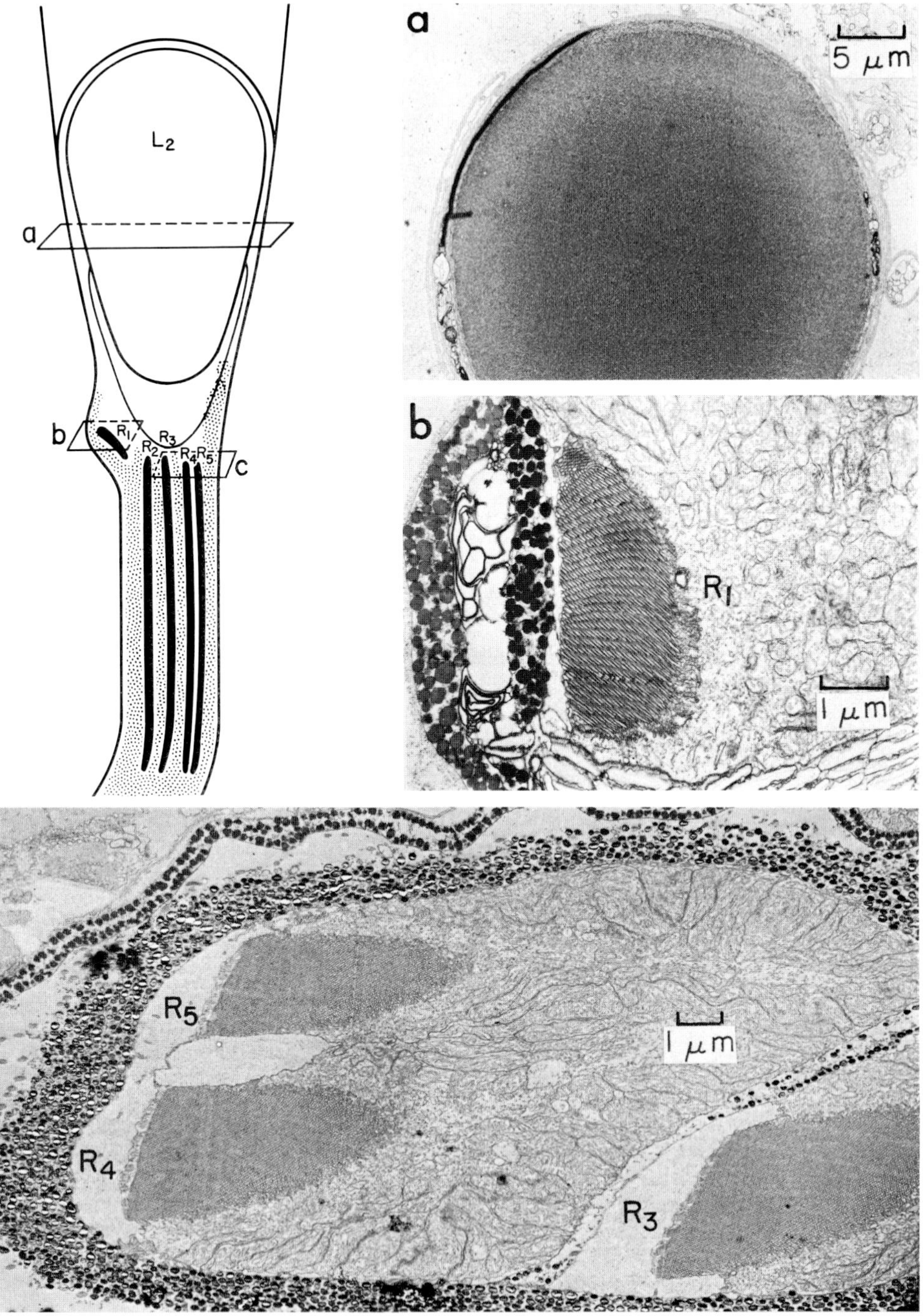


FIGURE 8.17 The *Copilia quadrata* eye structure, longitudinal view of crystalline cone and rhabdom. Electron micrographs (a) crystalline cone, cross section ($L_2$), (b) oblique section showing asymmetric rhabdomere ($R_1$) in rhabdom, (c) oblique section through the rhabdom showing three rhabdomeres ($R_3$–$R_5$) of the five rhabdomeres that form the rhabdom.

To see how this optical system would work, we estimated the refractive index of the anterior lens to be 1.425, which gave for this lens a focal length of 0.93 mm (i.e., the focal length measured by Exner, 1891). Using this focal length and the known position of the rhabdom, we found that the focal length of the posterior lens that would be necessary to place the image at the rhabdom level was 0.128 mm. The anterior lens alone has a relative speed or focal ratio of 5.5:1 but, when the posterior lens is taken into consideration, the focal ratio changes to 2.5, or an increase in light collecting efficiency of five times.

To determine how efficient this optical system would be, we constructed a holder for a 15 mm diameter and 25 mm focal length, Hastings triplet lens, and we mounted the lens 17.5 mm in front of the film plane of a 101.5 mm × 127 mm Burke and James commercial view camera. The image was then focused on the film plane, as formed by a combination of that lens with the regular camera lens (a 152.5 mm focal length lens with a focal ratio of 6.8:1), and photographs were taken. With the camera lens alone, an exposure of 1/25 second at *f*/18 was required to record an image. However, when the second lens was introduced, the exposure has to be reduced to 1/200 second to record an image with the same density on the negative. Calculations indicated that the lens speed was increased from *f*/18 to *f*/5.6, or an increase of more than eight times in image brightness. The photographs also showed that the size of the image formed by the combination of these lenses was reduced to about one-third, but this is much less than the gain in image brightness. A similar optical system has been described for a focal reduction camera used on the large telescope at Yerkes Observatory, Williams Bay, Wisconsin (Gascoigne, 1968).

Although the *Copilia* eye is considered primitive in that its field scanning mode is slow, it has adapted to low light levels by evolving a comparatively "advanced" optical system (Wolken and Florida, 1969).

## Molluscs

Next to the arthropods, the molluscs are by far the most numerous of the invertebrates. Included among the molluscs are clams, oysters, nautiluses, squids, and octopi. Almost every kind of imaging eye is found among them, from the simple pinhole eye of the *Nautilus* to the refracting type eye of the *Octopus*.

This then brings us to a third kind of optical system occurring in most vertebrates, including man, which is the refracting eye. In refracting eyes the image is formed by the refraction of light at one or more surfaces that separate media of different refractive indices. The refracting structures are the clear cornea and the lens. The refracting eye has the great advantage that image formation occurs through an integrative action, so that all rays falling on the eye from a given point of origin are brought to a point of focus on the retinal photoreceptors. The image produced there is inverted, and its size is inversely proportional to the distance of the object.

## THE *Octopus*

Of the mollusc cephalopods, it is the *Octopus* that attracts our immediate attention. The *Octopus* is a large muscular animal; it may grow to 10 feet in length and have eyes that are unusually large for the body size. It is of interest to us that its behavioral responses are primarily visual ones. Also, the brain behind the eye is one of the most highly developed of any of the invertebrates. It is enclosed in a cartilaginous "skull" and is divided into 14 main lobes that govern different sets of functions. The optic lobes are the largest of all; one set controls the jet apparatus, another set, the memory (Young, 1964). The *Octopus* is an ideal animal for studies of visual acuity, eye–brain relationships, and learning processes. Considering that the *Octopus* (like *Copilia*) is among the oldest species known, one is curious to see how its eyes are structured.

The *Octopus* uses both monocular vision and binocular vision. In monocular vision the eyes face in opposite directions with their long axes roughly parallel. In binocular vision the eyes are switched slightly forward relative to the body. Therefore the *Octopus* has a complete field of vision of 360°. The *Octopus* can learn to distinguish one shape from another, even when the only difference between the two is their orientation. The *Octopus* can also achieve color matching with its surroundings, producing the pattern that is seen by expansion and contraction of the chromatophores in its skin (Packard and Sanders, 1969). All of these visual responses are not completely understood.

The *Octopus* eye resembles, in physical organization, the vertebrate eye except that its simple lens is formed out of two halves joined together (Figure 8.18a), the image on the retina is not inverted as in the vertebrate eye, and the photoreceptors are directly exposed to the incident light (Hess, 1943; Ramsey, 1952). Histological studies indicated that the retinal photoreceptors are not rods and cones as in the vertebrate eye, but are retinula cells that form rhabdoms (Grenacher, 1879, 1886; Patten, 1887) similar to those just described for the arthropod compound eyes.

Grenacher (1886) recognized that the rhabdoms are formed from four retinula cells. This is borne out by more recent studies (Wolken, 1958b; Moody, 1964) of the retina of *Octopus vulgaris,* the closely related cuttlefish, *Sepia officinalis* (Figure 8.18b,c), and the squid (Zonana, 1961), showing that each rhabdom consists of four radially arranged rhabdomeres.

The retina of the *Octopus* is at least 1 cm$^2$ in area, which means that there

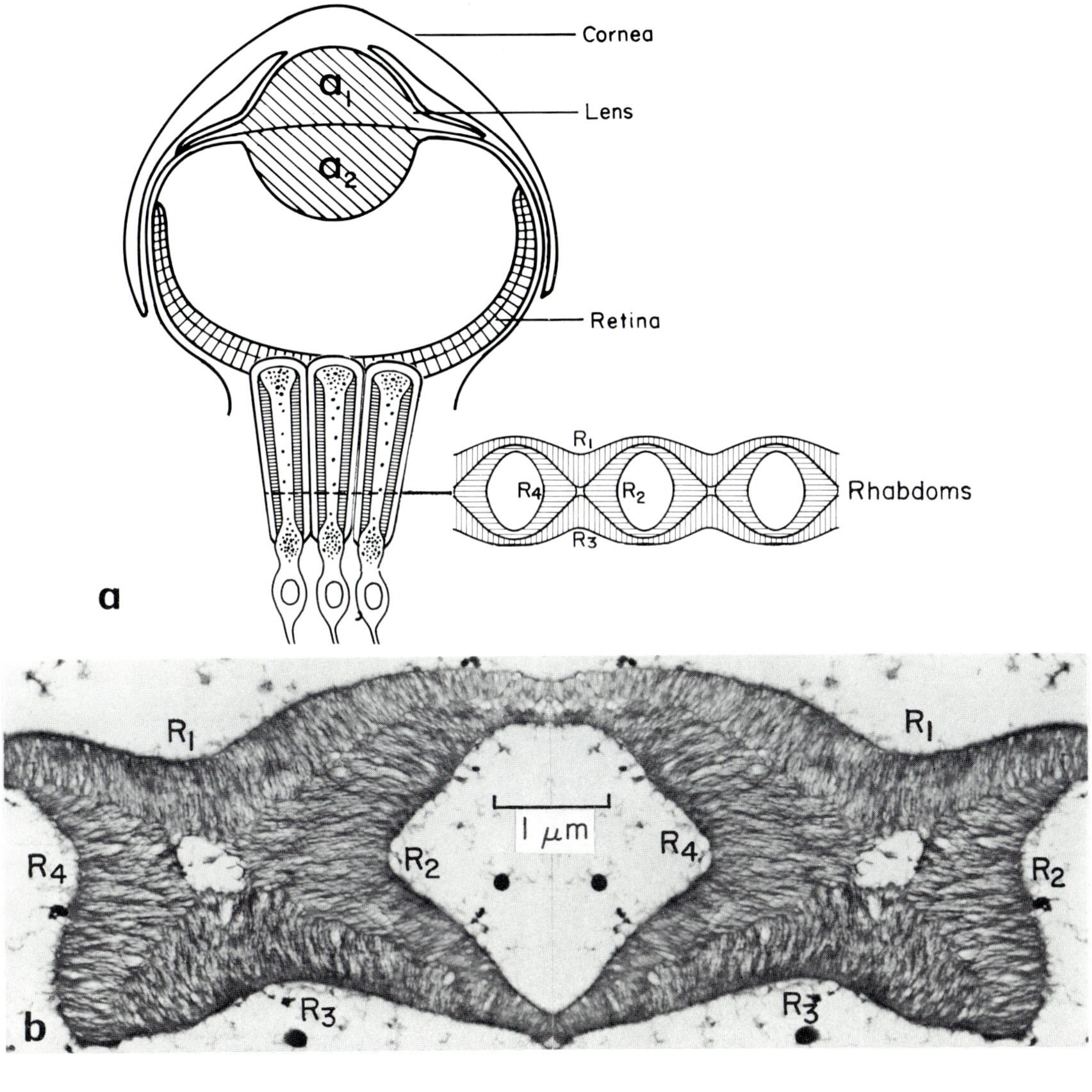


FIGURE 8.18 Mollusc cephalopod eye and structure of the retina (a). Note lens constructed of two halves ($a_1$–$a_2$) and that the retina is formed of rhabdoms. Each rhabdom has four rhabdomeres ($R_1$–$R_4$). (b) Electron micrograph of the rhabdoms.

would be about $2 \times 10^6$ rhabdomeres per eye. This number is roughly equivalent to the number of retinal rods of the vertebrate retina. However, the cephalopod retina differs fundamentally from that of vertebrates in that it contains no ganglion cells (Young, 1964).

In a cross section of the rhabdom of the *Sepia* eye, four sides of the rhabdomeres are isolated by pigment cells containing screening pigment granules. These appear to migrate depending on the light intensity. In numerous cross sections and longitudinal serial sections the rhabdomeres measure about 70 $\mu$m in length and from 1 to 1.5 $\mu$m in diameter. In all the cross sections the lamellar structure is observed, whereas in all oblique and longitudinal sections the microtubular structure is seen. Here, as in the arthropod rhabdomeres, the microtubules are about 500 Å in diameter and are separated by a double membrane 100 Å thick. The retinas of the *Octopus, Sepia,* and squid are similar to the ommatidia of insect and crustacea compound eyes. The retina consists of rhabdoms formed by rhabdomeres as schematized in Figure 8.18.

## Related Remarks

Despite the wide variety of invertebrate photoreceptor structures that has been revealed to us through electron microscopy, we will attempt to make some general observations about what we have seen.

### The Rhabdom Structure

As exemplified in the insects, crustacea, and molluscs just described, there are two structural arrangements for the rhabdomeres that form the rhabdom. One is the "open-type" rhabdom characteristic of *Drosophila* (Figure 8.6), the housefly *Musca domestica,* and the crustacean *Copilia* (Figure 8.17), in which the rhabdomeres project through a necklike part of the retinula cell into a comparatively large cavity. The other is a "closed-" or fused-type rhabdom as found in the cockroach, firefly, carpenter ant, hornet, honeybee, grasshopper, locust, dragonfly, moth, and butterfly (Figures 8.7 and 8.10–12). The closed-type rhabdom is found in the crustacea *Daphnia* and *Leptodora* (Figures 8.13–15), and in the mollusc cephalopods *Sepia, Octopus,* and squid (Figure 8.18). In the fused rhabdom the retinula cells are modified to form wedge-shaped rhabdomeres that lie in close proximity to one another around a narrow axial cavity.

The number of rhabdomeres that form the rhabdom vary; in the open-type rhabdom there are usually five to seven with an asymmetric rhabdomere lying in the same plane. In the closed-type rhabdom there are four to eight rhabdomeres forming a symmetrical arrangement with its asymmetric rhabdomere lying in another plane. All rhabdomeres contain microtubules of about 500 Å in diameter with a double-membraned wall about 100 Å thick. In general, the microtubules of adjacent rhabdomeres are oriented perpendicular to each other, whereas those of opposite rhabdomeres are parallel.

Electrophysiological measurements indicate that there are also two physiological types of eyes, a "slow-type" eye characterized by a negative monophasic potential which is dependent upon the state of dark-adaptation, and a "fast-type" eye in which the electroretinogram (ERG) is diphasic, the magnitude and the form of the potential being independent of the state of dark-adaptation (Autrum, 1950, 1958). It is of interest then, that except for the adult dragonfly, all the arthropods we have studied having closed-type rhabdoms possess a slow-type electrical response. On the other hand, all the dipterous and hymenopterous insects that have an open-type arrangement possess the fast-type electrical response. The slow-type eye is more characteristic of nocturnal insects and arthropods that navigate at low light levels. In the fast-type eye, the rhabdom occupies only a small part of the volume. In the slow-type eye, the rhabdom takes up a much larger part of the volume; for example, the volume of the rhabdom of the cockroach is about five times the volume of the *Drosophila* rhabdom (Wolken *et al.*, 1957a,b).

An interesting structural feature in certain insects is that the crystalline cone and the rhabdom are connected by a crystalline thread which can function as a waveguide (Horridge, 1969), as seen in the firefly *Photuris* (Figure 8.9). Døving and Miller (1969) demonstrated for the skipper butterfly, *Epargyreus clarus*, that only the light contained in the crystalline thread is effective for stimulating the retinula cells. Varela and Wiitanen (1970) then showed that in the worker honeybee the closed rhabdom and the surrounding zone act together as a waveguide. Stuermer (1970) found fiber optic elements in *Phacops*, a trilobite fossil. All of this indicates that the adaptation of fiber optics for visual detection evolved a long time ago.

### Polarized Light Analysis

One of the interesting phenomena exhibited by animals with compound eyes is that they are able to analyze the plane of vibration of polarized light. They use it as a compass for navigation, for the direction of polarization indicates the relative direction of the sun.

Sir John Lubbock (1882), an English banker, wondered how ants find their way to and from their nests while foraging for food. His observations led him to conclude that ants use the sun as a compass for orientation and navigation. Although this seemed to be a reasonable explanation, it was soon questioned—for how could insects find their way when the sun was hidden from view? The question was not really answered until von Frisch's (1949) behavioral studies of the bee suggested that, in addition to the direction of a point source, the sun, insects could utilize polarization information from a patch of blue sky. In fact there is an analyzer for polarized light in the bee's eye. von Frisch (1950, 1953) constructed a model to show that the rhabdom consists of eight triangular polarizing elements, each transmitting a quantity of light proportional to the degree of polarization. In this model, opposite pairs of rhabdomeres have their polarizers in parallel orientation (Figure 8.19). The arthropod and mollusc photoreceptor fine structure as revealed by electron microscopy, and the strikingly geometric arrangement of perpendicular and parallel microtubules that form the rhabdom, strongly support von Frisch's model (1950, 1967).

In contrast to this theory, it has been proposed that reflection patterns from the environment resolve polarized light into patterns of graded light intensity. According to this view, the compound eye merely discriminates intensity and is not a direct analyzer.

In our laboratory, ants (*Tapinoma sessile, Solenopsis saevissimae*), houseflies (*Musca domestica*), fireflies (*Photurus pennsylvanicus*), and Japanese beetles (*Popilla japonica*) were found to orient to plane polarized light under conditions which prevent the use of background reflections as cues for orientation. All the insects studied exhibited a compass reaction (the source of stimulation served as a fixed point by which an oriented path is maintained) $\pm 45°$ with respect to the plane of polarization of the incident light beam. All oriented at 0° and 90° except the Japanese beetle (Marak and Wolken, 1965).

The use of a white background, which reflects normally incident polarized light in a circular pattern, and a black background, which reflects elliptically, led to a consistent difference in the response curve. With the white background there was more orientation at 45° and less at 0° and 90°. This response difference is consistent with the hypothesis that there are more cues for simple phototactic response on the black background which reflects more light perpendicular, rather than parallel, to the plane of polarization.

Houseflies and fireflies appear to orient to polarized light only within a narrow range of intensities, which suggested a model that includes two photo-

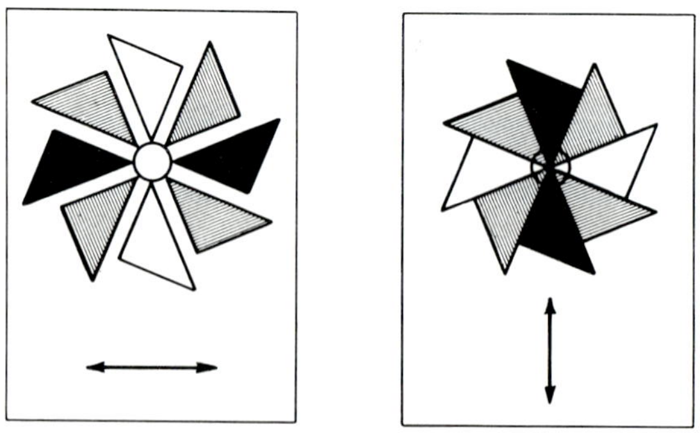

FIGURE 8.19 Model of rhabdom with its rhabdomeres and how they can act as an analyzer for polarized light (see K. von Frisch, 1950).

receptors with different absorption vectors. If these two receptors are oriented at 90° with respect to each other, and if the long axis of the visual pigment molecule is oriented in a single plane but free to vibrate in that plane, the receptors will absorb polarized light in a ratio of two to one when the electric vector is parallel to the absorption vector of one of the photoreceptors.

The reflectance pattern of nonpolarized light is circular, while the reflectance pattern of polarized light is elliptical, with the long axis perpendicular to the plane of polarization. In normal daylight, the long axis of the ellipse will point to the solar azimuth. This indicates that insect navigation can be accomplished extraocularly. Therefore, one crucial consideration must be experimentally determined before we can conclude that the rhabdom fine structure is in fact the analyzer for polarized light. We must establish whether more than one pigment is responsible for the organism's visual sensitivity. In Chapter XII we take up the nature of these visual pigments and their possible orientation within the rhabdomere microtubules to see if we can arrive at a molecular basis for the ability of these animals to analyze the direction of polarized light.

The structural arrangement of the arthropod and mollusc rhabdom suggests a number of interesting relationships in the evolution of the vertebrate eye and its retinal visual cells which we will now examine.

# IX
# Protochordates

Protochordates comprise three subphyla: Hemichordata, Urochordata, and Cephalochordata. These have traditionally been considered subphyla of the phylum Chordata, of which the fourth subphylum is Vertebrata. Some biologists consider the urochordates and cephalochordates more closely related to the vertebrates than are the hemichordates (Barrington, 1965). The relationship between the urochordates and cephalochordates and their relationship to the vertebrates is far from understood.

Possible phylogenetic relationships between the animal phyla are illustrated in Figure 9.1. The protochordates in this scheme are found to occupy an unusual position, for they have characteristics of invertebrates; that is, they do not possess a vertebrate column. But they do have certain structures, such as a notochord, a perforated pharynx, and a central nervous system, which are associated with the vertebrates.

The tunicates, a class of the subphylum Urochordata, are a group of remarkable animals which includes the ascidians, or sea squirts. They are known to be related to the vertebrates because they have motile tadpole-like larvae with definite chordate characteristics. It seems likely that ascidians and amphioxus, as well as the vertebrates arose from a primitive protochordate with unspecialized larvae (Barrington, 1965). These larvae might have been

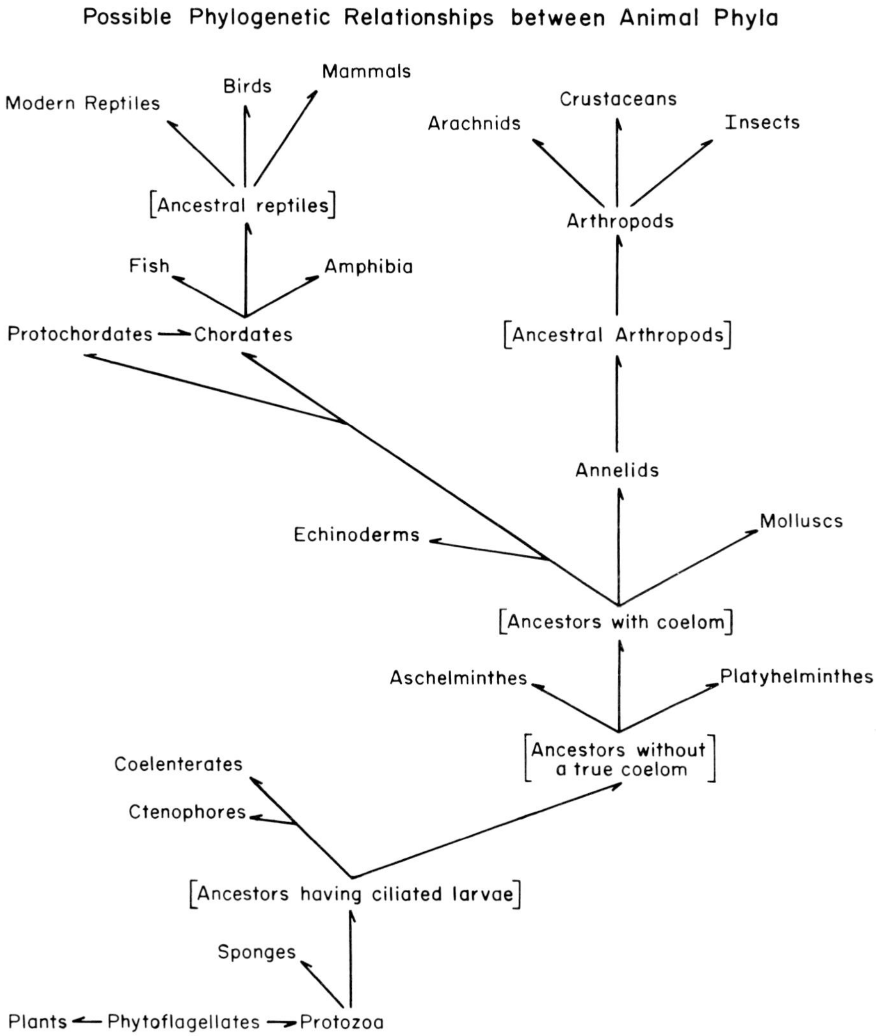


FIGURE 9.1 Possible phylogenetic relationship between animal phyla.

derived, as Garstang (1928) and later Berrill (1955) hypothesized, from single ciliated forms like the larvae of modern echinoderms.

Earlier theories regarding the evolution of the visual cells of the vertebrate eye suggested that they arose from the pigmented eyespot of amphioxus. For amphioxus appears to have two kinds of photosensitive systems. One type is described as pigmented nerve cells in the central nervous system which are believed to determine the direction in which the animal swims. The second type is a group of ependymal cells connected with nerve fibers, the infundibular organ shielded by a pigmented eyespot. As in the flatworm, when the animal is illuminated, it will turn in the opposite direction, i.e., away from the light source. It is these ependymal cells that may be homologous with the eyes of vertebrates. However, there is no experimental evidence to support the hypothesis that this is a photosensory structure. A summary of these earlier ideas on the evolution of the visual cells is discussed by Walls (1942) and Willmer (1960). More recent experimental data is discussed by Crescitelli (1972).

Ascidian larvae possess photoreceptor cells that do show some structural similarities to vertebrate visual cells. Therefore, it is of interest to see if there is more concrete evidence for the existence of an evolutionary development in the photoreceptor cell structure in the protochordates in line toward the vertebrate visual cells.

## *Ciona intestinalis*

The sea squirt *Ciona intestinalis* is a case of special interest in our discussion of evolution, for it is situated in the transition zone between invertebrates and vertebrates and exhibits features common to both.

Before proceeding with our discussion of the photoreceptors in *Ciona* siphons, it is of interest to recall a series of experiments with *Ciona* siphons which were carried out in the early 1900's by Paul Kammerer, an unusually interesting and gifted scientist. Kammerer held to the ideas of Lamarck and was convinced that useful characteristics acquired during the organism's lifetime could be passed on to the next generation. He set out to experimentally prove this, and he claimed that he did. I bring this up here because Kammerer considered *Ciona* crucial to the proof of his hypothesis.

*Ciona* siphons will regenerate after amputation. According to Kammerer (1915), if the two siphons are cut off, they will regenerate and with repeated amputations the siphons will become very long—the offspring of these individuals have longer siphons than normal. These results of supposed inheritance changes could not be repeated by other scientists. Although nutrition was found to be important to regeneration, little thought was given to the effect of light on the regeneration and behavior of the siphons.

Kammerer's experiments with the midwife toad *Alytes obstricans* received even greater notoriety at the time, and an even more violent controversy arose when the results could not be reproduced. However, some biologists were inclined to believe that Kammerer's experiments raised some fundamental questions regarding evolution. But others regarded him as a charlatan, for his sole remaining specimen of the midwife toad was found to have been doctored.

Koestler (1971) has recently raised questions about scientific research, and with it has brought our attention back to Kammerer for developing a line of research which he viewed as having far-reaching implications. No matter how sympathetically one views Kammerer's battle with the establishment, it should give scientists some pause before postulating theories based on misunderstood concepts, too little information, and incomplete experimental evidence.

With this bit of science history and forewarning let us return to *Ciona* and its photoreceptors in search of evolutionary clues.

The adult *Ciona* is a typical ascidian, shaped like a sack and enclosed within a gelatinous jacket, a tunic, made of a substance closely related to the cellulose of plants. *Ciona* has two siphons: the oral siphon into which water is drawn, and the "atrial opening" from which water is expelled (Figure 9.2a). The atrium leads into a very large cavity, the pharynx, which has a vast number of tiny slits in its side. These slits are ciliated so they act as pumps; they create the water current which passes in at the mouth, through the atrium, and then out the atrial opening. It is a matter of great interest to comparative zoologists that these slits are essentially homologous with the gill slits of fish.

A striking difference can be seen between the morphology of the motile larva and the sessile adult. It is the larval form, however, that has the characteristics of a vertebrate.

## The Larval Ocellus

In the larva, or tadpole, of *Ciona intestinalis* a relatively simple but organized ocellus has been described (Dilly, 1961, 1964, 1969). The ocellus is believed to serve as the photoreceptor to detect the direction of light in the environment of the tadpole, thereby permitting the larva to orient itself and move toward or away from the light stimulus. Thus, the tadpole is phototactic. The photoreceptor molecule responsible for such behavior has not been identified. The ocellus is made up of about 10 cells, including a lens cell, a large cell containing pigment granules, and from four to nine retinula cells. Dilly described the photoreceptor of the retinal cell as a pile of membrane discs, or lamellae, and a connecting cilium; however, he was unable to find synaptic connections. Dilly speculated that the ascidian retinal cell is probably comparable to the vertebrate retinal cell. However, Eakin (1963) questioned the homology of the ascidian larval photoreceptor cells with vertebrate photoreceptor cells.

Eakin and Kuda (1971) then reexamined by electron microscopy the receptor structure of the ocellus of two species of ascidian tadpoles, *Ciona intestinalis* and *Distaplia occidentalis*. They found that the ocellus is composed of one pigmented, cup-shaped cell with an estimated fifteen to twenty sensory cells and three lens cells. Each photoreceptor cell possessed an outer segment consisting of many lamellae and an inner segment bearing the basal body of the modified cilium (axonemes of $9 + 0$ doublets of microtubules), as schema-

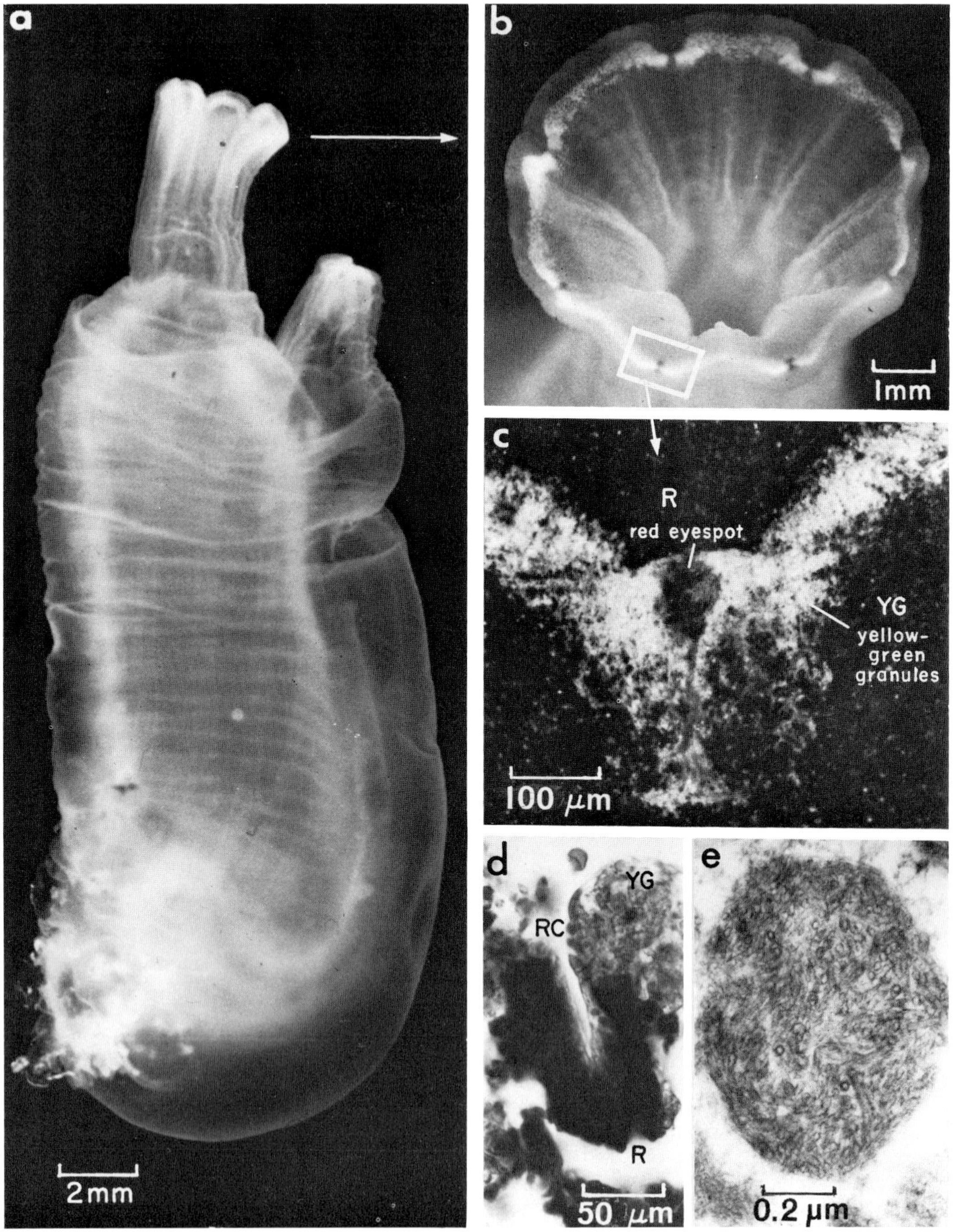


FIGURE 9.2 *Ciona intestinalis.* (a) Photograph of a living adult *Ciona* showing the two siphons and the tunic, (b) a view of *Ciona* siphons showing the red eyespots, (c) a fluorescence micrograph of an area in the siphons that contains the yellow–green pigment

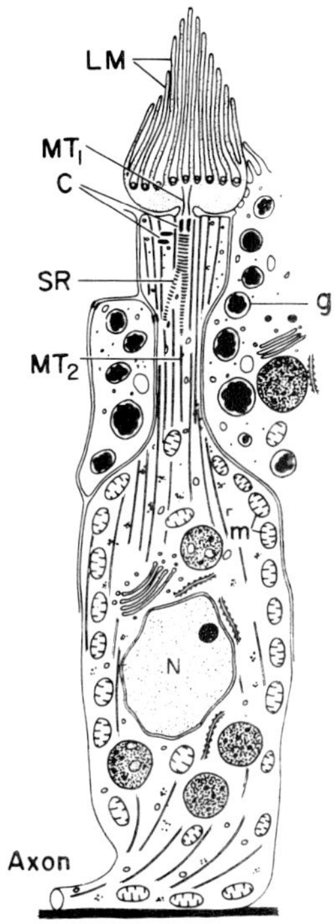


FIGURE 9.3 Diagram of photoreceptor cell in ocellus of *Ciona intestinalis* larva; LM, sensory lamellae derived from cilium; $MT_1$, microtubule (in axoneme of cilium); C, centrioles; SR, rootlet of cilium; $MT_2$, microtubules in cytoplasm; g, pigment granules; N, nucleus; m, mitochondria. From Eakin and Kuda (1971).

tized in Figure 9.3. They thereby established the ciliary nature of the tunicate photoreceptor, which would relate it to the vertebrate visual cell.

Now, from comparison of the morphology of the photoreceptor of the larva to that of the adult, it may be possible to ascertain the relative importance of the motile tadpole to the sessile adult in a line of evolutionary development within the chordates and toward the vertebrates. The problem is, where are the photoreceptors in the adult? Are the photoreceptors similar for the slow orientation and phototropism, as well as for the rapid contraction–retraction siphon responses? And what photoreceptor molecules do they contain?

## THE ADULT OCELLUS

The adult *Ciona* is sessile, but reacts to light by the closing and retraction of its siphons. The adult *Ciona* also exhibits positive phototropism, for the

granules and red eyespot, (d) histological stained section of an ocellus showing three main structural regions: receptor cells (RC), the cup of red pigment (R), and surrounding them, yellow-green pigment granules (YG), (e) electron micrograph of a yellow-green pigment body showing microtubules.

body grows in such a way that its long axis is oriented in a line with the direction of prevailing illumination, with the openings of the siphons turned to the light source (Millar, 1953). When Hecht (1918) studied the adult *Ciona,* he found that the most light-sensitive region was within 1 to 1.5 cm from the tip of the oral siphons. Hecht (1926) attributed such photosensitivity to two processes, one being photochemical in nature and another related process which was chemical. This implied that the adult should also possess a photoreceptor system.

Pigmented eyespots around the siphons have long been labeled "ocelli" (Figure 9.2), but their function as ocelli for photoreception has been doubted, for *Ciona* will continue to respond to light after the siphons have been amputated (Millar, 1953; Millot, 1957). To account for *Ciona* photosensitivity it was suggested that there is a *dermal light sense* associated with the surface of the adult (Steven, 1963; Millot, 1968).

Nevertheless, researchers have looked for structures that show a similar morphology to known photoreceptors in the ocellar regions of the adult *Ciona,* but light microscopy has failed to resolve any such structures. Hŭus (1937) suggested that the ocelli are vestigial structures. Previously Seeliger (1893), and then later von Haffner (1933), claimed that the pyramidal cells that have a mass of cilia along their outer edges could be photoreceptors. But, Millar (1953) found no ciliary-like processes extending from the cells, except that in some of the cells of the ocellus there was a vacuole in the cytoplasm between the cell nuclei and the cell surface. Das (1948) suggested that the site for the photoreceptor was the red pigmented region of the epithelium lining the siphons, with the red eyespots in the intersiphonal regions (Figure 9.2b–e).

Let us now reexamine the structure of the red eyespots, or ocelli. They are situated in the free border of the siphons, being recognizable in the adult as a red pigment in a yellow–green background (Figure 9.2b,c). The red eyespots lie at the bases of the notches between the lobes of the siphons. They can be resolved by microscopy as a cup of pigment around a clear central region, when viewed from the outside of the animal. When seen from above, the red area appears as a solid cylinder. These cylinders are arranged around the mouths of the siphons with their long axes lying radially in the plane of the siphon opening.

The red cylinders are about 100 $\mu$m long and 85 $\mu$m in diameter while the clear zone (diameter of the pigmented cup) is about 40 $\mu$m across. The cylinders are located in the gaps between the lips of the siphon folds, while between them, and extending along the free border of the siphon, is a yellow–green colored zone. This zone expands somewhat to enclose the red cylinder. The exact number of ocelli around a particular siphon varies but there are usually eight around the oral siphon and six around the atrial siphon. They are covered by the thin transparent tunic that encloses the animal.

Electron microscopy confirms that in addition to the layer of cells containing red pigment there is a deeper region containing the cells with yellow–green pigment and a more superficial layer of nonpigmented cells. The nonpigmented cells can be seen as a single layer of elongated epithelial cells, continuous with those forming the outer body wall. In the region of the ocelli the layer of epithelial cells forms a cup-shaped depression. Some of the cells have peg-shaped processes which extend to the tunic which covers them. Many of the processes contain numerous small vesicles. Deeper in the ocelli some of the epithelial cells contain cilia which project into the tunic. Away from the ocellus, the epithelial cells are approximately cuboidal in shape; their adjacent membranes show thickenings and they lie with their internal bases on a basement lamella. The tunic covering the ocellar region of the siphon is very similar to that described in the tadpole larva of *Ciona intestinalis* (Dilly, 1969).

Both the cilia and the processes projecting into the tunic could be sensory mechanoreceptors that underlie the epithelium, bringing about distortion of the tunic. We have already indicated that *Ciona* can respond to light in two ways, either with a short rapid contraction of the siphons followed by their retraction, or by a much longer phototropic response. A similar response of rapid contraction and retraction can be observed by bringing an object close to or touching the outsides of the siphons. Perhaps this illustrates a limited range of responses in *Ciona* and that receptors in the siphons are sensitive to both light and touch.

Many of the epithelial cells of the ocelli differ from those in other parts of the siphon in being more columnar. The deeper they lie in the cup-shaped depression of the epithelium, the more elongated they become. When the animal is contracted, the peg-shaped projections of the epithelial cells form a mass of interdigitating processes, but when the siphon is extended these processes are presumably drawn apart. This is perhaps a mechanism for controlling the amount of light reaching the receptors.

Like other epithelial cells in the region of the ocelli, some of the epithelial cells contain membrane-bound, electron-dense, oval masses. At higher resolution, these masses can be seen to contain a mass of granules and fibrils. Fedele (1923) described these to be photoreceptor cells in the epidermis of the siphons. They were distinguishable by the presence of pigment granules, but in other respects resembled tactile receptor cells.

## The Photoreceptor

What could be the photoreceptor? Deep in the ocelli some of the epithelial cells appear as if their free surfaces were invaginated or, alternatively, as if their lateral borders were extended into processes delimiting an apparent lumen. In each cell these processes enclose a single cilium adjacent to which arise microvilli (sensory lamellae) which extend along the length and tips of the processes. These are the structures that have been identified with the photoreceptor cell of the adult *Ciona* (Dilly and Wolken, 1973). The structure of these cells has similarities to various known photoreceptor cells (Wolken, 1971). Therefore, it is believed that these are probably the photoreceptor cells of *Ciona;* such a cell is schematized in Figure 9.4 (compare it with the tadpole photoreceptor, Figure 9.3). They usually occur in a group surrounded by other less specialized cells of the ocelli.

The single cilium found at the base of the lumen of the photoreceptor cells does not have a complex root system and is remarkable only for some transverse membranes just external to the point at which it joins the cell. These transverse membranes may cross the cilium or they can arise alternately on

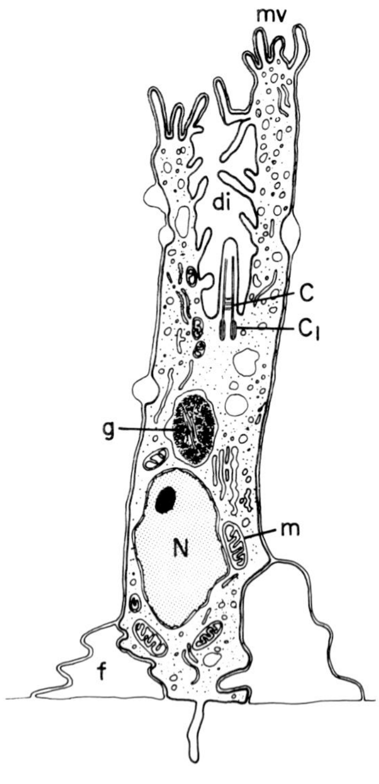


FIGURE 9.4 *Ciona intestinalis.* Diagram of adult photoreceptor cell; mv, microvilli arising from the cell membrane; C, microtubule in axoneme of cilium; $C_1$, centriole; g, pigment body; N, nucleus; m, mitochondria; f, nerve fibers; di, deep invagination, or lumen. From Dilly and Wolken (1973).

either side of the fibril. The microvilli are 1 μm to 2 μm in length and about 200 nm in diameter. Generally they are spaced at intervals, although occasionally several may lie immediately adjacent to each other. Along the lateral borders of adjacent photoreceptor cells there are thickenings of the plasma membranes and desmosomes. Between the bases of the receptor cells there are structures which can be identified with nerves.

## Pigment Cells

There is only a minute amount of red pigment compared with the yellow–green pigment in the adult *Ciona*. The cells that contain the red pigment are immediately beneath the epithelial cells of the ocelli and are separated from them by a narrow basement lamella. The pigment consists of small spherical granules from about 0.2 μm to 0.6 μm in diameter, bordered by a membrane. Each granule of pigment is homogeneous, electron-dense, and shows no internal structure. The granules are most concentrated in those regions of the pigmented cells which are closest to the invaginated layer of epithelial cells of the ocelli. It is this concentration of red pigment cells that gives the ocelli its color and is referred to as eyespots. It is not known if the granules move in response to changing light intensities, but no mechanism has been found that could account for such migrations. The red pigment may occur in cells which also contain the yellow–green pigment.

The cells which contain the yellow–green pigment are more or less confined to the small regions adjacent to the tops of the siphons. The pigment is found in oval or spherical bodies varying between about 0.8 μm to 4 μm in their long axes. The majority of these bodies are much larger than the red pigment granules, and, like the latter are enclosed by a membrane. They may occupy most of the cytoplasm or be confined to some small region of the cell; and they are not all the same size in a cell. They lie farther into the epidermis, may be several layers thick, and surround the red pigment region of the ocelli. Deep into the yellow–green pigment layers are groups of cells that contain both the longitudinal and circular muscle cells of the siphons. Sometimes the yellow–green pigment containing cells extend between these muscle cells and surround them on three sides.

Each yellow–green body seems to consist mainly of a mass of microtubules. The latter are not regularly arranged but lie together in a tightly packed tangle (Figure 9.2e). As yet we have not been able to determine whether this appearance is due to the plane of section passing through different parts of a single coiled microtubule or whether many microtubules are present. Some of the microtubules appear to branch. Individually they seem hollow, are about 33 nm in diameter, and possess a smooth wall about 5 nm thick.

Some of the cells that contain the yellow–green pigment also have cilia, and kinetosomes are occasionally observed between the pigmented bodies. Many of these cells contain large vesicles and channels of endoplasmic reticulum. Rather surprisingly, the microstructure of the yellow–green bodies resembles that of the prolamellar body found in etiolated algae and plants. These prolamellar bodies are the precursors of chloroplasts and give rise to the lamellae (Figures 5.5 and 5.6) of the chloroplast (Pickett-Heaps, 1968). A question then arises as to whether or not the yellow–green bodies that are in *Ciona* are of algal origin—for *Ciona* ingests algae and a symbiotic relationship could have resulted and could contribute to its photosensitivity. Such a relationship has been found in the marine flatworm *Convoluta roscoffensis*. However interesting, such a hypothesis for *Ciona* remains a matter of speculation and needs further investigation.

### The Pigments

The absorption spectrum of the extracted red eyespot pigment, in 80% acetone, resembles that of a carotenoid and probably functions as a screening pigment for the photoreceptor. The yellow–green pigment fluoresces in the green part of the spectrum. Extractions with phosphate buffer and various acetone concentrations yield absorption spectra with peaks near 360 nm and 460 nm, suggestive of a flavin or flavoprotein.

Even though the photoreceptor pigment molecule has not been identified the presence of a flavin would indicate, as in *Phycomyces* phototropism and *Euglena* phototaxis, that it could participate in the photoreceptor process and could be its photoreceptor molecule.

## Remarks Regarding the Ocelli

The red eyespots and the cells contained within are well suited to be a directional photoreceptor. If we assume that the ocelli respond to light, then it is possible to suggest a function for them. The long axes of the ocelli lie in the same plane as the opening of the siphon, and the ocelli are arranged as spokes around it. In the contracted siphons the group of peg processes will be interdigitated and obscure the receptor cells, but as the siphons extend, they will cause the walls of the ocelli to move. This will draw the interdigitating processes apart, thus exposing the sensitive endings to the light. When the open siphon is pointing directly at the light, no light will fall on any of the ocellar receptors. When the plane of the siphon is not at right angles to the light, then light will fall only on some of the ocelli. This would provide a mechanism to keep the siphon pointing toward the light, and thus away from the debris on the bottom. If the light shines horizontally on the animal, it will bend towards it, showing that this orientation response does not depend upon gravity. Since the siphon currents are involved in feeding, respiration, reproduction, and excretion, this photosensitivity would be of great advantage to the animal.

Earlier we mentioned the "dermal light sense" and a few remarks are in order here. Eyes are not the sole means of photoreception, for the general body surface of many eyeless and blind animals is remarkably light-sensitive. This diffuse photosensitivity over the whole or part of an animal's body was defined by Millot (1968) as the "dermal light sense." However, neither the photoreceptor structure nor the pigment responsible is known, so it is not certain that this behavior is dermal. Diffuse photosensitivity, though, is widely spread through all the major phyla, and more particularly in aquatic animals. For example, in the clam *Mya,* a sudden change of light intensity results in retraction of its siphon. In the earthworm, the surface of the anterior segments is light-sensitive. These animals possess large numbers of photosensory cells located beneath their skin. There is evidence to show that even deeper tissue cells can also be photosensitive, as in marine animals which exhibit responses if their nerve or ganglion cells are exposed to light (Chalazonitis, 1964). Why deeper tissue cells (nerve, brain, sex glands) in animals should be light-sensitive is little understood at present.

Looking back, we can see that there is some similarity between the photoreceptors we described in the adult *Ciona* (Figure 9.4) and the globular cilium in the ascidian tadpole (Figure 9.3). They are alike in that they both have microvilli; but, whereas those in the larva come from the cilium membrane, those in the adult come from the cell membrane.

In an attempt to find evolutionary significance in the differing types of photoreceptors, Eakin (1963) classified them as either *ciliary* or *rhabdomeric,* depending upon the origin of their photoreceptor membranes. Those where the photoreceptor membranes arose from the cilium (or flagellum), are of the vertebrate type, and those arising from the cell membrane, *rhabdomeric,* are of the invertebrate type. Such a rhabdomeric receptor, except in the origin of its photoreceptor membranes, may be very different morphologically from the typical highly organized rhabdomere of the arthropods and molluscs (see for example Figures 8.6, and 8.11–15). In *Ciona* it is preferable to consider the adult photoreceptor as a microvillous receptor, but according to Eakin, it could be classified as rhabdomeric. The photoreceptor in the tadpole larva is ciliary (Dilly, 1961, 1964, 1969; Eakin and Kuda, 1971). Eakin (1973) has correlated the rhabdomeric type photoreceptor with the annelid superphylum and the ciliary type photoreceptor with the echinoderm superphylum. But in *Ciona* we have both types.

Now that we have described both the larval and adult photoreceptors, we find that the adult photoreceptor is of the major invertebrate type (Dilly and

Wolken, 1973). The relative evolutionary significance of the adult and larval forms of *Ciona* suggests that the photoreceptor of the larva is on the main evolutionary path, and that the adult has been a secondary development.

Let us now turn to the visual cells of vertebrate eyes and see what structural relationships can be found between these and the photoreceptor structures that we have already described.

# X

# The Vertebrate Eye and the Retina

In the phylogenetic scheme that we presented in Figure 9.1, and according to the description given by Walls (1942), the lowest of the vertebrates are the cyclostomes. The cylostomes include the dogfish and lampreys. Just above the cyclostomes are the many types of true fish whose eyes are more fully differentiated than those of the cyclostomes. The oldest of the true fish are the elasmobranchs, whose modern representatives are the sharks and rays. From one of these types the first land animals, the ancient amphibians, were derived. They in turn gave rise to modern amphibians and to the reptiles. The reptiles differentiated into a large number of orders, only four of which have persisted to the present time. From one group of extinct reptiles came the birds. Certain skeletal characteristics, as well as other similarities, suggest that birds have a reptilian origin. For example, the avian eye has many features in common with the eyes of reptiles. Most birds are highly active diurnal animals and their retinas are characterized by numerous cones, single and double, whereas the rods are relatively few in number. Pigmented oil globules are numerous in the retina and they range from colorless to yellow, orange, and red. In nocturnal birds, the oil globules are weakly colored or colorless and there are a greater number of rods. From another group of extinct reptiles came the mammals. There are still many unknowns in this phylogeny. For example, according to Packard (1972), present-day cephalopods and fish show a resemblance even to the vertebrate refracting eye.

It is through the structure of the eye and the retinal photoreceptors that we have been trying to trace the evolutionary pathways. Here we will try to see what structural relationships can be found between the vertebrate mammalian eye and those of the fish and amphibians.

## The Eye

The eye and how it functions for vision have fascinated man for a long time. The optics of image formation were described by Johannes Kepler in 1611. How the eye accomplishes imaging was developed by René Descartes in 1664. At the beginning of the eighteenth century, William Molyneux (1709) of Dublin wrote the first treatise on optics in English, which contained several simple diagrams showing the projection of a real, inverted image in the human eye. With the development of the microscope came the first description of the morphology of the retina by van Leeuwenhoek in 1674. But it took almost two centuries before the retinal rods and cones were described by Schultze (1866), and another decade for Böll (1876) to recognize that the retina was photosensitive and bleached in the light. Soon afterward Kühne (1878) and co-workers began a series of studies to isolate the visual pigment. From these observations came the beginning of our understanding of the visual process. We say much more about this in our discussion of the visual pigments and visual excitation in Chapters XI and XII.

The gross anatomy of the human eye in cross section is shown schematically in Figure 10.1. The eyeball, which is approximately spherical, houses the complete optical apparatus. There are two refracting structures, the clear cornea (refractive index 1.336) and the crystalline lens (refractive index 1.437). A variable aperture, a diaphragm, is provided by a contractile, membranous partition, the iris, that regulates the size of the light opening, the pupil. In dim light, in order to admit more light through the lens, the pupil opens to an extent governed by the activity of the retina. The lens also acts as a filter by sharply cutting off the far edge of the ultraviolet region at about 360 nm. Excellent descriptions of the anatomy of the eye, its histology, and embryology are given by Walls (1942) and Polyak (1957), and its evolutionary development by Duke-Elder (1958) and Crescitelli (1972).

Image formation takes place by the refraction of light on one or more spherical surfaces that separate media of different refractive indexes. This causes the formation of inverted, reduced images on the surface of the retina. The combined optical and sensory apparatus is paired in two symmetrically constructed and oriented eyeballs. As a result, a large section of visual space is imaged on both retinas binocularly.

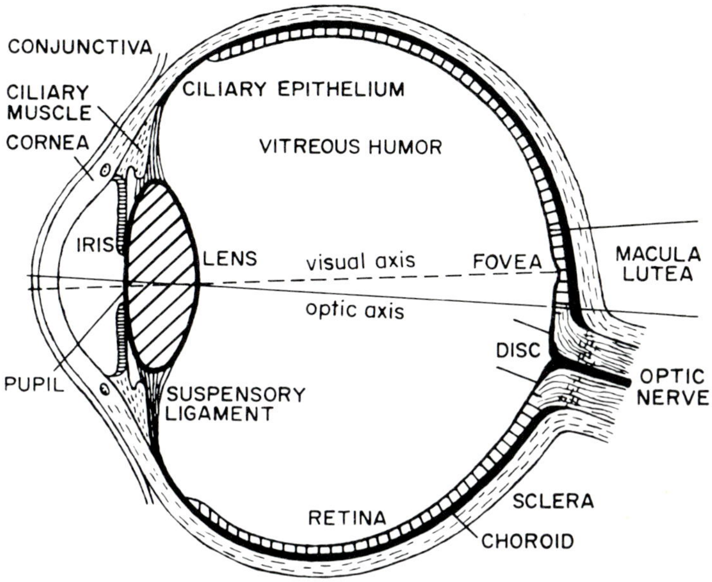


FIGURE 10.1 The structure of the human eye (see Figures 10.3 and 10.4 structure of the retina).

Refracting type eyes evolved a long way back, as we have already described in the arthropod, the trilobite fossil *Phacops* with its refracting type lenses (Towe, 1973), and in the cephalopod mollusc *Octopus* with a more developed refracting eye.

When I was in the Galapagos Islands, on the island of Santa Cruz, I was struck by the surf jumping fish, the "four-eyed" blenny, *Dialommus fuscus*, which frequents the rocks between tides. This fish is amphibious and its eyes can easily adjust to aerial and aquatic vision (Figure 10.2). Therefore, it must have two distinct optical systems. The cornea is partitioned and the lens is so shaped that it refracts the light on the lower part of the retina in the air, and on the upper part of the retina when in water. It seems, then, that both aerial and aquatic objects are focused simultaneously on different parts of the retina. This is an interesting evolutionary "adaptive" development, but too little is known about the optics of their eyes and the structure of their retinas.

## The Structural Development of the Eye

The development of the vertebrate eye consists principally of the conversion of the cells of the optic cup walls into cells of the retina. This involves the multiplication of cells and the differentiation of some of them into the

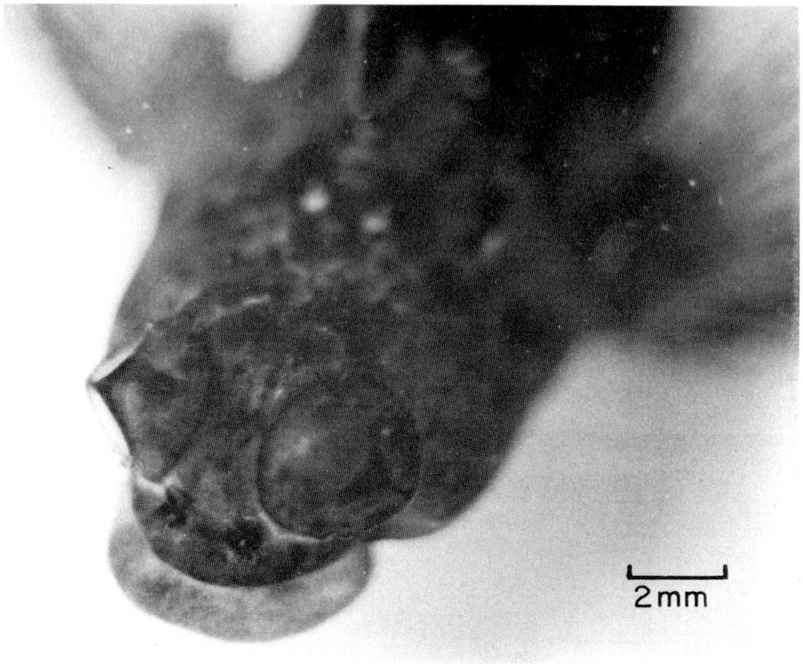


FIGURE 10.2 Four-eyed blenny, *Dialommus fuscus;* view of its eye, cornea, and lens.

light-sensitive retinal rods and cones and others into the nerve cells with which the rods and cones are connected and which in turn are connected with the brain. Although the original connection of the optic vesicle with the embryonic brain persists throughout this process (as the optic stalk), nervous connection of the retina with the brain is secondary and is formed by outgrowth of nerve fibers from the nerve cells of the retina through the optic stalk into the brain.

The retinas of all vertebrates develop from the walls of the neural tube, so that the retinal cells possess many features in common with neural cells. In invertebrates, epithelial cells give rise to the eye. But in the vertebrate eye, epithelial cells form the cornea and lens.

## The Retina

The vertebrate retina is a complex structure which is closely attached to the pigment epithelium. The retina consists of 10 cell layers (Figure 10.3). The photoreceptors, the rods and cones, are arranged in a single-layered mosaic or rhabdom-like array (Figure 10.4) connected with a highly developed system

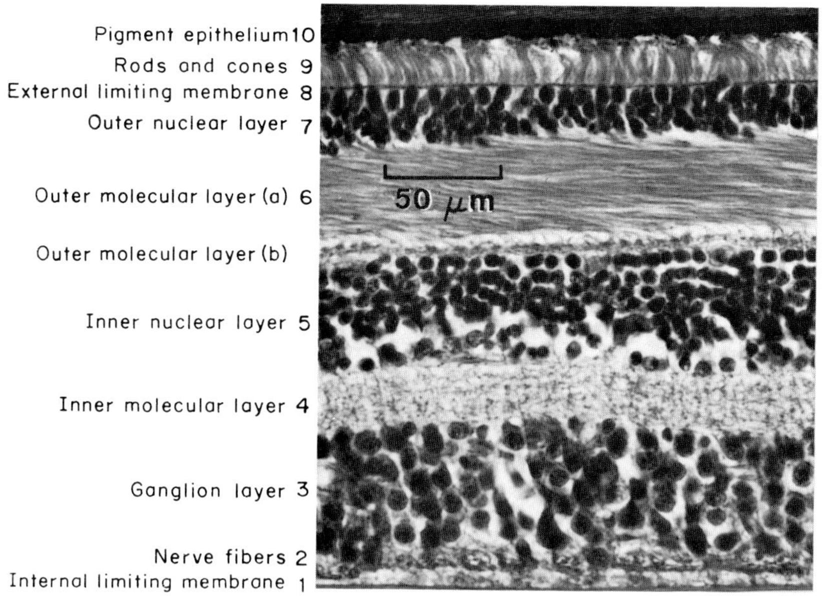


FIGURE 10.3 Retina of human eye; histological section showing the ten cell layers.

of connecting and interconnecting neurons. Other cell layers of the retina include the external limiting membrane, the external nuclear layer, the external molecular (plexiform) layer, the internal nuclear layer, the internal (plexiform) layer, the ganglion cell layer, the nerve fiber layer, and the internal limiting membrane. The nervous cell layers comprise the rod and cone cells, the bipolar cells, and the ganglion cells. It is estimated that in the human retina there are about $1 \times 10^8$ retinal rods and $7 \times 10^6$ retinal cones. Toward the center of the human retina there is a depression, the fovea, which is the fixation point of the eye, and where vision is most acute. It contains mostly cones, of which there are $4 \times 10^3$. The rods become more numerous as the distance from the fovea increases. The fovea and the region just around it, the macula lutea, are colored yellow; they contain a plant carotenoid, a xanthophyll (Figure 3.6).

The first four layers constitute the neuroepithelial layer and are the neurons of the first order. The remaining layers are considered the cerebral portion, where there exists a complex arrangement of nervous elements resembling in structure and function those of the central nervous system of which the retina represents an outlying portion. The fifth layer (internal nuclear layer), which contains the bipolar, horizontal, and amacrine cells, and the sixth layer comprise the neurons of the second order. The seventh and eighth layers make up the neurons of the third order, which pass centripetally to the primary

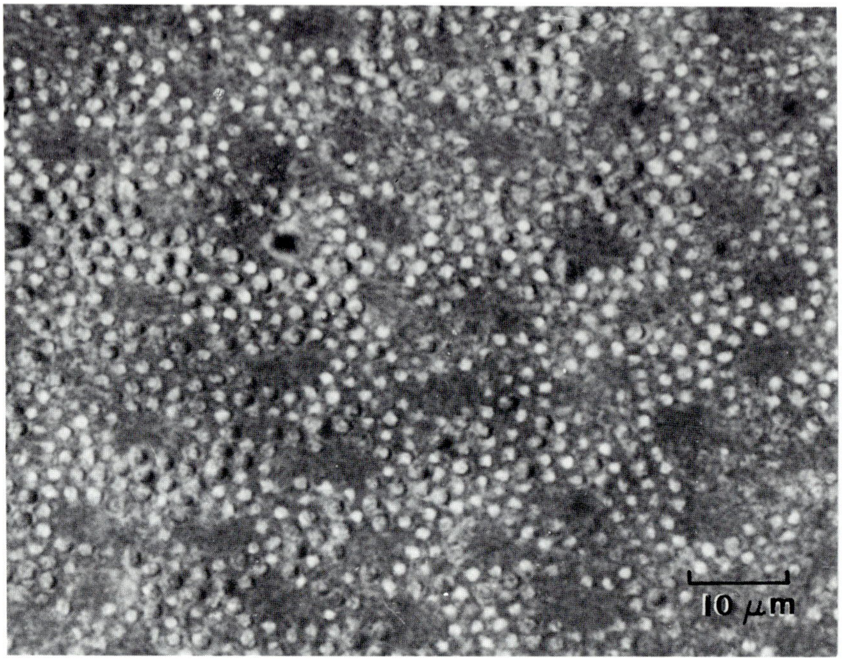


FIGURE 10.4 Retina of human eye; surface view of rods and cones. Not rhabdom-like array of rods.

optic center (the lateral geniculate) of the metathalamus (refer to description in Polyak, 1957).

Next to the retina is a sheet of cells filled with black pigment, the choroid coat, which absorbs extra light and prevents internally reflected light from blurring the image.

## The Retinal Rods and Cones

The rods and cones are differentiated specialized structures of the retinal cells. Each has an inner segment and what appears either as a rod- or cone-shaped outer segment which contains all of the photosensitive visual pigment (Figures 10.5 and 10.6). Over a century ago, Schultze (1866) described the general morphology of the retina of the frog and observed that when the retinal rods were exposed to light they disintegrated into platelets (Figure 10.5b,c). However, Schultze's observations had to await the development of the polarizing microscope and the electron microscope before they could be confirmed.

With the polarizing microscope, Schmidt (1935, 1937, 1938) was able to

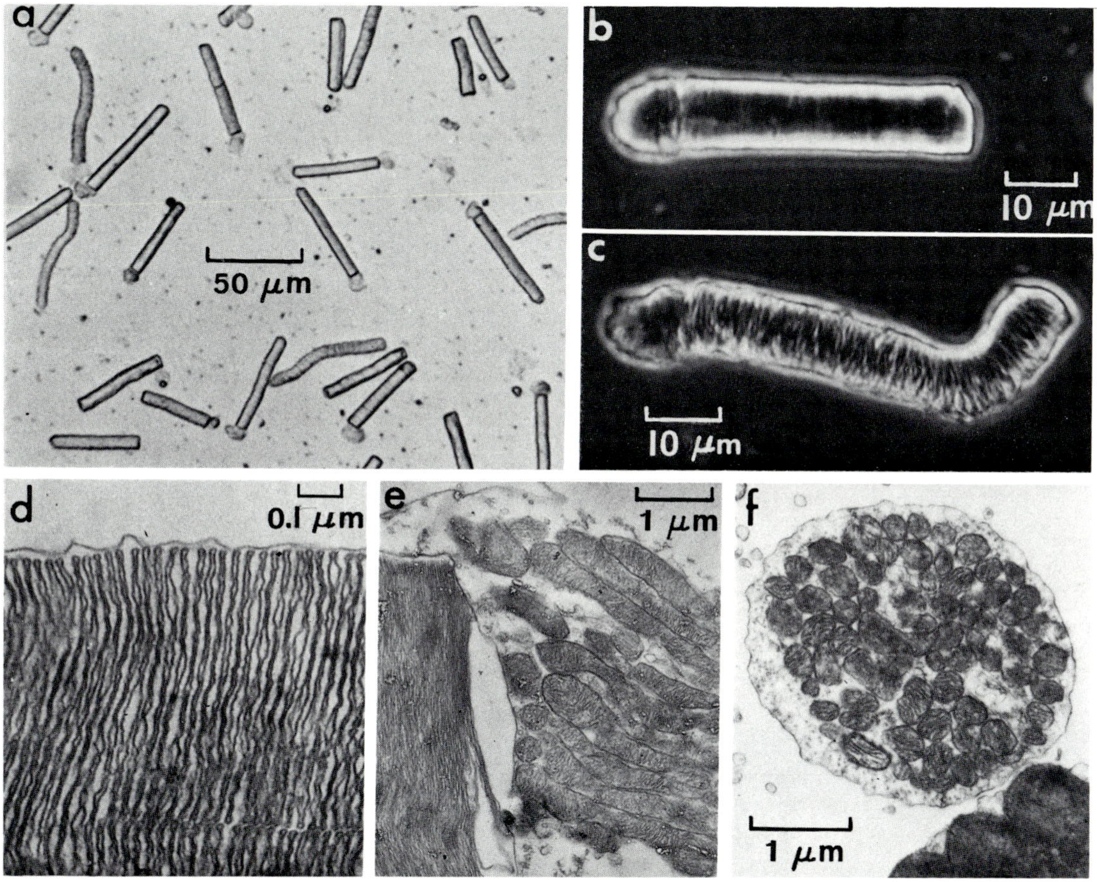


FIGURE 10.5 Frog retinal rod. (a) Freshly isolated rods from retina, (b) phase contrast of a single rod, (c) same rod, bleached at 525 nm. Electron micrographs, (d) outer segment, (e) and (f) inner segment.

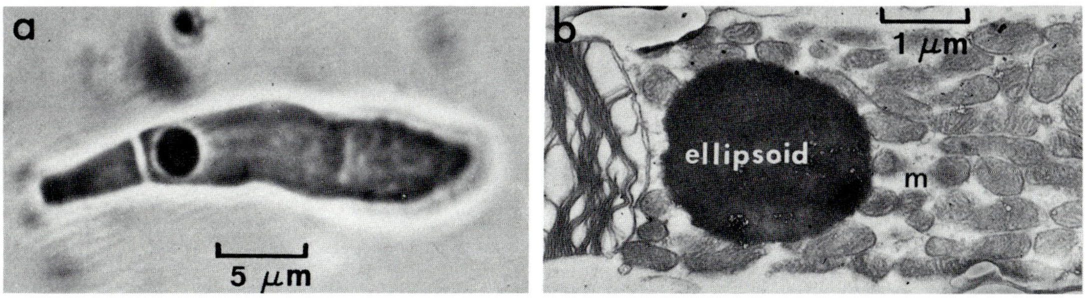


FIGURE 10.6 (a) Freshly isolated frog retinal cone; (b) electron micrograph of cone inner segment; note ellipsoid and mitochondria (m).

show that the rod outer segments were birefringent, indicating that they were highly ordered structures. He observed that when the rods were extracted with organic solvents to remove the lipids, there was a reversal in sign of the birefringence. To account for the observed optical changes, Schmidt suggested that the lipid molecules would lie parallel to the axis of the outer segment and that there would be nonlipid material arranged at right angles to the long axis. When the rods were treated with dilute acids or alkali, the birefringence disappeared, accompanied by a lengthening of the rod to as much as 10 times its original dimension. Further observations on the dichroism of the outer segment suggested that the visual pigment was present in the nonlipid regions of the rod. With the rods free from lipids, Schmidt was able to show a complete curve for birefringence as a function of the refractive index of the medium. These observations led him to postulate that the rod outer segments were laminated structures (lamellae) of lipids and proteins.

Sjöstrand (1949, 1953a,b), using the electron microscope, showed that the guinea pig retinal rod outer segments were lamellae with edges 75 Å in thickness. Electron microscopic studies that followed clearly established that all vertebrate retinal rod and cone outer segments (e.g., frog, perch, cattle, rab-

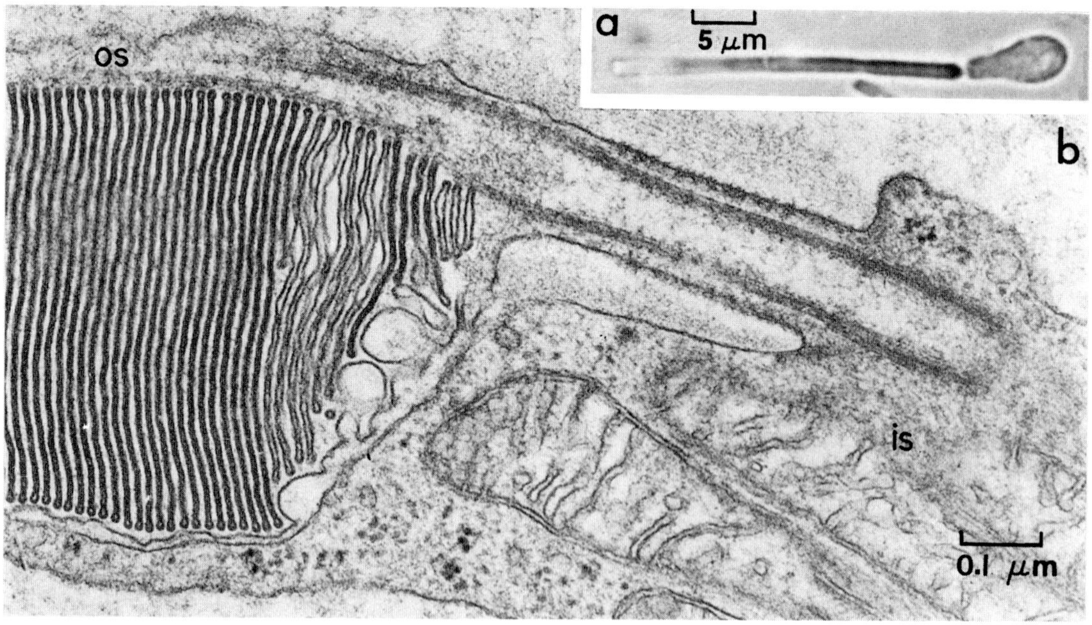


FIGURE 10.7 (a) Freshly isolated human retinal rod; (b) electron micrograph showing fine structure of outer (os) and inner (is) segments (courtesy of Dr. T. Kuwabara, National Institutes of Health, Bethesda, Maryland).

bit, rat, monkey, and human) are indeed double-membraned discs, lamellae of the order of 250 Å in thickness, separated by less dense material of approximately equal thickness. Electron micrographs of the frog rod (Figure 10.5), the human rod (Figure 10.7), and the bovine rod (Figure 10.8) illustrate this double-membraned lamellar structure. Each membrane of the lamellae is from 50 to 75 Å in thickness (De Robertis, 1956; De Robertis and Lassansky, 1961; Fernández-Morán, 1961; Cohen, 1961, 1963, 1964; Missotten, 1964; Dowling, 1965; Wolken, 1961a,b, 1963, 1966b).

The inner segments of the rods and cones are not lamellar but are densely packed with mitochondria, as seen in Figures 10.5e,f and 10.6b. The inner segment appears to be connected to the outer segment by a continuous membrane. Interconnection occurs by a cilium, or flagellum, that passes from the outer segment through the inner segment (Figures 10.7 and 10.8). Embryologically, the rods and cones are thought to have been derived from flagella, and in this respect it is interesting to note that communication between the segments is accomplished by a "cilium" (De Robertis, 1956). The fine structure of this cilium distinctly shows the characteristic nine fibrils found in the cilia and flagella of plant and animal cells. For example, compare the cross section of a *Euglena* flagella (Figure 7.1) and the *Ciona* photoreceptors (Figures 9.3 and 9.4) with the connecting cilium of the human (Figure 10.7) and bovine (Figure 10.8), retinal rod.

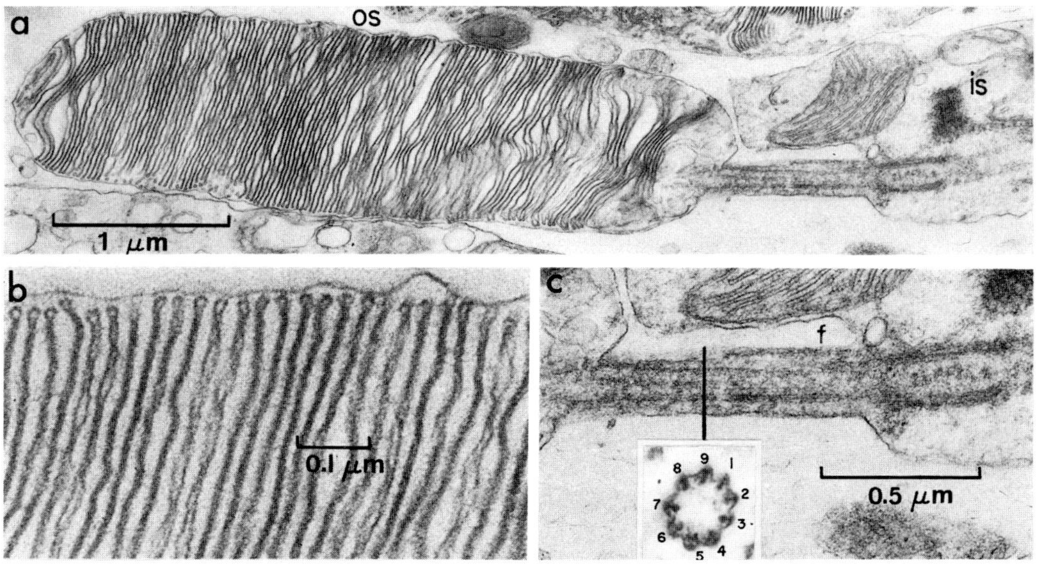


FIGURE 10.8 Cattle retinal rod; electron micrographs (a) outer segments (os) and connecting flagellum (f) to inner segment (is), (b) and (c) enlargements of areas in (a).

The connecting cilium of both these highly specialized outer and inner segments may be a crucial factor in the functional chemistry of the rods and cones. At one end, the outer segment, we find a highly ordered matrix containing photosensitive pigment designed for light capture. At the opposite end, the inner segment, we find a mass of tightly packed mitochondria whose enzymatic systems would provide the chemistry for oxidation–reduction reactions and hence, energy transfer.

## The Elasmobranch Retina

Although considerable advances have been made in the elucidation of the fine structure of the vertebrate retinal rods and cones, the evolution of the vertebrate visual receptor cells is still not clearly understood. A detailed study of the fine structure of shark retinal photoreceptors could provide additional information on the evolution of the elasmobranchs, but also of the amphibious and terrestrial vertebrates.

The elasmobranchs—sharks, dogfish, skates, and rays—are cartilaginous fish which are believed to be descendants of Devonian forms that evolved from placoderm-like ancestors. Therefore we can look for evidence that the elasmobranch eye and the ultrastructure of the retina developed from the structures of these more primitive creatures and foreshadow the development of more advanced vertebrate visual structures.

Early anatomical studies of the elasmobranch eye and retina were conducted by Franz (1931, 1934), but surprisingly, there is very little recent research directed to the shark eye (Gilbert, 1963). The numerous behavioristic studies of sharks give the impression that the shark is very dependent on vision, although its sense of olfaction is keenly developed. Most of the available information indicates that the visual cells in the elasmobranchs are undifferentiated and are neither rods nor cones. But the observations of Gruber *et al.* (1963) of the lemon shark indicated that the retina has rods and cones. The cones were characterized as being short, tapering outer segments with pyramidal inner segments containing ellipsoids. Also the visual cells were free of oil globules. However, the retinas of some elasmobranchs, for example the skate, do not have cones (Dowling and Ripps, 1970). Most of the visual cells of the shark retina resemble the retinal rods of vertebrates. But there are other visual cells that have a greater diameter, and although they have been identified as cones (Hamasaki and Gruber, 1965), they appear to be modified rods. They are not structurally like the cones of fish or amphibians (Figure 10.6). In addition, while sharks do distinguish changes in brightness, there is no evidence that they perceive color. Therefore, in the elasmobranchs only one type of visual cell, resembling a rod more than a cone, is indicated.

To find some structural relationships in the elasmobranch retina, an electron microscopic study was undertaken of the smooth dogfish, *Mustelus canis*. We found that the retina did not show any differentiation into what can be described as rods and cones. What appears as the rod outer segment is very long and narrow (Figure 10.9a,b). The inner segment, or ellipsoid, is attached to the outer segment by ciliary-like fibers. Thus each visual cell consists of an outer segment and an inner segment.

The *in situ* outer segment appears as a smooth cylinder of uniform thickness with a rounded distal end. It was observed to have a peculiar brilliancy and to be homogeneous and positively birefringent. The inner segment, or ellipsoid, is broader and shorter than the outer segment. At its junction with the outer segment the ellipsoid has a dark-staining fiber apparatus and a diphosome. The outer segment is about 1.5 $\mu$m in diameter and 20 $\mu$m in length and is composed of double membrane discs approximately 150 Å in thickness. The center of the disc is partially hollow. The membranes forming the discs are about 30 Å in thickness. It was noted that even in the regions in which the outer segments remain attached to the retina, they tend to have one or two swellings. In some regions the double membrane discs tend to become separated and may partially escape the enclosing membrane, thus leaving a free space in the outer segment. Such artifacts have been noted to occur very rarely in the case of the frog retina but are common in the trout retina.

The ellipsoids may be seen to contain three different regions. The distalmost third of the ellipsoid contains a dense aggregation of elongated mitochondria and a bundle of fibrils running in one side of the outer segment and joining it to the inner segment. These fibrils form the cilium with its basal body in the inner segment, as shown in the human and cattle retinal rod (Figures 10.7 and 10.8). The middle third of the ellipsoid is usually fibrous while the proximal third is granular or vesicular. In a few ellipsoids, the middle and proximal third (particularly the middle) appear to have a space reminiscent of an oil globule. Only one in five visual cells contains what appears as an oil globule. The proximal region is probably analogous to the contractile myoid in most fish and birds. When illuminated, the contractile myoid in these animals lengthens and pushes the outer segment deeper into the pigment epithelium toward the sclera. In dim light this region contracts, moving the outer segment closer to the external limiting membrane and thus exposing it to more light (Detwiler, 1943). However, little is known of the contractility of the myoid (proximal third of the ellipsoid) in the smooth dogfish or in the elasmobranchs in general.

The work of Hamasaki and Bridges (1965) on the electroretinogram (ERG) of three species of sharks lends some support to the view that the undifferentiated visual cells of the shark may be capable of functioning as rods and cones. While it may be possible to argue that the smooth dogfish's visual cells are indeed rods, we must consider the fact that their organization does

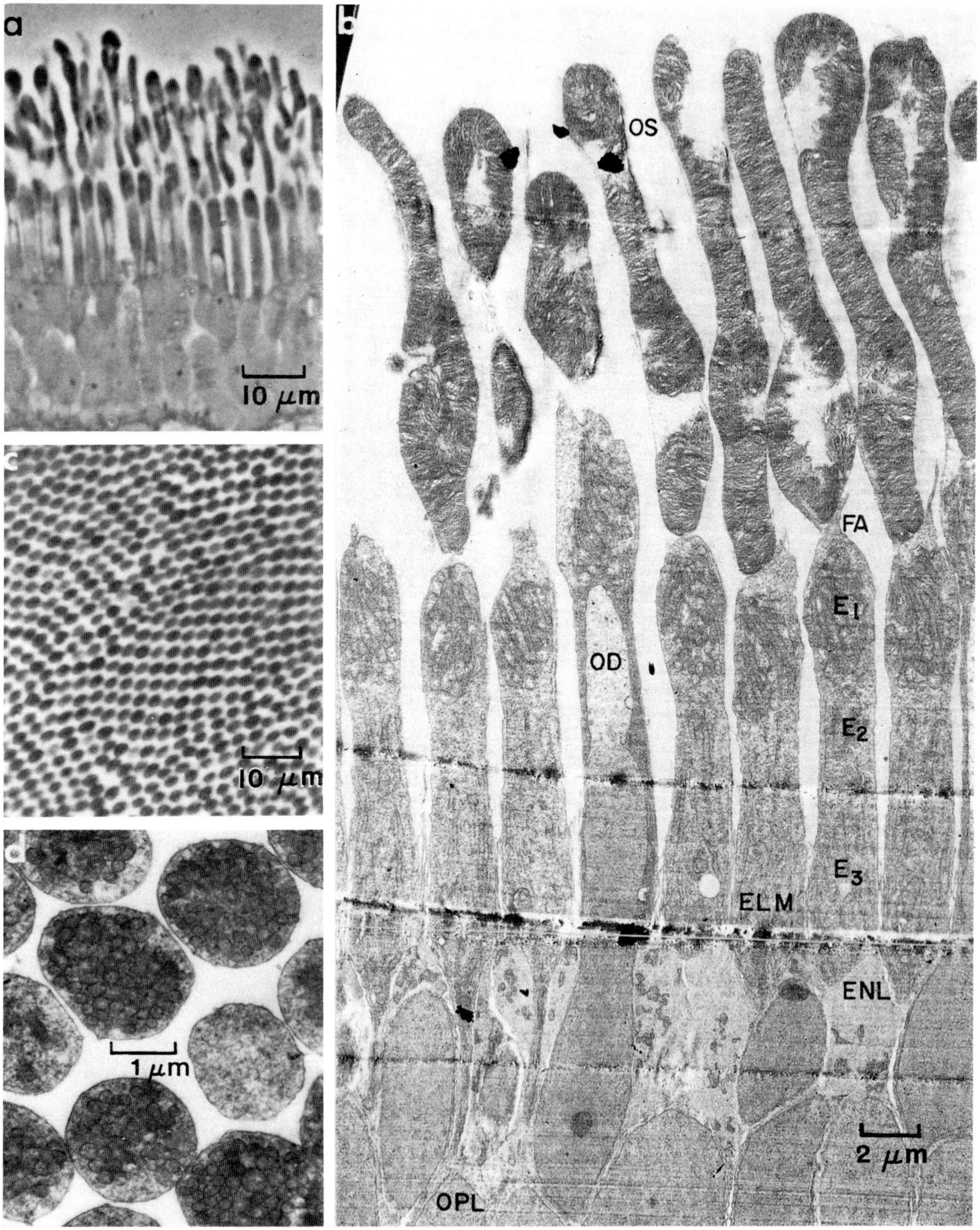


FIGURE 10.9 Smooth dogfish retina, *Mustelus canis*. (a) Light micrograph of a fixed and cut section of the retinal photoreceptors, (b) electron micrograph showing the outer and inner segments of rods, (c) light micrograph cross section of the retina area, (d)

not altogether resemble that of a rod retina. Thus, it appears more likely that the elasmobranchs, derived from the placoderms, represent a separate offshoot in the phylogeny of fish. In view of the fact that the sense of olfaction is so well developed in the sharks, one may postulate that their vision has evolved in a particular way so as to be especially useful in near vision. Once the prey is located by olfaction from a distance, the shark must depend on a fairly acute visual image. This it surely has, even with one type of visual cell. The visual cell resembles a rod morphologically, but due to the organization of its inner layers, is capable of functioning like a cone. The retina of the smooth dogfish has attained a very high level of organization comparable to that of the highly visual teleosts and mammals.

## The Amphibian Retinal Rod

Let us now reconsider the structure of the amphibian rod. The outer segment of the frog rod is an uncommonly large structure of about 60 μm in length and 6 μm in diameter (Figure 10.5). Thus it is easily observed with the light microscope. The retinal rods can be severed from the retina simply by shaking.

The isolated frog rods appear highly refractive and seem to be constructed of long, packed rodlets of about 1 μm in diameter. As soon as they are exposed to bright light, the entire outer segment swells, and as if subject to osmotic shock, begins to break transversely so that within seconds all structural identity is disrupted and its visual pigment is bleached (Wolken, 1961a).

A cross-sectional view of freshly fixed and sectioned frog rods shows a cylinder with scalloped edges and fissures extending into the rod, so that it is divided into 15 to 20 irregular wedges (Figure 10.10a). Longitudinal sections reveal that these wedges, or rodlets, produce further divisions within the structure of the rod, as shown schematically in Figure 10.10b,c.

In *Necturis,* the mud puppy, the rod outer segments are about twice the diameter of the frog rod and only one-half its length (12 μm in diameter and 30 μm long); they are the largest known among the amphibians (Wald *et al.,* 1963). Here, too, the outer segments are cut by a series of deep longitudinal grooves into an array of rodlets. There are from 20 to more than 30 such fissures, making the rod in cross section appear scalloped, as in the frog rod (Figure 10.10).

Thus the structure of the amphibian outer rod segment resembles roughly that of a closed-type rhabdom, which we have already observed in certain arthropods and molluscs (see for example Figures 8.10–8.12 and 8.15).

---

electron micrograph cross section through the photoreceptor inner segment. OS, outer segment; FA, fiber apparatus; OD, oil droplet; $E_1$, distal third ellipsoid; $E_2$, middle third ellipsoid; $E_3$, proximal third ellipsoid; ELM, external limiting membrane; ENL, external nuclear layer; OPL, outer plexiform layer.

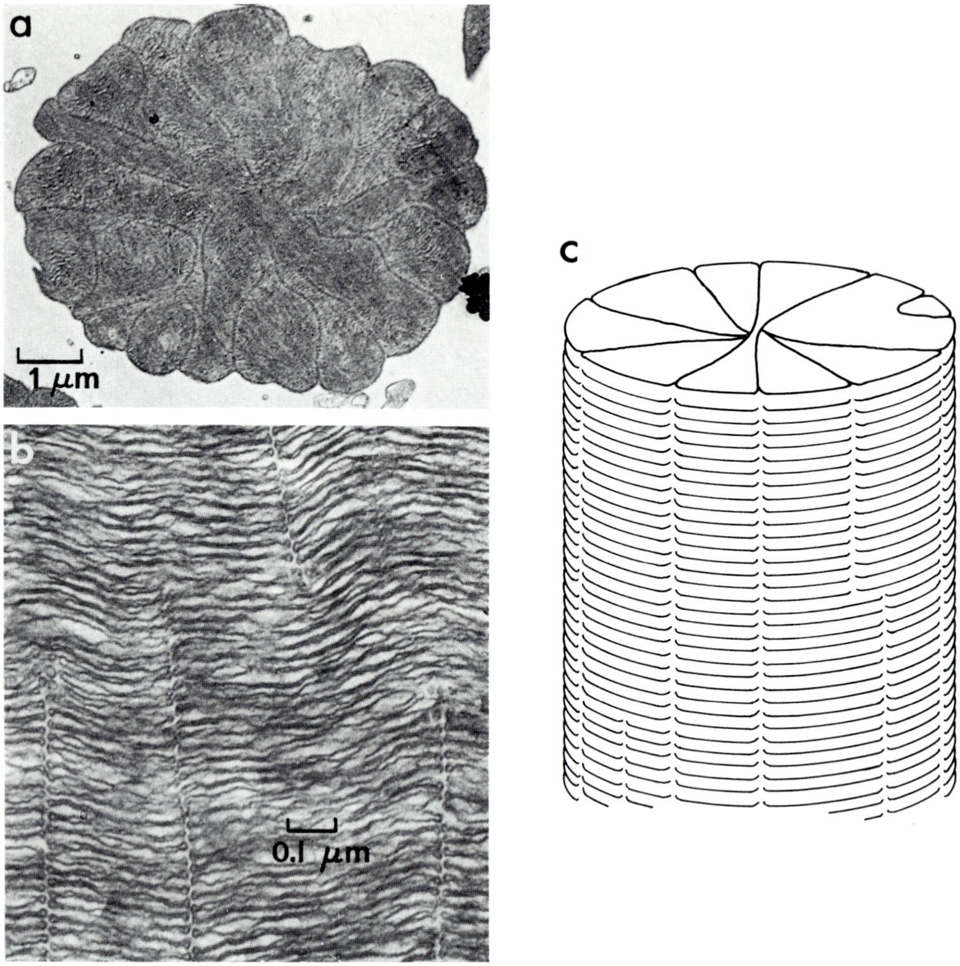


FIGURE 10.10 Frog retinal rod, *Rana pipiens*. (a) Cross section of outer segment, (b) longitudinal section, and (c) schematic of outer segment (note relationship to fused rhabdom of arthropods and molluscs, Figures 8.10–8.15).

Perhaps what we are seeing here in the amphibian rod structure is a clue to an evolutionary link with the arthropods and molluscs, for all other vertebrate retinal rods are units of only 1 μm in diameter and show no such rhabdom-like structure. On the other hand, the resemblance between the closed-type rhabdom and the amphibian retinal rod may be the result of independent development and may have no evolutionary significance.

# XI
# The Vertebrate Visual Pigments

The conclusion that we can draw from the structure of the invertebrate and vertebrate photoreceptors is that they all appear essentially to be modifications of a common plan. Now what about the visual pigments—do we find a common pigment molecule for all visual systems?

Let us now identify the vertebrate visual pigments and their photochemistry.

### Vitamin A and the Visual Pigment

We have already mentioned that carotenoids are easily and abundantly synthesized by plants (Figures 3.6–3.8). But animals cannot synthesize $C_{40}$-carotenoids and must obtain them in their nutrition from plants.

A finding of great importance was that it is not the ingested $C_{40}$-carotenoids, but their degraded $C_{20}$ derivative, vitamin A, that is necessary for animal life (Figure 3.9). The general function of vitamin A is to support growth and to maintain the tissue cells. When the stores of vitamin A in the liver and blood have been exhausted, the first symptom of vitamin A deficiency in man and in other animals is the rise of visual threshold known as night blindness.

To learn about the tissue function of vitamin A, vitamin A acid was substituted for vitamin A and was found to maintain growth in the rat, with a biopotency equivalent to vitamin A (van Dorp and Arens, 1947). Yet no matter how large the amount of vitamin A acid fed to the rat, no vitamin A was deposited in the liver. The rat was unable to reduce vitamin A acid to vitamin A, the form in which vitamin A is stored. This finding led Moore (1953) to suggest that although the tissue function of vitamin A seems to be fulfilled by vitamin A acid, it might not be able to serve as the precursor for the visual pigments, which need vitamin A for their synthesis.

The importance of vitamin A aldehyde, retinal, as an intermediate in vitamin A metabolism was established by Morton (1944) and Morton and Goodwin (1944). Then Glover *et al.* (1948) demonstrated that retinal$_1$ is rapidly converted to vitamin $A_1$ when it is administered orally, subcutaneously, or intraperitoneally. The conversion of retinal$_1$ to vitamin $A_1$ is a reduction which occurs in the gut and subcutaneous tissues. This was a plausible explanation for the displacement of the absorption maximum (depending on the organic solvent) from around 328 nm, that of vitamin $A_1$, to about 370 nm, that of retinal$_1$ (Figure 11.1). The shift in the absorption spectrum is explained by an increase in the number of conjugated bonds from 5 to 6; if the terminal —$CH_2OH$ group of vitamin $A_1$ (Figure 3.9) is replaced by a —CHO, the aldehyde group, this would provide the sixth conjugated bond (Figure 11.2).

Retinal has been synthesized from vitamin $A_1$ by a number of researchers (Morton and Goodwin, 1944; Hawkins and Hunter, 1944). Then it was demonstrated that retinal$_1$ could be obtained by the oxidation of $\beta$-carotene (Hunter and Williams, 1945). This is of special interest to us, for it provided chemical

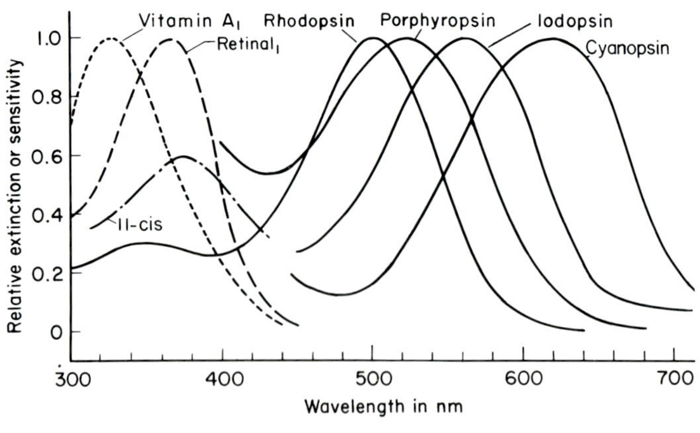


FIGURE 11.1 Visual pigments, absorption spectra. From Wald (1959) and Dartnall (1957).

proof of the conversion of $\beta$-carotene to vitamin A and demonstrated that retinal$_1$ was an intermediate product in the conversion of $\beta$-carotene to vitamin $A_1$ *in vivo*.

By their spectroscopic absorption characteristics, the chemical isomers vitamin $A_1$ and vitamin $A_2$, and hence retinal$_1$ and retinal$_2$, can be distinguished from one another.*

Vitamin $A_1$ can be identified by its absorption spectrum, for its maximum peak is at 325 nm, both in hexane and in ethanol. When reacted with the Carr-Price reagent ($SbCl_3$ in chloroform), the absorption peak is at 621 nm. Retinal$_1$ can also be identified by its absorption spectrum, depending on the solvent. For example, in hexane the absorption peak is at 368 nm, in ethanol it is at 383 nm, and when reacted with the Carr-Price reagent an absorption maximum at 664 nm is obtained (Hubbard *et al.*, 1971).

Vitamin $A_2$ and retinal$_2$ were found in the frog retina (Wald, 1938, 1939). It was later shown that retinal$_2$ could be converted to vitamin $A_2$ in the rat (Morton *et al.*, 1946, 1947). Vitamin $A_2$ is a chemical isomer and differs from vitamin $A_1$ by possessing an added double bond in the ring, at positions 3 and 4 (Figure 11.2). Retinal$_2$ is characterized as a deep yellow substance soluble in petroleum ether. It was crystallized by Salah and Morton (1948) and the maximum absorption peaks of *crystalline* retinal$_2$ were found to be at 385 nm in hexane, 388 nm in petroleum ether, and 401 nm in ethanol. Vitamin $A_2$ reacts with the Carr-Price reagent to give a maximum absorption peak around 693 nm. Retinal$_2$ also reacts with the Carr-Price reagent, producing an absorption maximum near 705 nm.

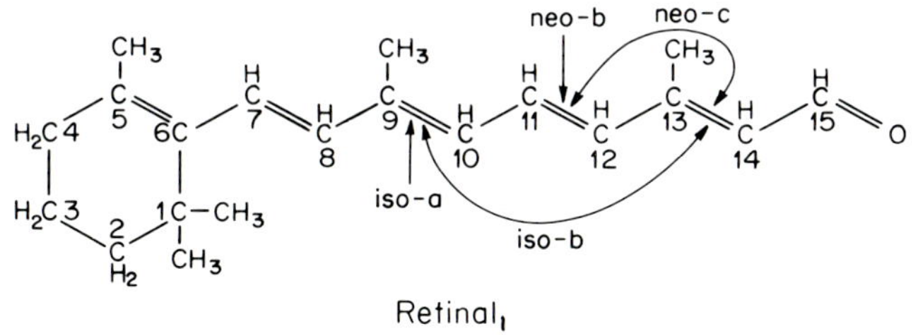


FIGURE 11.2 Retinal$_1$, indicating different geometric isomers (see Table 11.2, p. 202).

* In the revised nomenclature vitamin $A_1$ is retinol, vitamin A acid is retonic acid, and vitamin $A_1$ aldehyde or retinene, is retinal$_1$ or retinaldehyde; vitamin $A_2$ is 3-dehydroretinol; and retinene$_2$ is 3-dehydroretinal or retinal$_2$—all so named because of their discovery in the retina.

## Visual Pigments

The sensitivity of the eye, as one goes from dim to bright light, shifts toward the red end of the spectrum. This observation was first described in 1825 by the Czech physiologist, Johannes Purkinje. The basis of this change is a difference in spectral sensitivity between the retinal rods and cones. The retinal rods are active in vision at relatively low light levels, and their maximal sensitivity is in the blue–green at about 500 nm (Figure 11.3). The cones are active in bright light and color vision, and their maximum spectral sensitivity lies more toward the red part of the visible spectrum.

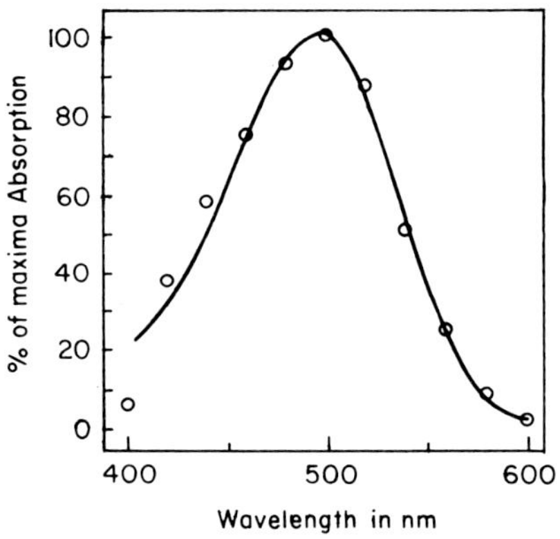


FIGURE 11.3 Comparison of the absorption spectrum of human rhodopsin to that of the action spectrum for the rod sensitivity of the eye; circles indicate rod sensitivity. From Crescitelli and Dartnall (1953).

To extract and isolate the visual pigments from the retina requires the use of special solubilizing agents. The procedures for extraction and the use of various solubilizing agents are described by Dartnall (1957) and Hubbard *et al.* (1971). An effective extractant, and one that is most commonly used, is an aqueous saturated solution of digitonin (a digitalis glycoside). After extraction, relatively pure rhodopsin can be obtained (Figure 11.4) by the use of column fractionation (Bowness, 1959; Heller, 1968, 1969).

The extracted visual pigments are identified primarily by their absorption spectra to be either *rhodopsin* ($retinal_1$ + rod opsin) or *porphyropsin* ($retinal_2$ + rod opsin) for the retinal rods, and *iodopsin* ($retinal_1$ + cone opsin) or *cyanopsin* ($retinal_2$ + cone opsin) for the retinal cones (Figures 11.1 and 11.5). Marine fish and land vertebrates characteristically possess the rhodopsin ($retinal_1$) system. The porphyropsin system ($retinal_2$) is found in freshwater fish. But fish that are neither freshwater nor marine and migrate between both environments can possess both $retinal_1$ and $retinal_2$ (Table 11.1).

TABLE 11.1 SOME VERTEBRATE VISUAL ABSORPTION MAXIMA[a]

| *Phylum* / *Class* / *Organism* | *Common name* | *Rhodopsin chromophore type* | |
|---|---|---|---|
| | | *retinal$_1$* | *retinal$_2$* |
| Chordata | | | |
| Cyclostomata | | | |
| *Petromyzon marinus* | Lamprey | 497 | 518 |
| Pisces | | | |
| *Salmo gairdneri* | Rainbow trout | 503 | 527 |
| *Salmo salar* L | Atlantic salmon | 503 | 527 |
| *Cyprinus carpio* L | Carp | | 523 |
| *Carassius auratus* L | Goldfish | | 522 |
| *Anguilla anguilla* L | Eel | 487 | 523 |
| *Conger conger* L | Conger eel | 487 | |
| *Gadus morrhua* Day | Cod | 499 | |
| *Gadus aeglifinus* L | Haddock | 494 | |
| *Morone labrax* L | Sea bass | 502 | 534 |
| *Scomber scombrus* L | Mackerel | 521, 487 | |
| *Pleuronectes microcephalus* | Lemon sole | 492, 501 | |
| *Pleuronectes flesus* | Flounder | 507, 510 | |
| Amphibia | | | |
| *Necturus maculosus* | Mud puppy | | 522 |
| *Xenopus laevis* | Clawed toad | 502 | 523 |
| *Rana temporaria* | Common frog | 502 | |
| *Rana pipiens* | Leopard frog | 502 | |
| Reptilia | | | |
| *Alligator mississippiensis* | Mississippi alligator | 499 | |
| *Crotalus viridis helleri* | Rattlesnake | 500 | |
| Aves | | | |
| *Pelicanus occidentalis* | Pelican | 502 | |
| *Anas boschas* | Duck | 502 | |
| *Gallus gallus* | Chicken | 500, 560 | |
| *Columba livia* | Pigeon | 502, 544 | |
| Mammalia | | | |
| *Oryctolagus cuniculus* | Rabbit | 502 | |
| *Sciurus carolinensis leucotis* | Gray squirrel | 502 | |
| *Cavia porcellus* | Guinea pig | 497 | |
| *Rattus rattus* | Rat | 498 | |
| *Mus musculus domesticus* | Mouse | 498 | |
| Primates | | | |
| *Macaca mulatta* | Rhesus monkey | 497 | |
| *Homo sapiens* | Man | 497, 493 | |

[a] Taken in part from Lythgoe (1972, pp. 605–621) and Wolken (1973, pp. 1268–1269).

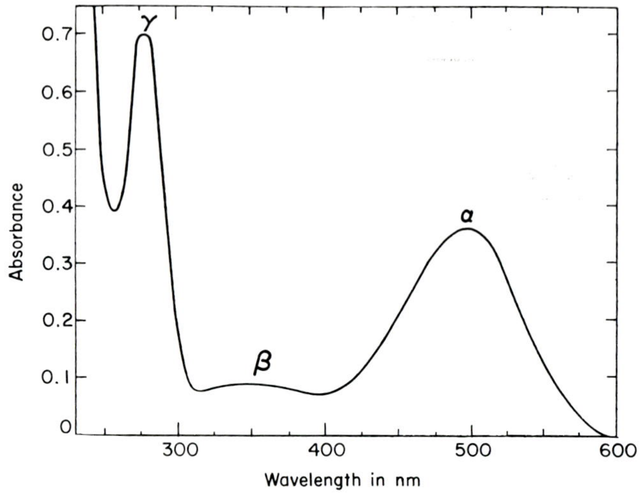


FIGURE 11.4 Cattle rhodopsin, purified on a chromatographic column, absorption spectrum. From Bowness (1959).

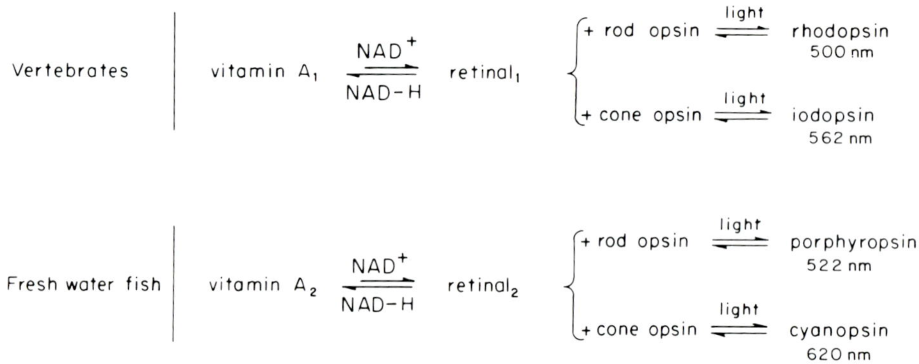


FIGURE 11.5 Visual pigments in the rods and cones, derived either from vitamin $A_1$ or vitamin $A_2$; indicating their absorption maxima (according to Wald, 1959).

## The Photochemistry of Rhodopsin

The mechanism of the visual process in the vertebrate retinal rod has been studied thoroughly by Wald (1953, 1954, 1955, 1956, 1959, 1961a,b), Dartnall (1957, 1962), and their associates. The isolated rod visual pigment, rhodopsin (Figure 11.4), bleaches upon exposure to light yielding retinal and its carrier protein *opsin* (Figure 11.6). Retinal is bound to opsin in a Schiff-base linkage. Ball *et al.* (1949) and Morton and Collins (1950a,b,c) have shown that retinal

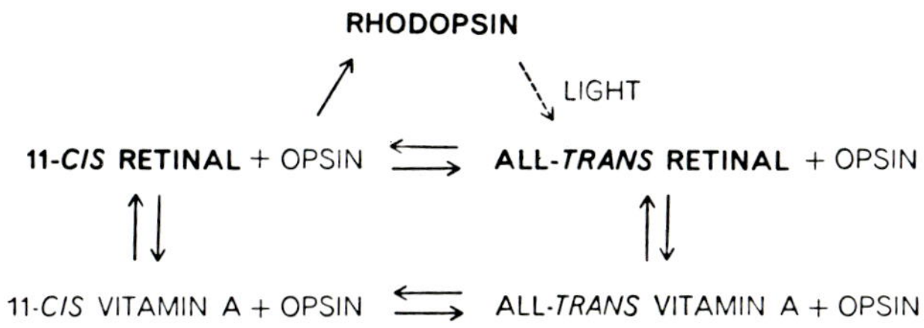


FIGURE 11.6 Photochemical events in visual excitation.

is attached to an amino group of opsin. Schiff bases of retinal with a variety of amino acids and amines involve the condensation of the aldehyde group of retinal with the amino group of amine or amino acid (Figure 11.7).

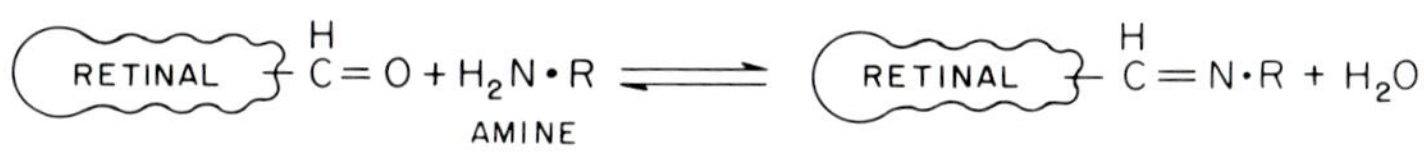


FIGURE 11.7 Schematic of a retinal–amine complex.

The bleaching of rhodopsin by light can be followed by spectroscopy (Figure 11.8). Curve 1 is unbleached rhodopsin and curves 2–7 show the displacement of the major rhodopsin peak at close to 500 nm to that of all-*trans*-retinal near 370 nm. Retinal recombines with opsin in the dark to form rhodopsin. The reaction is spontaneous, and therefore opsin may be looked upon as a retinal-trapping enzyme, removing free retinal from the mixture and causing the production of additional retinal from vitamin A to maintain the necessary equilibrium.

Vitamin A and retinal are polyene chains that exist in a number of different geometrical configurations corresponding to the possible *cis–trans* isomerizations around the double bonds of these molecules (Figure 11.2 and Table 11.2). For example, there are five possible geometric isomers of retinal, corresponding to rotation about the 9–10 carbon bond, the 11–12 bond, the 13–14 bond, the 9–10 and 13–14 bonds, and the 11–12 and 13–14 bonds. The retinal produced by bleaching rhodopsin has always been found to be in the all-*trans* form. However, to resynthesize rhodopsin from all-*trans*-retinal, it was found that retinal must be isomerized to the 11-*cis* form in order to recombine with opsin (Figure 11.6). The 9-*cis* retinal can also complex with opsin to form a series of photosensitive pigments, isorhodopsins, which have been found in small concentrations in the liver and the blood, but not in the eye.

The existence of the 11-*cis* isomer was considered improbable, because the steric interference between the methyl group at carbon 13 and the hydrogen

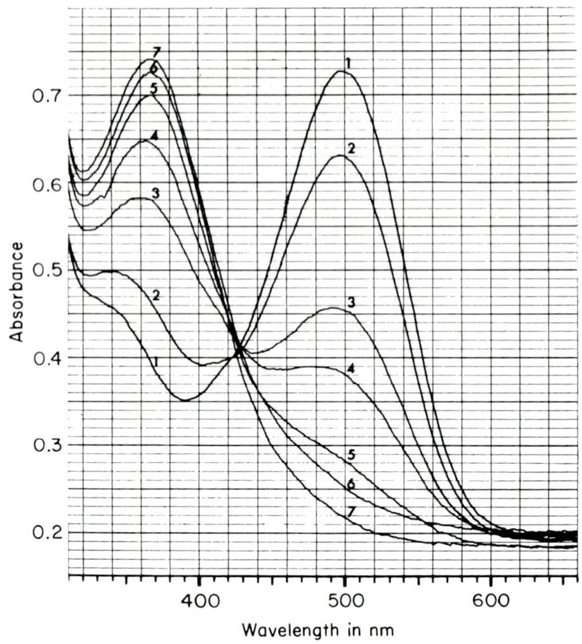


FIGURE 11.8 Frog rhodopsin (in 4% tergitol) absorption spectrum, curve 1. Light bleaching spectra, curves 2–7.

at position 10 would prevent the molecule from becoming entirely planar. Wald and his associates were surprised to find that the functional isomer of retinal was in fact the 11-*cis* configuration of retinal. Because of its hindered configuration, the 11-*cis* form is the least stable of the possible isomers; it is the most easily formed upon irradiation, and the most sensitive to temperature and light. Wald thinks that the instability of this form of retinal helps explain

TABLE 11.2 GEOMETRIC ISOMERS OF RETINAL

| *Isomerization around bonds* | *Nomenclature* | *Alternate name* |
|---|---|---|
| 9–10 | 9-*cis* | iso-*a* |
| 11–12 | 11-*cis* | neo-*b* |
| 13–14 | 13-*cis* | neo-*a* |
| 9–10, 13–14 | 9, 13-di-*cis* | iso-*b* |
| 11–12, 13–14 | 11, 13-di-*cis* | neo-*c* |

its presence in the rhodopsin molecule. What could be more appropriate than a molecule that is very unstable in the light, but extremely stable in the dark?

The bleaching of rhodopsin to release retinal from opsin does not proceed directly (Figure 11.9), but in a series of intermediate steps (Abrahamson and Ostroy, 1967; Guzzo and Pool, 1969). In the process, the 11-*cis*-retinal is isomerized to all-*trans*-retinal, resulting in a release of energy due to its transformation to a more stable form. (For review, see Menger, 1975).

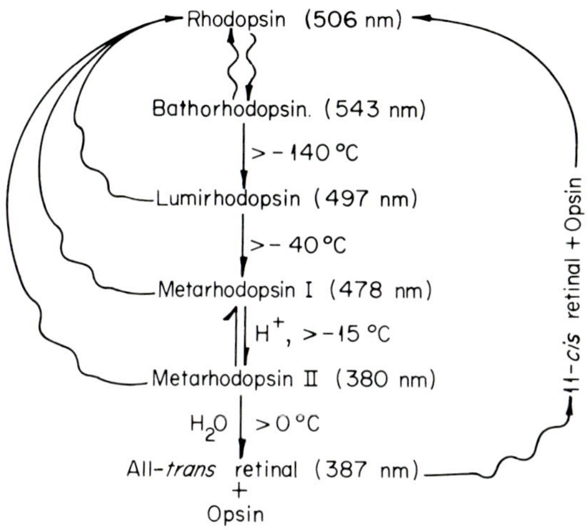


FIGURE 11.9 Rhodopsin (bovine); photochemical events denoted by wavy lines, thermal (dark) events by straight lines (see Figure 11.6).

Therefore, the only apparent action of the absorption of light by rhodopsin is the isomerization of the 11-*cis*-retinal to the all-*trans*-retinal (Figures 11.6 and 11.9), resulting in a release of energy and excitation; all other reactions can be considered dark reactions.

## The Cone Pigments and Color Vision

We can now ask questions about the cone pigments and color vision. It was noted earlier that chemical isolation has yielded only one photosensitive cone pigment, either *iodopsin* ($\lambda$ max 560 nm) if $retinal_1$ is present, or *cyanopsin* ($\lambda$ max 620 nm) if $retinal_2$ is present (Figures 11.1 and 11.5).

How do these two pigments fit into an account of color vision in vertebrates? It may be recalled that there are two general theories of color vision. The first is the tricolor theory, which arose from the work of Young (1802,

1807), von Helmholtz (1852, 1867), and Maxwell (1861, 1890), and the other is the theory of Hering (1885). The tricolor theory asserts that there are three different pigments in the retinal cones with maximum absorption in the blue, green, and red regions of the spectrum. According to this theory, the brain computes yellow and white from green and red at high light intensities and white from blue at low intensities. In contrast, Hering's theory postulates that there are the following six basic responses, which occur in pairs: blue–yellow, red–green, and black–white. Excitation leading to any single response suppresses the action of the other member of the pair.

To test these theories, the cones *in situ* have been isolated and their absorption spectra obtained by microspectrophotometry. For example, frog cones were found to have general absorption throughout the whole of the visible spectrum with maxima near 430, 480, 540, 610, and 680 nm (Wolken, 1966b). For the carp, cone peaks in the regions of 420–430, 490–500, 520–540, 560–580, 620–640, and 670–680 nm were found (Hanaoka and Fujimoto, 1957). In the goldfish, which belongs to the carp family, Marks (1963) found cones with absorption peaks at 455, 530, and 624 nm, and Liebman and Entine (1964) found absorption peaks at 460, 540, and 640 nm. For human and monkey foveal cones, absorption peaks were found at 445, 535, and 570 nm (Marks *et al.*, 1964; MacNichol, 1964). Wald (1964b), Brown and Wald (1964), and Wald and Brown (1965) found for human cones absorption peaks near 450, 525, and 555 nm, which compare well with their psychophysical data for the spectral sensitivity of the human eye, with peaks at 430, 540, and 575 nm. These studies indicate that for color vision there are at least three different spectrally absorbing cone pigments consistent with the psychophysical data, one for sensing blue, one for green, and one for red (Figure 11.10).

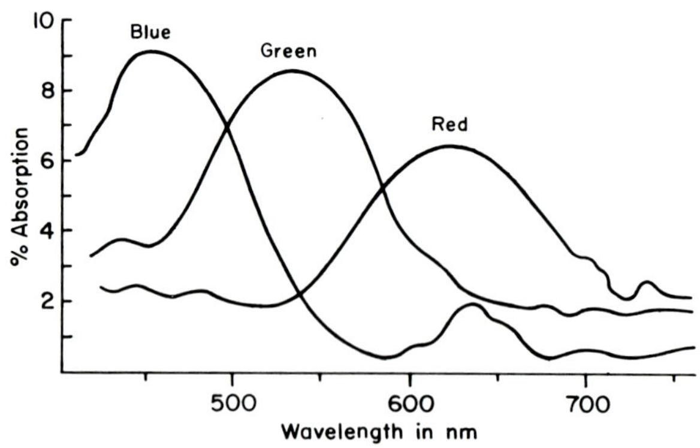


FIGURE 11.10 Absorption spectra obtained from three different cones of the goldfish, smoothed out spectra. From Liebman and Entine (1964).

## The Oil Globules

In the retinas of birds, amphibians, lizards, and snakes there are pigmented oil globules (Figure 11.11). These globules are believed to have evolved a long time ago, since they are found in the retinas of sturgeon, belonging to the ancestral fish chondrosteans. However, most fish, with the exception of the lungfish, do not have them. The globules are situated between the inner and outer segments of the cone, so that light must pass through them before reaching the photosensitive pigments in the outer segment. The globules are thus in a position to act as color filters; but their function in color vision is unknown.

Nearly a century ago, Krause (1863) suggested that these oil globules, by differentially transmitting light to the outer segments, affect the spectral response of the cone, and thus can provide the basis of color differentiation in the animals possessing them. The ophthalmologist, Siegfried Garten, published a paper in Germany in 1907 indicating a system of color photography based

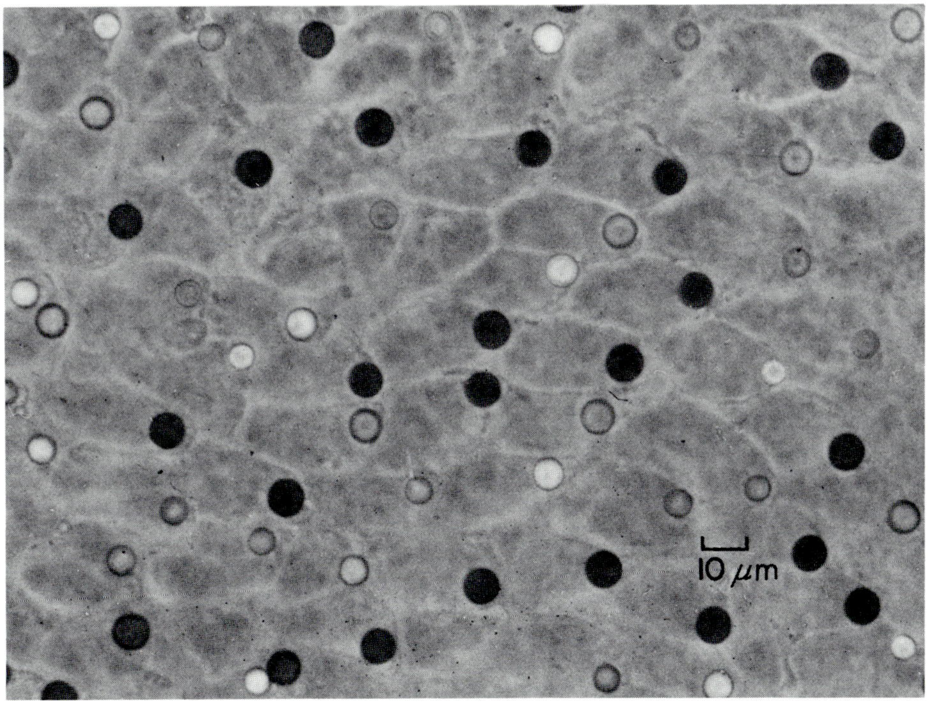


FIGURE 11.11 Oil globules in swamp turtle (*Pseudemys scripta elegans*) retina.

upon this principle. In 1906 the French chemist, Louis Lumiérè, brought out an autochrome process for color photography (see Wall, 1925). Lumiérè used suspensions of starch grains dyed red, green, and blue, mixed in roughly equal proportions and spread over the surface of a photographic plate. The granules were squashed flat and the interstices filled with carbon black. Each colored granule served as a color filter for the photosensitive emulsion which lay beneath it.

Therefore, the different colored oil globules have continued to be suspected of acting as color filters for the retinal cones (Walls, 1942; Fujimoto *et al.*, 1957). Since a red globule absorbs light somewhere in the blue–green region, the location of the globule between the inner cone segment and the light source makes a color theory reasonable. However, there is no direct correlation between the color of the globule and the color discrimination of the animal possessing it.

Cones predominate in the retinas of chicken, pigeon, and turtle eyes. These animals necessarily function only at high light intensity. In a freshly excised chicken retina, the oil globules are pigmented yellow–green, orange, and red, and the globules range from 3 to 5 $\mu$m in diameter.

Wald and Zussman (1938) and Wald (1948), using organic solvents and chromatography, isolated three different colored fractions from the chicken retina. These fractions resembled the *in situ* oil globule colors and were identified as carotenoids: lutein, zeaxanthin, astaxanthin, and in addition, a new carotenoid, gallaxanthin.

The absorption spectra of the *in situ* chicken yellow–green, orange, and red globules, obtained by microspectrophotometry, show absorption maxima in three different regions of the spectrum (Strother and Wolken, 1960). The absorption spectra do not show sharp peaks, but very broad bands are evident; the width of the band increases as the maximum absorbing wavelength increases. The yellow–green globules have a general absorption in the region 390–440 nm. The orange globules have a broad absorption in the region 440–480 nm, and the red globules have an absorption from 480–560 nm (Figure 11.12).

Observations on bleaching indicate that the yellow–green globules are the most unstable of the three. When stored in the dark at 4°C, they become colorless. When all the colored globules are mixed together, a reddish orange color results, which, when exposed to white light bleaches first to orange, then to yellow, and then becomes colorless. If one assumes that the globules are acting merely as color filters, then the yellow–green and orange globules are relatively inefficient. For example, the chicken cone pigment, iodopsin, has its absorption peak around 550 nm, and only the red globule has appreciable absorption in this region.

These colored retinal oil globules have been postulated as the basis for the observed increase in spectral sensitivity of these animals to red light as com-

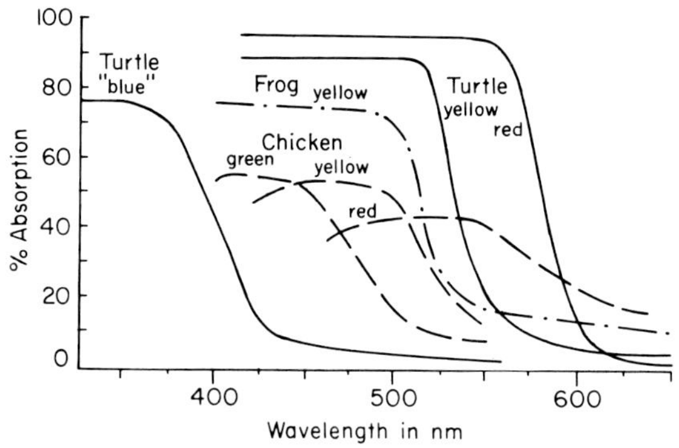


FIGURE 11.12 Spectra of various colored oil globules acting as retinal filters in turtle (——), chicken (– – –), and frog (– · – ·).

pared to animals without them (Walls, 1942; Fox, 1953). The effect of colored globules on the spectral sensitivity of the pigeon retina has been discussed by Donner (1953), using electrophysiological data. He observed three "modulator" curves for the pigeon, with maxima at 470–490, 540–550, and 600–620 nm. Comparison with three similar curves obtained by Granit (1941) for the frog retina revealed a shift to longer wavelengths for the pigeon. On the basis of the observed shift, Donner (1953) calculated what the absorption spectra of the oil globules should be in the pigeon. Donner's calculated values and the spectra obtained by Strother (1963) could account for the red shift in the pigeon spectral sensitivity.

Spectral sensitivity for the swamp turtle, using electroretinogram techniques, shows maxima at 575, 620, and 645 nm. Absorption spectra recorded by microspectrophotometry for their colored oil globules show absorption maxima for the red globules from 555 to 565 nm, for the yellow globules from 510 to 515 nm, and for the "colorless" globules from 370 to 380 nm (Figure 11.12), confirming the assumption that the spectral response of the turtle eye corresponds to the absorption curve of its cone pigment, cyanopsin. Strother (1963) has obtained spectral absorption peaks at 625 and 650 nm, which is in agreement with two of the spectral responses obtained from the electroretinograms.

In the amphibians, the frog retina has yellow oil globules (Figure 11.12). The frog also possesses "green" rods in addition to the "red" rods, as well as cones. The green rod absorption peak is near 540 nm and the red rod absorption peak is near 610 nm, whereas the cones have their maximum absorption near 570 nm. These spectral absorption peaks taken together would cover the spectral range necessary for color vision.

In mammals, the primate fovea which contains mostly cones, and the region surrounding it, the *macula lutea*, are colored yellow. This pigment is the plant carotenoid, xanthophyll (Figure 3.6). This, then, would provide a color filter for the three different spectral absorption peaks of the cones. The yellow oil globule pigment serves as a yellow filter and has basically the same function, that is, to improve the visual acuity and contrast for all animals that possess it.

### Retinal Rod Molecular Structure

The central problem in understanding photoexcitation and energy transfer from the cell to the more highly evolved and specialized sensory organelles is to decipher the structure of the membranes in which the photoreceptor molecule resides. With the information we have on the photoreceptor structure and the visual pigments and their photochemistry, let us see if we can develop a molecular structure which corresponds to the function of the photoreceptor for visual excitation.

It is apparent from Figures 10.5, 10.7, and 10.8 that the rod outer segments are comprised of double-membraned lamellae with a repeated unit of about 250 Å. Although the precise location of the rhodopsin, protein, and lipid molecules within these structures is not known, we can hypothesize a molecular model of the retinal rod. In order to do this, the geometry of individual rods, their length and diameter, and the number of lamellae and their dimensions were determined from numerous electron micrographs (Wolken, 1956, 1961a). With these data and the rhodopsin concentration, we can calculate the surface area that each rhodopsin molecule would occupy in the lamellae.

By extraction of a known number of retinal rods and from the absorption spectrum (Figure 11.4) and the molar extinction coefficient for rhodopsin at 500 nm, the concentration of rhodopsin per rod can be estimated. We can also determine the rhodopsin concentration in a single rod outer segment by microspectrophotometry. The spectrum from a single frog rod is shown in Figure 11.13. Using the microspectrophotometry data, we found that there were $3 \times 10^9$ rhodopsin molecules per frog rod. This compares very closely with the estimates obtained for the rhodopsin concentration in the extracted frog rod which is $3.8 \times 10^9$ molecules per rod (Table 11.3).

In order to calculate the area that the rhodopsin molecules would occupy in the retinal rod outer segment lamellae, several assumptions were necessary. These were that the retinal rod outer segment lamellae are double membranes of lipid and protein (see Figure 4.6) or a lipoprotein; that there is a monomolecular layer of rhodopsin molecules associated with the protein (or lipoprotein) membrane; and also that these double membranes, or rod discs, are separated by aqueous proteins. These assumptions are supported by chemical analyses which show that, in general, the visual pigment rhodopsin accounts for 4–10%, protein for 40–50%, and lipids for 20–40% of the total outer rod segment weight.

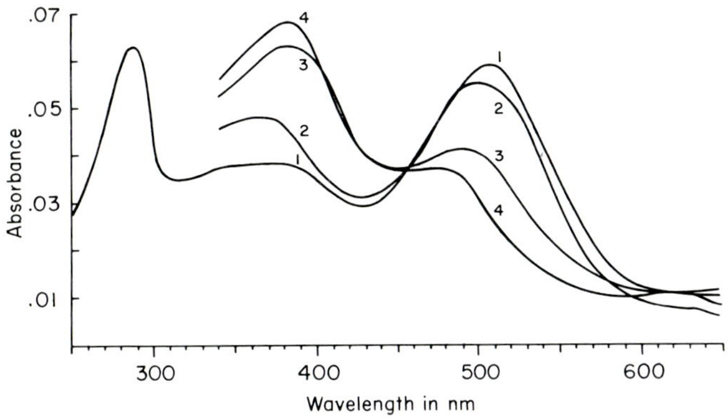


FIGURE 11.13 Absorption spectra of an isolated frog retinal rod, curve 1; curves 2–4, upon irradiation with green light (500 nm) at 15 second intervals. Note the shift in spectra from rhodospin, curve 1, to that of retinal, curve 4. The absorption peak near 280 nm for opsin does not shift (compare to Figure 11.4). Obtained with microspectrophotometer.

The cross-sectional area that would be associated with each rhodopsin molecule is expressed by $A = \pi D^2/4P$ where $D$ is the diameter of the retinal rod and $P$ is the number of rhodopsin molecules in a single monolayer. In our calculation, $P$ is replaced by $N/2n$, where $N$ is the rhodopsin concentration in molecules per retinal rod and $n$ is the number of double-membraned lamellae per rod. So the expression for the maximum cross-sectional area for each rhodopsin molecule is given by $A = \pi D^2 n/2N$. Using this equation, the cross-sectional areas calculated for cattle and frog rhodopsin were 2500 and 2620 Å$^2$, respectively (Table 11.3), which means the diameter of the rhodopsin molecule should be of the order of 50 Å. This value agrees well with theoretical calculations which assert that a rhodopsin molecule, if symmetrical, would have a diameter of the order of 40 Å (Wald, 1954). Thus, our figure for the available area implies that there would be sufficient space for all of the rhodopsin molecules to cover all of the lamellar surfaces of a single rod.

The molecular structure for a rod outer segment is schematized in Figure 11.14. A small area is enlarged to show the molecular packing of rhodopsin in the lamellar membranes. Such a structural model provides sufficient space on the lamellar surfaces to accommodate all the rhodopsin molecules. This membrane structure maximizes the surface area available for light capture and visual excitation. In fact, this model may be very close to reality, for X-ray diffraction studies of the frog rod outer segments seem to support such a model (Blaurock and Wilkins, 1969; Blasie, 1972).

The rhodopsin molecular weight $M$ has also been calculated from the retinal rod structure. Where $D$ is the diameter, $T$ the thickness of the lamellar mem-

TABLE 11.3 RETINAL ROD STRUCTURAL DATA[a]

| *Animal* | *Average diameter, D* (μm) | *Thickness of dense lamellae, T* (Å) | *Number of dense layers per rod, n* | *Number of rhodopsin molecules per rod, N* | *Calculated cross-sectional area of rhodopsin* (Å²) | *Calculated diameter of rhodopsin molecule, d* (Å) | *Calculated molecular weight, M*[b] |
|---|---|---|---|---|---|---|---|
| Frog | 5.0 | 150 | 1000 | $3.8 \times 10^{9}$[c] | 2620 | 51 | 60,000 |
| Cattle | 1.0 | 200 | 800 | $4.2 \times 10^{6}$ | 2500 | 50 | 40,000[d,e] |

[a] Taken from Wolken (1961a).
[b] Heller (1969) finds that the molecular weight of cattle and frog rhodopsin is 28,000.
[c] Microspectrophotometry of frog rod gives $3.0 \times 10^{9}$ rhodopsin molecules.
[d] Calculation based on a lipoprotein, density 1.1, gives a molecular weight of 32,000.
[e] Abrahamson and Fager (1973) indicate molecular weight of 35,000–37,000.

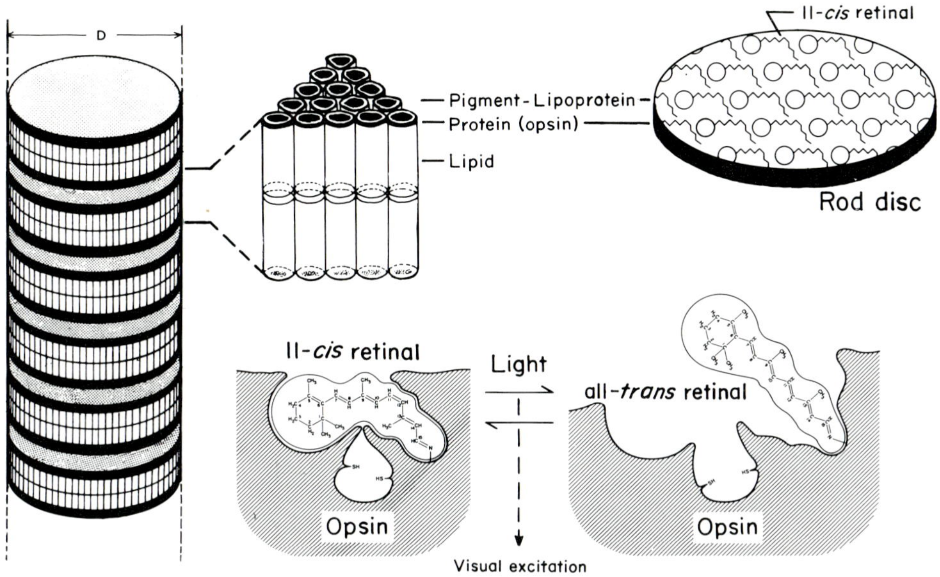


FIGURE 11.14 Model for molecular structure of retinal rod, showing the light bleaching of rhodopsin in the lamellae of the outer segment from 11-*cis* retinal to the all-*trans*-retinal.

branes, $s$ the density (taken as 1.3 for a protein or 1.1 for a lipoprotein), $L$ Avogadro's number, $n$ the number of lamellar membranes, and $N$ the number of rhodopsin molecules, the molecular weight is then obtained by $M = \pi D^2 TsLn/4N$. The molecular weights calculated from this equation (density = 1.3) for frog and cattle rhodopsin were found to be 60,000 and 40,000, respectively (Table 11.3). This compares well with the molecular weight of 40,000 calculated by Hubbard (1954) for cattle rhodopsin and recent calculations by Abrahamson and Fager (1973) of molecular weights from 35,000 to 37,000. If, in calculating $M$, the density of a lipoprotein (1.1) is used, the molecular weight would be reduced by 20%, resulting in a molecular weight of 32,000 for cattle rhodopsin and 54,000 for frog rhodopsin (Wolken, 1966b).

We can now go back to the photoreceptor structures of the invertebrates and find out whether their visual pigments use a similar photochemistry to the vertebrate photoreceptors.

# XII
# The Invertebrate Visual Pigments

There is good reason to believe that throughout the animal world the kinds of pigment molecules and receptor structures involved in visual excitation have been similar. If this is true, then we must expect certain features of vertebrate visual chemistry to repeat themselves in the invertebrates. Even though only a relatively few invertebrate visual pigments have been isolated and identified, sufficient experimental data have now accumulated to indicate that like the vertebrate visual system, the invertebrate visual system also depends on vitamin A and retinal as the visual pigment chromophore.

## Molluscs

### Cephalopods

The cephalopod mollusc visual pigments were among the first to be isolated and studied. The extraction was accomplished with 1–2% aqueous digitonin, a method used for extracting vertebrate rhodopsin. The visual pigment obtained from the eyes of the squid, *Loligo pealleii,* with a maximum absorption at 493 nm, was identified as a rhodopsin (Bliss, 1943, 1948; St. George and

Wald, 1949). However, in contrast to the vertebrate rhodopsin system, the squid visual process only involves the transformation of rhodopsin to metarhodopsin (Figure 11.9). The transformation of metarhodopsin to retinal and opsin does not occur normally, but can be brought about by subjecting metarhodopsin to elevated temperatures. The regeneration of rhodopsin takes place directly from the metarhodopsin. Furthermore, this reaction cannot take place in the dark, as with vertebrate rhodopsin, but requires light. In the light then, a continuous equilibrium is set up between rhodopsin and metarhodopsin (Hubbard and St. George, 1958).

$$\text{Rhodopsin (11-}cis\text{)} \xrightleftharpoons{\text{light}} \text{metarhodopsin (all-}trans\text{-retinal)}$$

Squid metarhodopsin can exist in two forms, depending upon the pH of the solution. In acid solution its maximum absorption is near 500 nm, whereas in alkaline solution, its maximum absorption is at 380 nm. These two forms are referred to as acid and alkaline metarhodopsin. They are interconvertible (in the dark only) simply by changing the pH of the solution. The reaction hinges on the addition of a single hydrogen ion to the alkaline metarhodopsin.

When the metarhodopsin is denatured with acid, the product, called "indicator yellow," is also pH sensitive, having λ maximum at about 440 nm in acid solution and 370 nm in alkaline solution. These are rather complicated reactions, but they explain the earlier observation of Krukenberg (1882) that squid rhodopsin either in the retina or in solution does not "bleach" on exposure to light. Bliss (1943, 1948) found that bleaching did occur when formaldehyde was added to his extracts. This is analogous to the denaturation of metarhodopsin by acid.

Since, in alkaline solution, metarhodopsin absorbs light at 380 nm, the composition of the rhodopsin–metarhodopsin equilibrium mixture can be shifted in either direction, depending upon the wavelength of light used. Near-ultraviolet light, which is absorbed more strongly by alkaline metarhodopsin, results in a larger quantity of rhodopsin in the equilibrium mixture, while orange light, absorbed more strongly by rhodopsin, produces a larger quantity of metarhodopsin. Irradiation with white light maintains the equilibrium, and irradiation of an acid solution produces a similar mixture, since both components absorb light in the same region of the spectrum.

The isomers of retinal involved in the visual process are all-*trans*-retinal, the chromophore of metarhodopsin, and 11-*cis*-retinal, the chromophore of rhodopsin. It has been suggested that the squid opsin molecule can accommodate both the all-*trans* and 11-*cis* forms of retinal, and therefore free retinal is produced only when the system is exposed to conditions that would denature opsin, and not as part of the system's normal function. That is, metarhodopsin, the

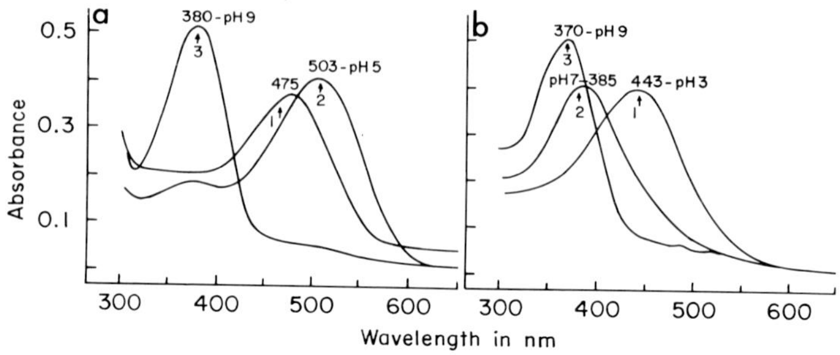


FIGURE 12.1 *Octopus* rhodopsin. (a) Curve 1, rhodopsin, $\lambda_{max}$ 475 nm; curve 2, metarhodopsin formed by irradiation, $\lambda_{max}$ 503 nm, pH 5; curve 3, metarhodopsin, $\lambda_{max}$ 380 nm, pH 9; (b) curve 1, indicator yellow, formed by denaturing metarhodopsin, $\lambda_{max}$ 443 nm, in acid, pH 3; curve 2, a mixture of free retinal, $\lambda_{max}$ 385 nm, pH 7.0; curve 3, in alkaline solution, pH 9, $\lambda_{max}$ 370 nm. From Brown and Brown (1958).

stable combination of opsin and all-*trans*-retinal, is produced, and upon further irradiation the retinal is isomerized to the 11-*cis* form, producing rhodopsin. As in the vertebrates this interconversion is blocked by low temperatures (—65°C), which prohibit the necessary rearrangement of the opsin molecule.

Visual pigments have also been extracted in the same manner from the cuttlefish, *Sepia officinalis,* with a maximum absorption at 492 nm, and from the octopus, *Octopus vulgaris,* with a maximum absorption at 475 nm (Hubbard and St. George, 1958; Brown and Brown, 1958). This is illustrated in the spectra of Figure 12.1a. Octopus rhodopsin has a maximum absorption at 475 nm (curve 1); upon irradiation, metarhodopsin is formed (curve 2). Metarhodopsin, being a pH indicator, absorbs maximally at 503 nm in acid solution (curve 2) and in alkaline solution at 380 nm (curve 3). When metarhodopsin is denatured by acid, pH 3, it forms "indicator yellow" (Figure 12.1b) with a maximum absorption at 443 nm (curve 1), in neutral solution (curve 2) it forms a mixture of free retinal ($\lambda$ max 385 nm) and denatured opsin, and in alkaline solution (curve 3) a mixture of retinal and opsin (Brown and Brown, 1958).

Therefore, in the photochemistry of the cephalopod rhodopsin, free retinal is not liberated in the process. However, these mollusc visual pigments exhibit a basic similarity to vertebrate rhodopsins in that 11-*cis*-retinal is complexed with a protein opsin to form rhodopsin.

Beside rhodopsin, there is a second light-sensitive pigment which has been extracted from the retinas of cephalopods named *retinochrome* (Hara and Hara, 1972). The retinochromes, like the rhodopsins, contain retinal as the chromophore. The $\lambda$ maxima of the retinochromes range from 490 to 522 nm, whereas those of the cephalopod rhodopsins range from 475 to 500 nm. Thus, the retinochrome is shifted 15 to 20 nm toward longer wavelengths than the corresponding rhodopsins and may account for the dual sensitivity of these animals.

## Arthropods

### Crustacea

Among the crustacea, a visual pigment was extracted from the lobster *Homarus americanus,* with a maximum absorption near 515 nm (Wald and Hubbard, 1957). This value was in agreement with the spectral sensitivity as determined from its electroretinogram (ERG). On irradiation, the lobster visual pigment yields a stable metarhodopsin, absorbing at 490 nm and exhibiting behavior similar to that described for the cephalopod metarhodopsins. Here, too, $retinal_1$ is present in the 11-*cis* isomeric form. In addition, considerable vitamin A was found in the lobster eyes, but not in their bodies.

The visual pigment of the euphausiid shrimp, *Euphausia pacifica,* was found to have maximal absorption at 462 nm (Kampa, 1955), but some doubt was expressed about this absorption maximum (Dartnall, 1962). In the euphausiid *Meganyctiphanes norvegica,* the absorption maximum was also found to be near 462 nm. $Retinal_1$ was identified in the 11-*cis* configuration and vitamin $A_1$ was also found in their eyes (Wald and Brown, 1957).

In the northern crayfish, *Procambarus,* the maximum spectral sensitivity was found to be near 570 nm. This is shifted further toward the red part of the spectrum than any of the known invertebrate rhodopsins (Kennedy and Bruno, 1961). From the crayfish, *Orconectes virilis,* two photosensitive pigments with absorption maxima at 510 and 562 nm were extracted (Wald, 1968). The maximum near 510 nm is close to the value for vertebrate rhodopsin, and the peak near 562 nm is close to the vertebrate cone pigment, iodopsin (Figure 11.1).

The spider crab, *Libinia emarginata,* like the crayfish, has two visual pigments which are differentiated not for color vision but for use in dim light and bright light, as are the vertebrate rods and cones (Wald, 1968). Microspectrophotometry of the rhabdoms revealed only one pigment with a maximum absorption at 493 nm. In the common prawn, *Palaemonetes vulgaris,* two different visual receptors seem to be present, with maximal sensitivities at 540 nm and in the near-ultraviolet at 390 nm (Wald and Seldin, 1968). Microspectrophotometry of the rhabdom also confirms two peaks, one near 496 nm and another near 555 nm (Goldsmith *et al.,* 1968).

For *Leptodora,* the behavioral action spectrum indicates a broad band of absorption, from 500 to 550 nm, and microspectrophotometry of the rhabdom area shows a peak near 510 nm (Wolken, 1971). As in other crustacea, these spectra may indicate that there are two different spectrally absorbing visual pigments (Wolken and Gallik, 1965).

### Insects

The insects are the most dominant and numerous of the arthropods and exhibit great diversity in behavior. Many insects are diurnal, well-adapted for navigating in bright light, and are able to distinguish colors; on the other hand, many insects are nocturnal and color blind (Wigglesworth, 1964). However, until recently very little was known about insect vision and their visual pigments (Wulff, 1956; Mazokhin-Porshnyakov, 1969).

This is understandable, for to extract and isolate the visual pigments from insect eyes is extremely difficult; it requires that thousands of insects be collected, dark-adapted, decapitated, and their eyes dissected, all in the cold and in red light. Thus, to avoid these tedious procedures, researchers have preferred to gather most of their information from behavioral studies in which the spectral sensitivity, or action spectrum, could be related to the absorption of their visual pigments. Electrophysiology has been used as a tool to obtain the spectral sensitivity from electroretinograms (ERG). From behavioral and electrophysiological data, two receptor pigments have been inferred, one with an absorption maximum in the visible around 500 nm and another in the ultraviolet, maximal near 365 nm. The absorption maximum around 500 nm suggested that insects also possess a rhodopsin visual pigment.

#### *Drosophila*

Let us first examine *Drosophila melanogaster,* whose various eye-color mutants have been of great importance in biochemical genetics. The eye-color pigments are contained in four types of cells: (1) the preretinal primary cells found in the crystalline cone, (2) the secondary cells which surround the retinula cells, (3) the basal cells located at the basement membrane, and (4) the postretinal cells in the outer optic ganglion (Nolte, 1950).

It was suggested that a link between the eye pigmentation and neural function could be the tyrosine pathway common to the synthesis of melanins and catecholamines, which are used as chemical transmitters in various synapses (Hotta and Benzer, 1969). The brown pigments are derived from tryptophan by way of kynurenine and 3-hydroxykynurenine (Kikkawa, 1941; Butenandt, 1952). It was postulated that the tryptophan-derived pigments are composed of metallic salt complexes, that the pigments are produced by the function of an enzyme having a specific metal as part of its prosthetic group, and that the enzyme function is controlled by a particular gene (Kikkawa *et al.,* 1955).

In three different *Drosophila* eye-color mutants, scarlet (st), white (w), and wild-type red (Canton), it was shown, using the extraction method of Kikkawa *et al.* (1955), that the scarlet eye pigmentation is influenced by an iron- and/or molybdenum-bearing complex and that the white eye is controlled by a nickel complex (Wolken *et al.*, 1957b). The pigments from the wild-type red eyes were found to be pteridines or pteridine derivatives (Forrest and Mitchell, 1954). These pigments have not been implicated in the visual pigment chemistry but are associated with the insect screening pigments.

It was demonstrated that *Drosophila* possesses color vision at high light intensities, but at low light intensities there is a "Purkinje shift" from photopic to scotopic vision, similar to the vertebrate shift from cone to rod vision (Fingerman, 1952; Fingerman and Brown, 1952, 1953). This shift suggests that there are two spectrally different absorbing pigments in the photoreceptors of the *Drosophila* eye. The phototactic response curves obtained by Fingerman and Brown (1952) indicated that the basic photosensitivity curve for *Drosophila* is that of the white-eyed mutant, and that the other response curves ot the other mutants differ from this only because of the screening effects of their eye-color pigments.

For the three eye-color mutants, scarlet, white, and wild-type red-eyed *Drosophila,* the action spectrum shows a single maximum near 508 nm, as shown in Figure 12.2 (Wolken *et al.*, 1957b). If these action spectra are indicative of the absorption spectrum of a visual pigment, then the maximum is very close to vertebrate rhodopsin, and implies that *Drosophila* should have a rhodopsin-like visual pigment.

To determine if this is so, the eyes from these eye-color mutants were extracted, using the procedure for extraction of vertebrate rhodopsin. Spectra of these extracted pigments showed a single maximum in the visible near 482 nm for the scarlet and wild-type red-eyed mutants, but not for the white-eyed mutant. However, all the extracts had a maximum in the ultraviolet at 342 nm and strong absorption in the region of 260–290 nm (Wolken *et al.*, 1957b).

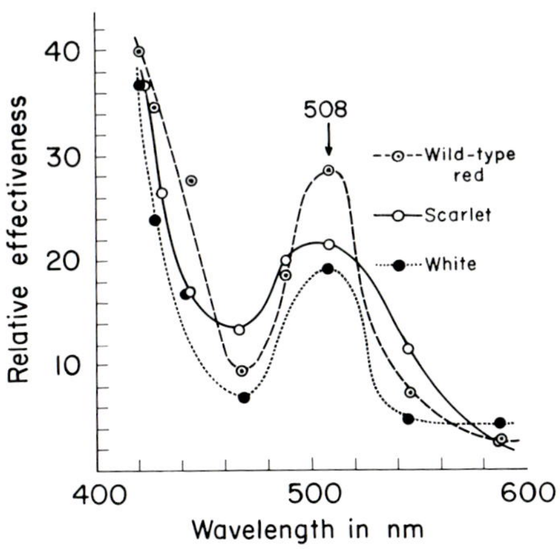


FIGURE 12.2 Action spectra of *Drosophila melanogaster* for three different eye-color mutants; wild-type red - -⊙- -, scarlet —○—, and white ····●····.

Also, the extracts did not bleach like a visual pigment. These spectra would suggest a carotenoid pigment complex, but no chemical identification was obtained.

Previously, Wald and Allen (1946) isolated from *Drosophila* a pigment that had a maximum at 436 nm and was not photosensitive, but no positive tests for vitamin A or retinal were obtained. Later, using a larger number of flies, Wald and Burg (1957) still were unable to identify for certain whether vitamin A was present. Most probably these extracted pigments are related to the ommochrome and pteridine screening pigments and are not visual pigments. Therefore, in *Drosophila,* the most actively studied of the insects, the isolation and identification of vitamin A and retinal has not been confirmed.

### *The Bee*

The identification of the visual pigment of the honeybee, *Apis mellifera,* has been of great interest since von Frisch (1914) demonstrated that it could be trained to distinguish red, yellow, and green from blue and violet. Kühn (1927) showed that blue–green and near ultraviolet could also be distinguished by the bee. These behavioral studies were confirmed and extended by Bertholf (1931), Hertz (1939), Daumer (1956), and Kuwabara (1957), who concluded that the primary ranges of bee sensitivity are the near ultraviolet 300–400 nm, blue 400–500 nm, and yellow 500–600 nm. Using electrophysiological methods, Goldsmith (1958a,b; 1960) showed that there should be several different absorbing visual pigments, one near 440 nm, another near 535 nm, and an ultraviolet sensitivity near 345 nm. Autrum and von Zwehl (1964) demonstrated that there are three different receptors for the worker bee with spectral sensitivity peaks at 340, 430, and 530 nm.

To obtain the visual pigment from honeybees, the insects were dark-adapted, decapitated, and after grinding, extracted with phosphate buffer, pH 7. The extract was further purified by chromatography and a pigment was isolated which by difference spectra (obtained by several minutes exposure to a yellow light) was found to have a maximum near 440 nm and a negative maximum near 370 nm (Figure 12.3). This latter peak was attributed by Goldsmith (1958a,b) to the formation of retinal$_1$, for it was one of the peaks he found in the electrophysiological measurements.

Goldsmith (1958a) was able to show that retinal$_1$, presumably the 11-*cis*-retinal$_1$, was indeed present in the honeybee head. He estimated that there was a maximum of 0.22 $\mu$g of retinal per gram of fresh heads, which corresponded to $3 \times 10^{-6}$ $\mu M$ per eye (Goldsmith, 1958b). No retinal could be extracted from the bodies. Furthermore, unlike all other visual pigments so far isolated, that of the bee could be brought into solution by grinding the heads of dark-adapted bees with 10 times their weight of neutral phosphate buffer.

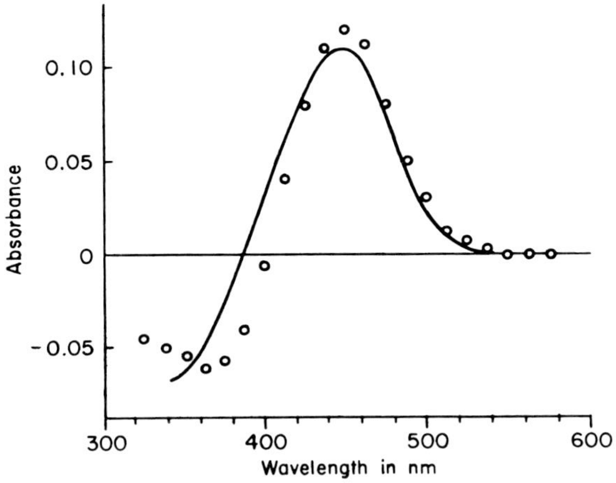


FIGURE 12.3 The honeybee, *Apis mellifera,* difference spectrum of extracted photosensitive pigment compared to that obtained by Goldsmith (1958b) (points on curve).

The finding that retinal$_1$ occurs in the head of the honeybee raised many questions about other insect visual pigments. Retinal is formed in the vertebrate eye by the oxidation of vitamin A, which is apparently not required by insects, for there has been no adequate evidence that vitamin A occurs in the bodies of insects. Goldsmith and Warner (1964) showed that vitamin $A_1$ is confined to the honeybee (*Apis mellifera*) head, that the vitamin $A_1$ is converted to retinal$_1$ on dark-adaptation, and that during light-adaptation the amount of vitamin $A_1$ increases.

## *The Housefly*

A search for visual pigments in the housefly, *Musca domestica,* was pursued using a similar procedure to that used for the honeybee. From the housefly heads a light-sensitive yellow pigment was extracted with phosphate buffer. The yellow pigment was separated from a number of other pigments by chromatography on a column of calcium phosphate mixed with Celite (Bowness and Wolken, 1959). The isolated pigment appeared to have many of the properties of the honeybee pigment, although its absorption maximum was shifted slightly down from 440 nm to 437 nm (Figure 12.4). The difference spectrum also showed a maximum at 437 nm (Figure 12.5a). When the photosensitive yellow pigment was bleached by light at pH 6.5, there was a shift toward longer wavelengths in the 500–540 nm region, and the main absorption peak was gradually replaced by a plateau of absorption around 440–460 nm.

It is of interest to note a number of similarities in spectroscopic properties between the housefly pigment and the vertebrate visual pigments. First, there

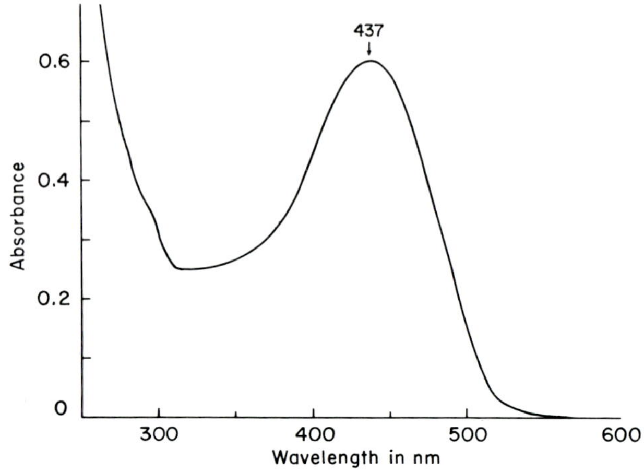


FIGURE 12.4 Absorption spectrum of the housefly, *Musca domestica,* photosensitive eye pigment. Eluate P with 0.2 *M* phosphate, pH 6.5 (Bowness and Wolken, 1959).

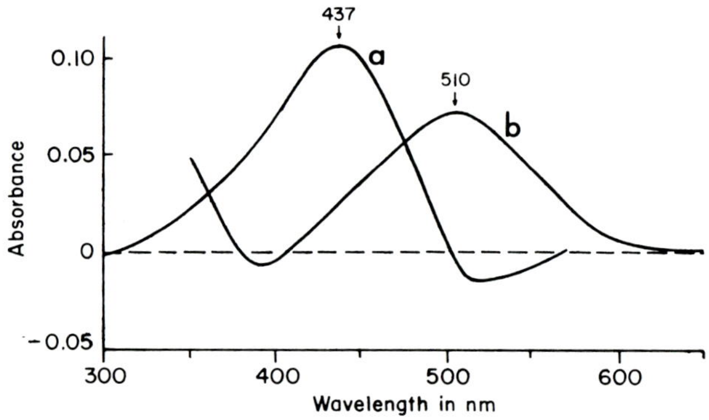


FIGURE 12.5 The housefly, *Musca domestica,* photosensitive eye pigment (Fig. 12.4) difference spectrum (a), compared to the difference spectrum of digitonin extract (b).

are the pH indicator properties shown by the light-sensitive housefly pigment and by its bleached products. On bleaching in the light at pH 6.5 a solution with plateaus of absorption at 440–460 nm and 350–360 nm is produced. Addition of a strong acid to this solution shifts the absorption peak to 470–475 nm. In alkaline solution there is a plateau at 360 nm only. Absorption maxima in these three wavelength regions are given by retinylideneamines and indicator-yellow under similar pH conditions (Ball *et al.,* 1949; Collins, 1954), although the 440 nm form of retinylideneamine is not stable except at pH 1 (Morton and Pitt, 1955). At pH 12 the housefly pigment produces an absorption maximum at 380 nm. A peak at this wavelength is obtained from squid metarhodopsin at pH 9 (see Figure 12.1a) and from cattle metarhodopsin at pH 13 (Hubbard and St. George, 1958; Hubbard and Kropf, 1959). Second, the bleaching process of the housefly pigment, as with vertebrate rhodopsin, appears to involve more than one stage. Third, the ultraviolet absorption spectrum of the pigment has a peak at 290 nm in alkaline solution. Protein may be part of this light-sensitive housefly pigment, for most proteins show an absorption peak at 275–280 nm in neutral or acid solution and at 290 nm in alkaline solution (Beaven and Holiday, 1952). Also, the heat bleaching of the pigment at 100°C gave a coagular precipitate which, when dissolved in 0.2 *N* sodium hydroxide, showed a peak at about 290 nm. Fourth, a precipitate containing about 10.5% nitrogen was obtained from the light-sensitive pigment solution upon addition of sulfosalicylic acid.

To see whether another photosensitive pigment could be extracted whose absorption peak would be closer to the behavioral action spectrum, aqueous digitonin (1.8%) extraction and the purification procedures described for cattle rhodopsin were used (Bowness, 1959). The flies were dark-adapted prior to dissection and their heads ground and extracted. All operations were carried out in the dark and in the cold. Under these conditions a pigment was eluted from the chromatographic column whose difference spectrum had an absorption maximum near 510 nm (Figure 12.5b). This absorption peak is closer to the spectral sensitivity peak for the housefly than the absorption peak for phosphate buffer-eluted, photosensitive yellow pigment (Figure 12.4).

In order to identify whether the chromophore retinal is part of the visual pigment of the housefly eye, three different extraction procedures were used. In the first method the heads were completely ground with anhydrous sodium sulfate and acetone. The mixture was centrifuged at 3000 rpm and the supernatant poured off and retained. The residue was reextracted with acetone, and after centrifugation it was combined with the first extract. This was repeated until all the color was removed.

The second method was to grind the heads with 0.2 *M* phosphate buffer (pH 6.5) and to centrifuge this mixture at 12,000 rpm for 20 minutes. The residue was extracted with acetone, while the supernatant was fractionated by saturation with ammonium sulfate at 45% and 60%. The ammonium sulfate precipitates were then extracted with acetone.

The third method was to grind the heads with 50% sucrose in 0.06 *M* phosphate buffer (pH 6.5) and then to centrifuge the sucrose mixture at 3000 rpm for 10 minutes. The residue was extracted with acetone, while the supernatant suspension was diluted with phosphate buffer to a sucrose concentration of 12.5% and centrifuged at 14,000 rpm for 20 minutes. The sedimented material was then extracted with acetone.

The initial acetone extract from each of the above preparations was evaporated to dryness, and the residue was dissolved in petroleum ether. The petro-

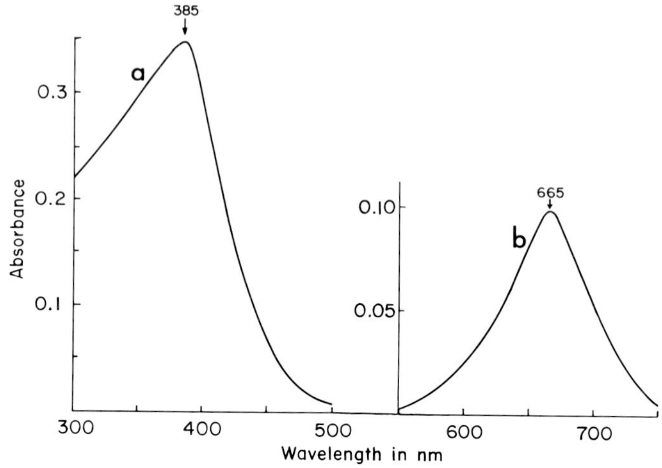


FIGURE 12.6 The housefly, *Musca domestica*. (a) Absorption spectrum of the retinal containing fraction in chloroform, (b) absorption spectrum of fraction (a) reacted with the Carr–Price reagent.

leum ether extract was then dried over anhydrous sodium sulfate, redissolved in petroleum ether, and chromatographed on alumina. The first column was eluted with 40% acetone in petroleum ether (v/v). The eluate was evaporated to dryness, redissolved in petroleum ether, and applied to a second column; fractions were eluted with 4, 17.5, 25, and 40% acetone in petroleum ether (v/v). The effluent fractions were evaporated to dryness, then stored in a vacuum desiccator in the dark. The acetone extracts of the precipitate from the ammonium sulfate fractionation (procedure 2) and the extract of the material which sedimented in 12.5% sucrose solution (procedure 3) were evaporated to dryness and dissolved in chloroform.

The results of these chromatograms showed a separation of one or more of the carotenoids present. Of the various carotenoids obtained, retinal$_1$ was identified. The fraction eluted with 25% acetone in petroleum ether had an absorption maximum at 385 nm in chloroform (Figure 12.6a), and when reacted with the Carr–Price reagent, the absorption maximum shifted to 665 nm (Figure 12.6b). This was a positive test for the presence of retinal$_1$. The amount of retinal$_1$ in the housefly heads was estimated from the extinction coefficient of the Carr–Price reaction (Ball *et al.*, 1949) to be 0.31 $\mu$g of retinal per gram weight of fresh heads. From the average number of 24,500 rhabdomeres per eye, this corresponds to $3.7 \times 10^7$ molecules of retinal per rhabdomere (Wolken *et al.*, 1960).

The presence of retinal$_1$ in the extract indicates that vitamin A$_1$ should also be present, since in the visual cycle retinal$_1$ is reduced to vitamin A$_1$.

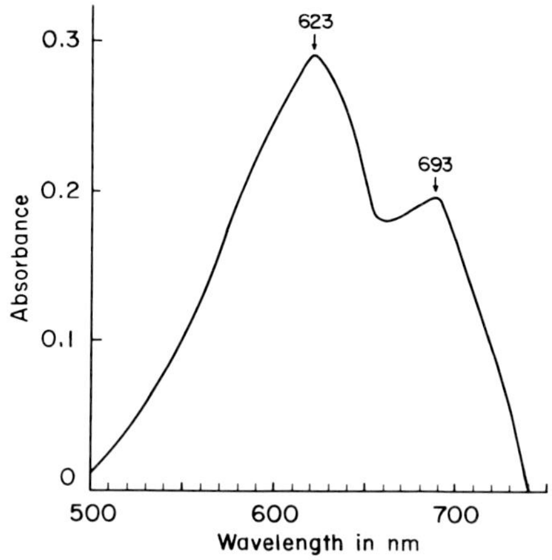


FIGURE 12.7 The housefly, *Musca domestica*. Eye pigment fraction eluted with 17.5% acetone in petroleum ether (v/v) reacted with Carr–Price reagent. Maxima at 623 and 693 nm correspond to that of pure vitamin $A_1$ and vitamin $A_2$, all-*trans*, at 621 nm and 693 nm.

Vitamin $A_1$ has a principal absorption maximum at 328 nm in chloroform and with the Carr–Price reagent gives an absorption peak at 620 nm. Fractions were obtained which gave maxima near 620 nm and near 696 nm (Figure 12.7), the maxima for vitamins $A_1$ and $A_2$; in some cases these were found in the same fraction (Wolken *et al.*, 1960). The identification of vitamin $A_2$ is particularly crucial, since it is so far known to occur only in freshwater fish and in some amphibians. This would be the first instance of its presence outside these vertebrate groups.

## *The Cockroach*

Of the nocturnal insects, it is the cockroach that attracts our interest, for it is considered more primitive than the honeybee and the housefly.

Electrophysiological studies of the cockroach eye (Walther, 1958; Walther and Dodt, 1959) suggested that the cockroach eye contains two types of photoreceptors; the lower half of the eye containing one type with an absorption maximum at 507 nm, and the upper half containing both types, the one at 507 nm and another between 314 and 369 nm. Mote and Goldsmith (1970), using intracellular recordings, found in the white-eyed mutant of *Periplaneta* two receptor cells: green receptor cells with a spectral sensitivity of about 507 nm, and ultraviolet receptor cells with a maximum spectral sensitivity at 365 nm, which is in agreement with Walther and Dodt (1959).

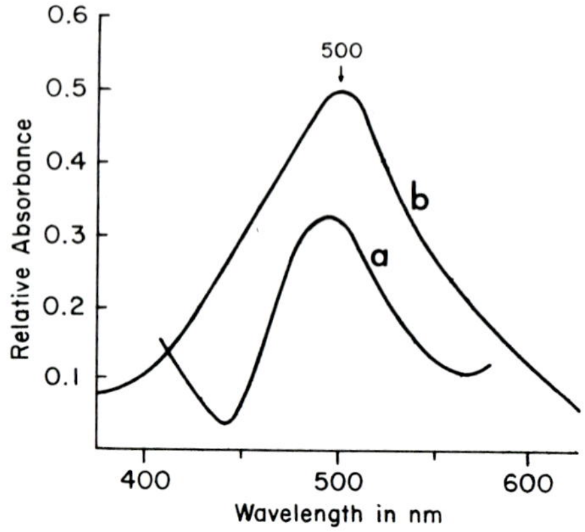


FIGURE 12.8 (a) Cockroach, *Blaberus giganteus,* absorption spectrum of rhabdom obtained by microspectrophotometry, (b) cockroach, *Blatta orientalis,* absorption spectrum of rhabdomere fraction extracted with 1.8% digitonin in 15 *M* phosphate buffer, pH 7.0.

Cockroach eyes are large in comparison with those of other insects and are therefore more easily dissected in the dark. To see whether their visual pigment could be spectrally identified *in situ,* sections were made through the rhabdom area of the eye of *Blaberus giganteus.* These sections were scanned with the microspectrophotometer which revealed an absorption spectrum with a major peak around 495 nm (Figure 12.8a), close to that of rhodopsin.

To chemically identify rhodopsin as the visual pigment, large numbers of eyes from *Blatta orientalis* and from *Periplaneta americana* were dissected and the eyes ground in 45% sucrose–phosphate buffer solution. This was followed by differential and high speed centrifugation to isolate a rhabdomere fraction. The rhabdomere fraction so obtained was then extracted with 1.8% aqueous digitonin. The pigment extract had a maximum absorption peak near 500 nm and was photosensitive (Figure 12.8b). The difference spectrum of the light bleached against the unbleached pigment extract (Figure 12.9a) also showed maximum absorption near 500 nm, corresponding to a rhodopsin, and a negative maximum at 375 nm, corresponding to retinal (Wolken and Scheer, 1963). These absorption peaks are close to those found by Mote and Goldsmith (1970) for the *Periplaneta* white-eyed mutant.

To determine whether retinal was really a part of this photosensitive pigment, extraction and column chromatography were employed (Wolken and Scheer, 1963). The chromatographic fractions eluted from the alumina columns in acetone gave an absorption spectrum indicative of retinal$_1$. When the fraction eluted with 17.5% acetone in petroleum ether was reacted with the Carr–Price reagent, the absorption maximum was found to be at 664 nm, which is characteristic of retinal$_1$ (Figure 12.9b).

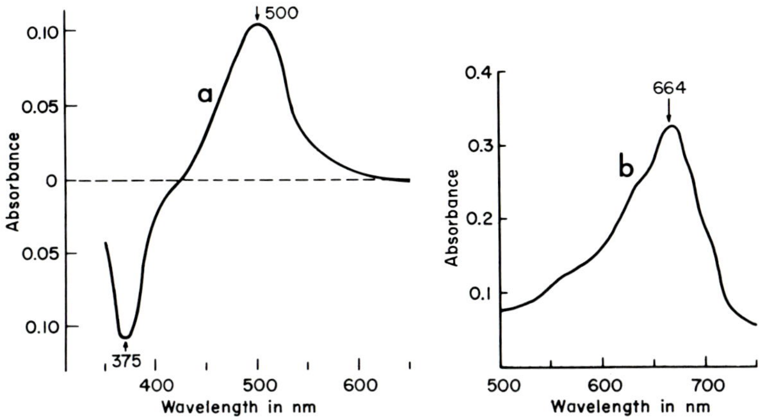


FIGURE 12.9 (a) Cockroach, *Blatta orientalis,* difference spectrum of visual pigment, (b) absorption spectrum of the eye pigment fraction chromatographed on alumina column, eluted with 17.5% acetone in petroleum ether, reacted with Carr–Price reagent.

The amount of retinal per gram of fresh cockroach eyes was estimated from the Carr–Price reaction spectrum to be 0.08 $\mu$g. On the basis of 28,000 rhabdomeres per eye (Wolken and Gupta, 1961). The amount of retinal or rhodopsin per rhabdomere was estimated to be $4.3 \times 10^7$ molecules.

Only one photosensitive pigment was isolated from the compound eyes of the cockroach (Wolken and Scheer, 1963). It is possible that the structure and arrangement of the lens, which can play a significant role in determining the near ultraviolet sensitivity of the eye (Wald, 1945; Dodt and Walther, 1958; Buriain and Ziv, 1959), together with the rhodopsin bleaching characteristics (Wald and Brown, 1958) could account for the dual sensitivity of the cockroach eye. This possibility is supported by the findings that the corneal lens fine structure permits it to function as an interference filter (Miller and Bernard, 1968).

### *Moths*

In the nocturnal moth, *Manduca sexta,* microspectrophotometry revealed three differently absorbing pigments with maxima in the visible part of the spectrum at 450 nm, 490 nm, and 530 nm, and an absorption peak in the ultraviolet at 350 nm (Carlson, 1972). In another moth, *Deilephila elepenor,* maxima were found at 350, 440–460, and 540–550 nm. If the ultraviolet sensitive pigment is discounted, then these receptor peaks come close to the absorption peaks found for different cones in the vertebrate retina.

*Arachnids*

The arachnids, or spiders, are of special evolutionary interest, for they are considered ancient in comparison to other insects. It was thought for a long time that spiders, like the bees, see color (Peckman and Peckman, 1887). More recently, Kästner (1950) made a thorough study of the jumping spider, *Evarcha falca,* using behavioral methods, and concluded that these spiders do have color vision. In the wolf spider, though, only one visual pigment for the median eye could be identified, with a λ maximum around 505–510 nm. However, their greatest sensitivity, like most insects, was in the ultraviolet around 380 nm. This ultraviolet sensitivity is believed to be due to another pigment, for example a flavin, which is not necessarily a primary visual pigment.

Closely related to the arachnids is the king crab or horseshoe crab, *Limulus polyphemus,* belonging to the class Merostomata. *Limulus* is a living fossil that continues to inhabit the shallow waters along the Atlantic coast. Its visual pigment has been isolated and identified as a rhodopsin, with an absorption maximum at 520 nm. The rhodopsin bleaches upon irradiation to yield $\text{retinal}_1$ and opsin (Hubbard and Wald, 1960).

## Color Vision

Behavioral studies have shown that the arthropod eye can sense dim light like the rods and bright light like the cones of the vertebrate eye. Also, certain insects and crustacea can distinguish colors. This implies that they should have different spectrally absorbing visual pigments as do the cones. Spectral sensitivity measurements do indicate that there should be several absorbing pigments. In examining the arthropod visual pigment, two different spectrally absorbing pigments were found which resemble the vertebrate retinal rods and cones (Figure 11.1). For example, the lobster visual pigment shows absorption peaks at 480 and 515 nm, the crayfish near 510 and 562 nm, the honeybee near 430 and 530 nm, the housefly near 440 and 510 nm, and the moth near 450 and 545 nm. In addition, all exhibit strong absorption in the ultraviolet in the neighborhood of 340–390 nm.

To clarify the dual spectral sensitivity and spectrally different absorption peaks, Autrum and Burkhardt (1961) and Burkhardt (1962) measured spectral sensitivity for the blowfly *Calliphora erythrocephala,* using microelectrodes placed in single retinula cells. They found three different spectral sensitivities in the visible with maxima at 470, 490, and 520 nm. Each was coupled with an ultraviolet band around 350 nm. In similar experiments with drone bees, Autrum and von Zwehl (1962) found two different receptors with maxima at about 340 and 447 nm. The 447 nm peak compares well with the observed spectral sensitivity maximum for drones at 430 nm (Autrum and von Zwehl, 1964) and to the extracted photosensitive honeybee pigment, maximal near 440 nm (Goldsmith, 1958a,b).

In the blowfly (*Calliphora erythrocephala*), the eye-color mutant "chalky" lacks all eye screening pigments, and Langer and Thorell (1966) were able to obtain spectra of the rhabdom by microspectrophotometry. The *Calliphora* rhabdom consists of seven rhabdomeres of the open type (see Figure 8.6c); Langer and Thorell found that for six of the rhabdomeres the absorption maximum was about 510 nm and for the seventh or asymmetric rhabdomere the absorption maximum was about 470 nm. These spectral peaks come close to Burkhardt's maxima for spectral sensitivity, and presumably these absorption peaks are associated with two differently absorbing visual pigments. To find that there are two differently absorbing pigments in different rhabdomeres of the rhabdom raises some interesting evolutionary questions.

## Visual Screening Pigments

The various eye colors seen in the arthropods come from differently colored pigment granules which surround the ommatidia. These pigment granules regulate the light that reaches the rhabdom. It was thought that the arthropod pigment granules would have similarities to the pigmented oil globules found in the retinas of birds, turtles, lizards, and snakes (Wolken, 1971). But there is no evidence to indicate that they are chemically identical or have similar function. The pigmented oil globules are all carotenoids (Wald, 1948; Wolken, 1966b).

In the isolation of the visual pigment of the housefly, the light yellow pigment that drained through the column with 0.025 *M* phosphate buffer appears to be of the melanin type. Of the other fractions eluted from the column, two contained light-stable carotenoids. One, eluted with 4% acetone in petroleum ether, was yellow in color, and its absorption spectrum in chloroform showed maxima at 426, 455, and 486 nm, which correspond to the plant pigment, xanthophyll (Figure 3.6), found in the fovea of the mammalian retina. The other was a red pigment, which required 2 *M* KOH or 1 *M* acetate–acetic acid buffer, pH 4.8, for elution from the column, and exhibited a shift in absorption maximum from about 490 to 440 nm in changing from alkaline to acid conditions (Bowness and Wolken, 1959). This is similar to the shift shown by rhodommatin, a red ommochrome-type pigment (Figure 12.10) obtained from insects by Butenandt *et al.* (1954).

In the compound eye of the housefly, *Musca domestica*, yellow pigment granules are found at the top of the ommatidium close to the corneal lens and crystalline cone, whereas the red pigment granules are found closer to the rhab-

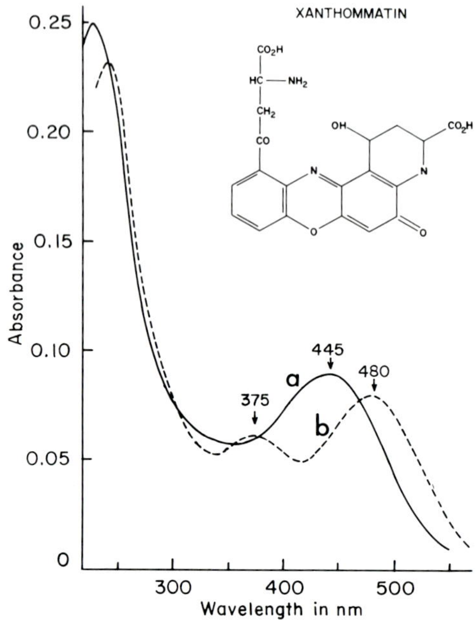


Figure 12.10 Absorption spectrum of xanthommatin, extracted from *Calliphora*. (a) pH 7.0 (——) and (b) in 5 *N* HCl, pH 2.0 (- - - - - -). From Butenandt and Neubert (1955).

dom. Microspectrophotometry of the yellow and red granules *in situ* shows that the yellow pigment in these granules has a broad absorption band maximal at about 440 nm, and that the red pigment also has a broad absorption band with its maximum nearer to 530 nm, with another smaller peak close to 390 nm (Strother and Casella, 1972).

In the blowfly, *Calliphora erythrocephala,* the yellow pigment granules have a single peak at about 445 nm and the red pigment granules show peaks near 380 and 540 nm (Langer, 1967). The yellow pigment absorption spectrum resembles that of an oxidized xanthommatin-protein complex and the red pigment is similar to the rhodommatin pigment (Butenandt *et al.,* 1954; Butenandt and Neubert, 1955; Burkhardt, 1962).

These spectra are also interesting from a physiological standpoint, for no obvious correlation between them and the overall spectral sensitivity for these insects is evident. Accordingly, Langer and Thorell (1966) made direct microspectrophotometric measurements of the *Calliphora* rhabdomeres within a rhabdom. They found two different spectra; one had two absorption peaks with maxima at about 380 and 510 nm, and closely resembled the red screening pigment spectrum, and the other showed only a single peak near 470 nm and its shape closely resembled the yellow pigment spectrum (Strother and Casella, 1972). These results would indicate then that the screening pigments function to absorb light for the visual pigments. In fact, a photosensitive yellow pigment with an absorption peak near 440 nm was extracted from housefly eyes (Bowness and Wolken; 1959) and from the honeybee (Goldsmith, 1958a,b), which closely matches the yellow screening pigment absorption peak (Figures 12.3 and 12.4). Data obtained by Langer (1967) and Strother and Casella (1972) indicate that for the blowfly and housefly the combined effect of both yellow and red screening pigments is to effectively screen the separate ommatidia from light leakage from 320 nm to about 640 nm. However, it is possible that the yellow and red screening pigments are acting separately to screen two different visual pigments, namely, the one absorbing at 440 nm and the other near 510 nm.

These spectra suggest that those regions where the pigments permit passage of light coincide with the spectral regions where the insect is most strongly sensitive. Thus, ultraviolet light of 350–400 nm and red light beyond 600 nm are transmitted and not absorbed by the screening pigment granules. This transmitted light is then available to the photoreceptors and results in a greater sensitivity for the insect. The arthropod pigment granules spectra obtained by microspectrophotometry have been recently reviewed by Strother and Casella (1972).

The screening pigment granules have for the most part been identified as ommochromes, pterines, and pteridines (Ziegler–Günder, 1956; Grossbach, 1957; Ziegler, 1964, 1965). Ommachrome pigments are yellow to dark red and fall into the class of ommatines and ommines (Linzen, 1959). The ommachromes of the xanthommatin type are photosensitive, they can be oxidized and reduced, and they are pH sensitive (Figure 12.10). Yoshida *et al.* (1967) suggested that these ommochromes could function as do the quinones and the cytochromes in the electron energy transfer.

The name pterines is derived from "lepidopterin," since it was first found in the wings of butterflies by Sir Frederich Gowland Hopkins in 1889. These pigments account for the yellow colorations in the wings of insects. The wing pattern disguises are, then, effective mimics to attract prey or ward off predators. This kind of adaptive selection by insects is seen particularly in the butterfiies, moths, and wasps. Comparison between the pteridine and alloxazine structure of flavins shows a close molecular relationship (see Figures 3.13 and 12.10). Also, certain pterine complexes with protein may serve as coenzymes analogous to the flavoproteins. The pterine pigments are usually chemically represented as the yellow xanthopterin (2-amino-4,6-dihydroxypyrimidopyrazine ring) which upon ultraviolet excitation, fluoresces blue (Forrest and Mitchell,. 1954). The pterines are stable *in vivo,* but photosensitive *in vitro.* Although they participate in some metabolic processes in the eye, they do not appear to be a primary photoreceptor molecule in visual excitation (Goldsmith, 1972).

## Polarized Light Analysis

The detection of polarized light by arthropods and molluscs suggests that the analyzer mechanism is based on the dichroism of the photosensitive pigment within the rhabdom (De Vries, 1956; Moody, 1964). The basis for such a hypothesis goes back to the earlier studies of Schmidt (1934, 1935, 1938) on vertebrate retinal rod outer segments. He found that the absorption of polarized light is independent of the orientation of the electric vector for light traversing the rod parallel to its long axis, but is dichroic when irradiation is transverse. This dichroism was studied in more detail for marine retinas (Denton, 1959) and for frog retinal rods (Liebman, 1962). An interesting feature of these studies of dichroism in vertebrate rods is that when the dichroic ratio is plotted against wavelength, it corresponds to the rhodopsin absorption spectrum. When rhodopsin is bleached, the dichroism is lost. This would mean that at certain wavelengths, polarized light absorption would be good, while at other wavelengths it would not.

Burkhardt and Wendler (1960) recorded action potentials of single retinula cells in the compound eye of the blowfly, *Calliphora,* with intracellular electrodes. They found that rotating the plane of polarization resulted in a 50% difference in amplitude between maximum and minimum responses. This effect was obtained with blue and white light, but was not observed when red light was used as a stimulus. This finding that polarized blue and white light but not red light had an effect on the receptor potential may find its explanation in the dichroic absorption of oriented photopigments in the rhabdomere.

If, in fact, the rhabdom constitutes a dichroic analyzer, its properties would depend on the arrangement of the microtubules in individual rhabdomeres and on the orientation of the rhodopsin molecules, with their major axes parallel to the tubule direction and hence, perpendicular to the normally incident illumination. Microspectrophotometry of the crayfish, *Orconectes virilis,* rhabdoms illuminated transversely does show that their photosensitive absorption exhibits a dichroic ratio of 2 *in situ.* The major absorption axis matches the axial direction of the closely parallel microtubules of the rhabdomere. Since these microtubules are regularly oriented transversely in about 24 layers, with the axes of the microtubules perpendicular to each other in alternate layers, transverse illumination of a properly oriented rhabdom displays alternate dichroic and isotropic bands. The explanation for the polarized light analyzer action in the rhabdom is that the absorbing dipoles of the rhodopsin molecules, as in the vertebrate retinal rods, lie parallel to the membrane surface (see Figure 11.14), but are otherwise randomly oriented (Waterman *et al.*, 1969).

TABLE 12.1 SOME INVERTEBRATE VISUAL ABSORPTION MAXIMA[a]

| *Phyla*<br>*Class*<br>*Organism* | *Common name* | *Maxima (nm)* |
|---|---|---|
| Mollusca | | |
| Cephalopoda | | |
| *Eledone moschata* | Octopus | 470 |
| *Loligo peallii* | Squid | 493 |
| *Octopus ocellatus* | Octopus | 477 |
| *Octopus vulgaris* | Octopus | 475 |
| *Sepia esculenta* | Cuttlefish | 486 |
| *Sepia officinalis* | Cuttlefish | 492 |
| *Sepiella japonica* | Cuttlefish | 500 |
| *Todarodes pacificus* | Squid | 480 |
| Arthropoda | | |
| Crustacea | | |
| *Callinectes sapidus* | Blue crab | 480 |
| *Euphasia pacifica* | Shrimp | 462 |
| *Homarus americanus* | Lobster | 515 |
| *Leptodora kindtii* | Water flea | 510, 550 |
| *Orconectes virilis* | Northern crayfish | 510, 562 |
| *Palaemonetes vulgaris* | Prawn | 390, 496, 540, 555 |
| *Porcellio scaber* | Wood louse | 480 |
| *Procambaras clarkii* | Swamp crayfish | 570 |
| Insecta | | |
| *Apis mellifera* | Honeybee | 440, 510 |
| *Blaberus giganteus* | Cockroach | 495 |
| *Blatta orientalis* | Cockroach | 500 |
| *Calliphora erythrocephala* | Blowfly | 470, 510 |
| *Deilephila elepenor* | Moth | 440–460, 540–550 |
| *Musca domestica* | Housefly | 437, 510 |
| *Manduca sexta* | Moth | 350, 450, 490, 530 |
| *Periplaneta americana* | Cockroach | 500 |
| Arachnida | | |
| *Evarcha falca* | Jumping spider | 350, 440, 540 |
| Merostomata | | |
| *Limulus polyphemus* | Horseshoe crab (king crab) | 520 |

[a] Taken in part from Wolken (1971, p. 138; 1973, pp. 1269–1270).

The point to be made here is that all studies of arthropod and mollusc eyes indicate that their visual pigment is a rhodopsin and the chromophore is 11-*cis*-retinal$_1$, as in the vertebrate visual systems (Wolken, 1971; Goldsmith, 1972). But unlike vertebrate rhodopsin, the photochemistry of cephalopod rhodopsin involves only the transformation of rhodopsin to metarhodopsin (Figure 11.9).

TABLE 12.2 CONCENTRATION OF VISUAL PIGMENT, RHODOPSIN, IN ARTHROPODS AND VERTEBRATES[a]

| | | *Rhabdomere* | |
|---|---|---|---|
| *Arthropods* | *Genus and species* | *Average volume in cm³* | *Concentration of rhodopsin molecules* |
| Cockroach | *Periplaneta americana* | $3.1 \times 10^{-10}$ | $5.9 \times 10^{7}$ |
| Cockroach | *Blatta orientalis* | $3.0 \times 10^{-10}$ | $4.3 \times 10^{7}$ |
| Housefly | *Musca domestica* | $6.8 \times 10^{-11}$ | $3.7 \times 10^{7}$ |
| Honeybee | *Apis mellifera* | $2.6 \times 10^{-11}$ | $2.6 \times 10^{7}$ |

| | | *Retinal rod* | |
|---|---|---|---|
| *Vertebrates* | | *Average volume in cm³* | *Concentration of rhodospin molecules* |
| Amphibian | | | |
| Frog | *Rana pipiens* | $1.5 \times 10^{-9}$ | $3.0 \times 10^{9}$ |
| Cattle | | $7.5 \times 10^{-12}$ | $1.0 \times 10^{6}$ |
| Man | | $1.6 \times 10^{-10}$ | $1.0 \times 10^{7}$ |

[a] From Wolken (1971, p. 150).

Some invertebrate visual pigment maxima which we have mentioned are listed in Table 12.1. The concentrations of rhodopsin that have been determined for arthropod and mollusc photoreceptors are of the order of $10^7$ molecules per rhabdomere (Table 12.2), which is comparable to the concentration of rhodopsin found in the retinal rods of vertebrates.

# XIII

# The Visual Photoreceptors and Pigments in Evolution

The evolution of organisms is usually traced through evidence of progressive changes leading from one morphological form to another. Comparisons are made between those structures which serve similar functions. This approach to the visual photoreceptors was pursued through a comparative structural analysis along phylogenetic lines. Based on our original hypothesis, the development of the photoreceptors should be found at the molecular level, for we can begin to see as we approach molecular dimensions that morphology becomes one with chemical structure and with the functioning molecules themselves.

The results of our structural investigations, from protozoan flagellates to the vertebrate visual photoreceptors, are summarized schematically in Figure 13.1. One can see how once a photoreceptor, the eyespot-flagellum (A), evolved, more complex visual imaging systems of ocelli and retinula cells (B,C), of compound eyes and rhabdoms (D), refracting eyes and rhabdoms of mollusc cephalopods (E) and vertebrate eyes and retinas (F), could have developed.

## Visual Photoreceptors

In the arthropods and molluscs all kinds of imaging systems are found, from pinhole eyes and ocelli to compound eyes and refracting eyes.

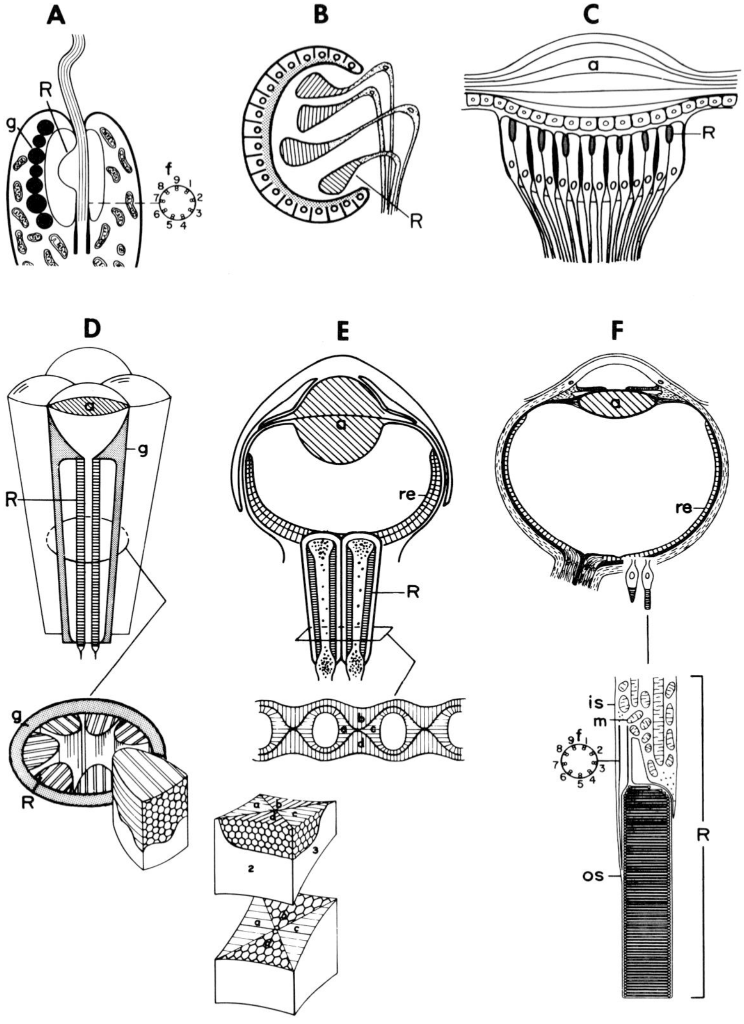


FIGURE 13.1 Phylogenetic development of various kinds of eyes and their photoreceptor structures. (A) Eyespot-flagellum of protozoan flagellates; (B) photosensory cell, ocellus of flatworms; (C) insect ocellus; (D) compound eye of arthropods; (E) mollusc cephalopod eye (a,b,c,d are rhabdomeres forming rhabdoms); (F) vertebrate eye; a, lens, f, flagellum; g, pigment granules; re, retina; R, photoreceptor (retinal rod or rhabdomere); m, mitochondria; os, outer segment; is, inner segment.

The compound eyes of arthropods are composed of numerous ommatidia. Each ommatidium consists of a corneal lens, a crystalline cone, and retinula cells (Figure 8.5). Each retinula cell has a differentiated photoreceptor structure, a rhabdomere, whose structure and function is similar to that of the vertebrate retinal rod. The rhabdomeres within the ommatidium form the rhabdom, which serves as the "retina" for these organisms. Compound eyes are found in extinct arthropods, for example, the trilobites of the Paleozoic era. Their compound eyes are well preserved and appear to be essentially the same as the present-day arthropod compound eyes (Figures 8.3 and 8.4). But their corneal lenses have been found to be calcite and to be doubly refracting. So we see in the trilobites that an advanced optical system evolved early in the history of animal evolution (Towe, 1973).

Another structural observation in certain insects was that the crystalline cone and the rhabdom are connected by crystalline threads, as seen in the firefly *Photuris* (Figure 8.9). It was suggested that the crystalline threads function as waveguides (Horridge, 1969), and Døving and Miller (1969) showed that for the skipper butterfly, *Epargyreus clarus,* only the light contained in the crystalline thread is effective for stimulating the retinula cells. Also, in the compound eye of the worker honeybee the closed rhabdom and the surrounding zone were found to act together as a waveguide (Varela and Wiitanen, 1970).

In the marine crustacean, *Copilia,* we observed a single ommatidium with a corneal lens ($L_1$) and at some distance away, the crystalline cone lens ($L_2$), attached to a scanning rhabdom of five rhabdomeres (Figures 8.16 and 8.17). Gregory (1967) believes that *Copilia* could be a surviving form of a prototype, single-channel scanning eye. In the freshwater *Daphnia* compound eye there are about 20 ommatidia that are in continuous oscillation, suggesting that *Daphnia* is part way up the evolutionary scale from the single-channel eye of *Copilia* to the fully developed compound eye of arthropods.

In the mollusc cephalopod the eye is a refracting type eye similar to the vertebrate eye, but its retina resembles that of the compound eye with retinula cell rhabdomeres forming rhabdoms (Figure 8.18).

In the insects and crustacea that have been described there are two geometrical arrangements for the rhabdomeres that form the rhabdom. One is the "open-" type rhabdom as seen in *Drosophila* and *Copilia* (Figures 8.6 and 8.17), in which the rhabdomeres project through a neck-like part of their retinula cell into a cavity. The other is the "closed-" type or fused rhabdom seen in the cockroach, firefly, and hornet, as well as in the honeybee, grasshopper, locust, dragonfly, moth, and butterfly (Figures 8.7 and 8.10–8.12), and in the crustacean waterfleas, *Daphnia* and *Leptodora* (Figures 8.13 and 8.15). The closed-type rhabdom is also found in the mollusc cephalopods *Octopus, Sepia,* and squid (Figure 8.18). The amphibian retinal rod resembles a closed-type rhabdom (Figure 10.10), and the vertebrate retina with rods and cones form rhabdom-like structures (Figure 10.4). Whether this is an evolutionary development for greater light capture or arose independently is not known.

The number of rhabdomeres that form the rhabdom vary; in the open-type rhabdom there are usually five to seven with an asymmetric rhabdomere lying in the same plane (Figure 8.6). In the closed-type rhabdom there are four

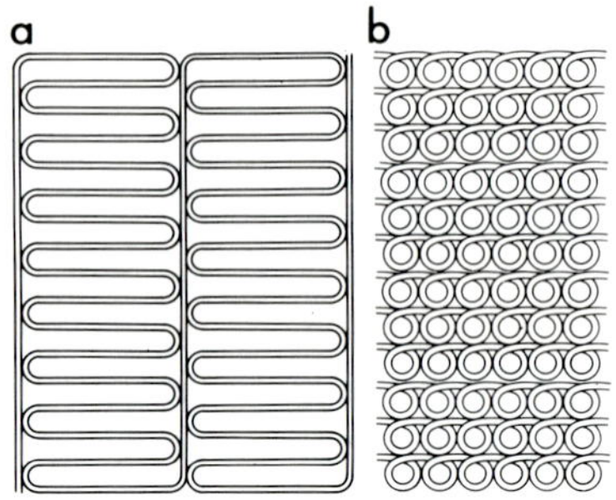


FIGURE 13.2 Schematic diagram of how membranes can form (a) lamellae or (b) microtubules.

to eight rhabdomeres forming a symmetrical arrangement, with an asymmetric rhabdomere lying in another plane (Figures 8.10–8.15). The number of rhabdomeres in the arthropod eye ranges from only a few to the order of thousands. But in refracting type eyes there are more than $10^6$ retinal photoreceptors.

The fine structure of all rhabdomeres is that of microtubules (or microvilli) of about 500 Å in diameter whose double-walled membranes are about 100 Å in thickness. In the rhabdom the microtubules of two adjacent rhabdomeres are oriented perpendicularly to each other, whereas those of the opposite rhabdomeres are parallel (Figures 8.7 and 8.11). Such a geometric arrangement would permit the rhabdom to act as an analyzer for polarized light (Figure 8.19).

We can now say that in the course of evolution, invertebrates developed every known type of optical system for image formation.

In comparison, the vertebrate retinal rod and cone outer segments are lamellae (Figures 10.7, 10.8, and 10.10). The lamellae are double-membraned structures of the order of 250 Å; each membrane is about 50 Å in thickness. Even though the vertebrate rods and cones are described as lamellae and the invertebrate rhabdomeres as microtubules, they are both membraneous processes of the retinal or retinula cells. These two types of membrane geometry, microtubules for the invertebrates and lamellae for the vertebrates, depend on how the photoreceptor membranes evolved. How these two types of membrane structures can be developed to produce the maximum surface area for the visual pigments is illustrated in Figure 13.2.

## Photoreceptor Origins

The origin of the photoreceptor must, however, go back further than these specialized visual receptor organelles to the cell membrane itself. The cell membrane structure of proteins and lipids, as we have indicated (Figure 4.6), has a number of advantages for receptor molecules. Membranes can provide large surface areas for the visual pigment and hence light capture; they bring molecules sufficiently close for interaction and energy transfer, and one part of a reaction site can be separated from another.

As cells evolved, cilia and flagella developed which provided greater mobility and more rapid locomotion in an aqueous environment. We have shown that cilia and flagella consist of nine contractile filaments which emerge from the roots of the basal body (Figures 9.3, 9.4, 10.7 and 10.8). The basal body also bears a remarkable resemblance to the centriole of the mitotic apparatus. As the cell synthesized pigments, they aggregated into pigment granules and could have become attached to the flagellum. In the protozoan flagellate *Euglena* we have shown that the paraflagellar body contains a photosensitive pigment which is associated with the flagellum, and that this provides a photoreceptor system for sensing light in the environment (Figures 7.1 and 7.8). Since all cilia and flagella are similarly structured and are found in sensory cells throughout all animal phyla, for example the cilium that penetrates from the inner segment to the outer segment of the vertebrate retinal rod (Figures 10.7 and 10.8), it is conceivable that a cilium or flagellum could have been the origin of the sensory receptor systems—and therefore they continued to be a basic structure incorporated into all photoreceptor systems.

Accordingly, a possible line of photoreceptor evolution was suggested by Eakin (1965, 1968, 1972), who developed a hypothesis based on structural observations. He asserts that the photoreceptors are derived from either a cilium or flagellum which became modified in various ways or from the cell membrane which became evaginated to form lamellae or microtubules.

In examining the structure of eyes from the simple eye to ocelli to compound eyes and finally to refracting eyes, there seems to be no smooth phylogenetic development. For each of these kinds of eyes has been found to differ widely, even among species of the same phylum. However, as we approach molecular dimensions we see that there is a basic molecular structure of packed membranes for all photoreceptors. This structure has similarities to the lattice structure of crystals (Figure 6.13) which has led to the idea that the photoreceptors may bear a close relationship to the physics of the *solid state*. This kind of molecular organization is by no means unique to the visual photoreceptors, for it has counterparts in the chloroplasts of photosynthetic plant cells (Figures 5.5 and 5.6).

## The Visual Pigments in Evolution

All visual pigments that have so far been identified are based on either vitamin $A_1$ or $A_2$ and their respective aldehydes, $retinal_1$ or $retinal_2$. The only chemical difference between these visual systems is the addition of a double bond at position 3–4 in the ring of the molecule.

The visual pigment identified for all vertebrate and invertebrate eyes is a rhodopsin in which the chromophore is retinal$_1$. Retinal in rhodopsin is in

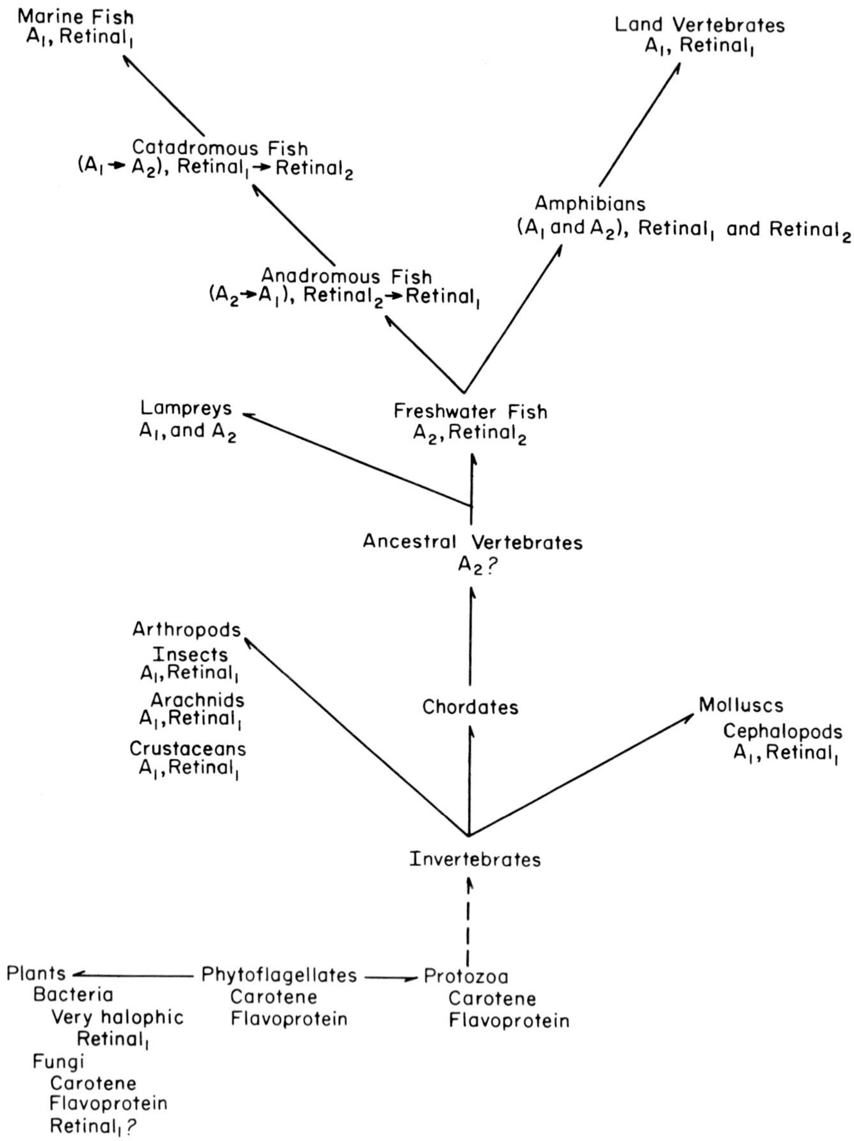


FIGURE 13.3 Phylogenetic relationship of carotene, vitamin A, and retinal in photoreceptors. Modified from Wald (1952, 1960, 1970) and Wolken (1971).

the 11-*cis* form and upon absorption of light it changes to the all-*trans* form. All retinal rods of the vertebrate eye and rhabdomeres of the invertebrate eye contain from $10^6$ to $10^9$ rhodopsin molecules. The number of rhodopsin molecules is directly related to the number of lamellae or microtubules of these photoreceptors. From the geometry of the photoreceptors, the number of lamellae and the rhodopsin concentration, the diameter of the rhodopsin molecule was estimated to be about 40 Å.

The finding that retinal$_1$ occurs in the heads of certain insects raised questions about the insect visual pigment. Retinal is formed in the vertebrate eye by the oxidation of vitamin A, which is apparently not required by insects, for there is no adequate evidence that vitamin A occurs in the bodies of insects other than in their heads. From housefly heads fractions were obtained indicating for the Carr–Price reaction maxima for vitamin $A_1$ and vitamin $A_2$. The identification of vitamin $A_2$ is particularly crucial since it is so far known to occur only in freshwater fish and in some amphibians, and this would be the first instance of its presence outside the vertebrate phyla.

Turning now to the vertebrates, Wald (1939) found that all the freshwater fish he has examined possess only porphyropsin systems. He also discovered that those fish which spawn in fresh water and live normally in the sea, anadromous fish, possess primarily the porphyropsin system, whereas those that live mostly in fresh water but spawn in the sea, catadromous fish, possess primarily the rhodopsin system. The so-called euryhaline fish, which are capable of existence in both environments, possess either predominantly or exclusively the photopigment and type of vitamin A which are ordinarily associated with the environment in which the fish is spawned.

It became apparent to Wald (1942) that differentiation of vitamin $A_1$ or $A_2$ usage was not primarily environmental, but rather genetic. Wald proposed that vitamin $A_2$ may represent a comparatively recent evolutionary development associated with the remigration of marine fish into fresh water. Wald (1942) then found that the porphyropsin (vitamin $A_2$) system predominates in the anadromous lamprey, a cyclostome, which is a primitive living vertebrate. This suggested to him that the ancestral vertebrate visual pigment was based on vitamin $A_2$ rather than vitamin $A_1$.

We can now begin to understand how the carotenoids, in the form of vitamin A and retinal, fit into some universal scheme for all animals possessing eyes (Figure 13.3).

## Opsins

Retinal in rhodopsin is complexed through a Schiff-base linkage to its protein, opsin, forming rhodopsin. The absorption spectra and other properties, such as solubilities of the rhodopsins from different species, indicate that the various forms of the protein opsin differ. For example, photosensitive pigments have been extracted from the bee and housefly in phosphate buffer, whereas all other visual pigments require solubilizing agents to bring them into solution. The molecular weights for different rhodopsins (Table 11.3) may also indicate species differences in the protein opsin. At present we know too little about the chemistry of opsin, its amino acid composition and sequence, to draw any evolutionary conclusions.

A dramatic example of adaptation through modification of protein structure is the relationship between the absorption spectra of the visual pigments of marine fish and the depths of their habitats (Anfinsen, 1963). Clarke and Denton (1962) observed that the increasing blueness of light with depth in the ocean suggests the possibility of a shift in the eyes of deepsea fish toward absorption at the blue end of the spectrum. This is in fact confirmed by evidence that the absorption spectrum of the extracted rhodopsin from surface fish is around 500 nm and shifts to around 480 nm for fish that live at depths from 200–400 meters. These shifts can be attributed to differences in the associated opsin. Since opsins do not themselves absorb light in the spectral range 480–500 nm, the shifts in the positions of the visual pigment absorption maxima must be attributed to the effect of the opsins on the spectral properties of rhodopsin. We have already seen evidence for this in the correlation observed between the habitat of the various species of fish and the spectral properties of their visual pigments.

The study of the structural modifications in opsin should be especially useful in elucidating the evolution of fish. Evidence for evolution is scarce, since large spans of evolutionary time may pass without too extensive a change in a particular protein molecule. The changes in the opsin molecule may not involve extensive alterations in sequence and folding. On the other hand, if alterations have been extensive, a very complex set of interactions between opsin and retinal will have to be explored. The questions are intriguing, and the study of the chemistry of opsins from different species is necessary before any real contribution to our understanding of the evolution of visual pigments at the molecular level can be established.

## How Universal Is Retinal?

It is of interest to turn again for evolutionary clues regarding the biosynthesis of visual pigments to the very primitive photoreceptor systems found in the bacteria, fungi, and protozoa. We are now interested in the question of how these primitive light receptor systems are related to the animal visual pigment, rhodopsin. Until recently retinal was not found in organisms which do not possess eyes, and so it was believed that animals could synthesize retinal only from its precursors, vitamin A and β-carotene.

In the phototaxis of *Euglena* we showed that the action spectrum indicates that a flavin-complex is the photoreceptor molecule and that the carotenoids associated with the eyespot act as a filter for the photoreceptor molecule (Figures 7.6 and 7.7). In phototropism of fungi and plants there is also experimental evidence to implicate the flavins and carotenoids in the photoreceptor process. The dilemma of whether flavins or carotenoids are the primary photoreceptors in phototactic and phototropic behavior has not been clearly resolved.

Most fungi are phototropic, and in the fungus *Phycomyces* it was suspected that the photoreceptor molecule could be a carotenoid and that retinal, as in the visual pigment, would be found. In *Phycomyces* wild type, small quantities of retinal were extracted from the sporangiophores (Meissner and Delbrück, 1968). This suggested that a rhodopsin-like molecule could be the photoreceptor pigment in *Phycomyces*. However, this possibility was ruled out because no retinal was detectable in β-carotene deficient mutants, which are still phototropic. It was believed that the retinal found in the sporangiophores was an oxidation product of β-carotene. Whether *Phycomyces* synthesizes retinal or whether retinal was converted from β-carotene in the process of extraction needed confirmation.

The most likely place to search for retinal was in the previously described crystals which abound in *Phycomyces* (Figure 6.7). To test whether retinal was either adsorbed on the crystal surfaces or was part of the crystal itself, we took the isolated octahedral crystals from Stages I–III, fixed them in 4% gluteraldehyde and obtained their absorption spectrum using the microspectrophotometer. The absorption spectrum showed a maximum near 520 nm (Figure 13.4a). This absorption peak was previously found for one of the crystal layers in the centrifuged sporangiophore (Figure 6.10) and in the recrystallized rhomboids. Another test was made with freshly isolated crystals. The crystals were dried and reacted with the Carr–Price reagent. The absorption spectrum of this reaction gave a peak at 664 nm (Figure 13.4b), typical of that of $retinal_1$.

Whether retinal is synthesized from β-carotene or is a metabolic waste product which is removed by the crystals awaits further studies. We have already indicated that the photoreceptor molecule responsible for *Phycomyces* phototropism is probably a flavin complex. Therefore, retinal, if present in these crystals, would not be expected to function in these fungi as it does for visual excitation in the retinal photoreceptors of the eye. However, it could function in some photochemical process related to the energetics of the cell, for example, with the light-growth response.

But even more interesting are the recent investigations concerning the extremely halophilic bacteria that grow in a marine environment with an unus-

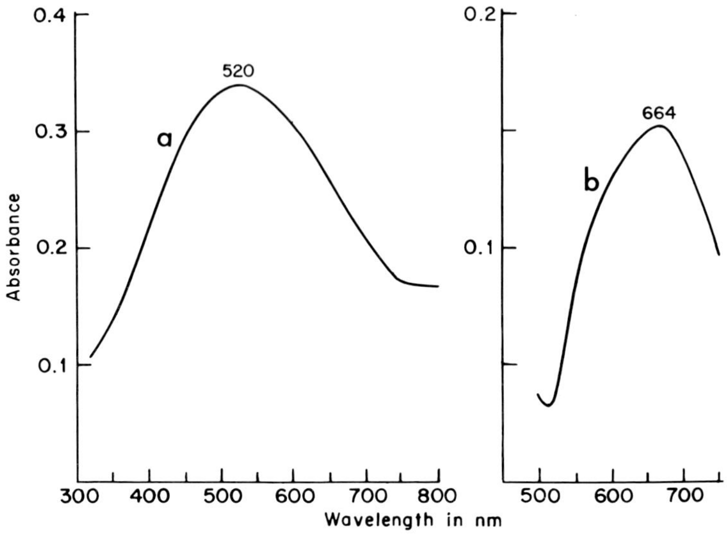


FIGURE 13.4 Isolated crystals from *Phycomyces blakesleeanus*; (a) absorption spectrum of a crystal fixed in 4% gluteraldehyde, embedded and sectioned, 4 μm thick, area scanned 18 $\mu m^2$, (b) suspension of crystals, dried and reacted with Carr–Price reagent, area scanned 12 $\mu m^2$. (Obtained with microspectrophotometer.)

ually high salt (NaCl) concentration of 25%, that of the Dead Sea in the Middle East and the Great Salt Lake in Utah. These findings destroyed the old myth that the Dead Sea is sterile. However, only two types of bacteria, *Halobacterium* and *Halococcus*, grow at these salt concentrations and the adaptability to such an environment is unique and indicates an equally unique biochemistry.

The bacterium *Halobacterium halobium* growth medium is given in Gochnauer and Kushner (1969). They grow at high temperatures near 44°C and in continuous direct sunlight. They synthesize large quantities of carotenoids which give the culture a red to purple color. In our laboratory cultures of *Halobacterium halobium* changed within days from pink to red to purple (Figure 13.5a). Absorption spectra of the bacteria by microspectrophotometry show a broad absorption with maxima at 421, 462, 503, and 535 nm (Figure 13.6a). Ruptured bacteria in distilled water have absorption peaks at 415, 475, 507, and 544 nm; extracts of the bacteria with 60% acetone have absorption peaks at 465, 496, and 532 nm (Figure 13.6b). These absorption spectra are typical of $C_{50}$-bacterioruberins and the $C_{40}$-carotenes which are the major pigments. Many of these have been identified by Gochnauer *et al.* (1972).

A surprising finding for these bacteria was that the ruptured membrane (Figure 13.5b,c,d), described as the "purple membrane," possesses the visual

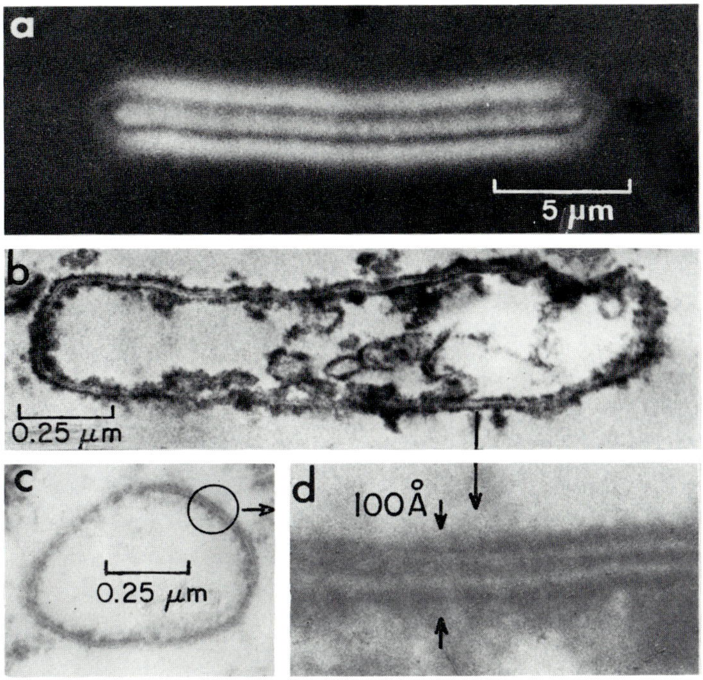


FIGURE 13.5 *Halobacterium halobium.* (a) Photomicrograph (phase contrast) of living cell, (b) electron micrograph of longitudinal cut through a fixed and embedded cell, (c) cross section cut through fixed and embedded cell, and (d) enlargement of cell membrane.

pigment rhodopsin (Oesterhelt and Stoeckenius, 1971; Blaurock and Stoeckenius, 1971). The purple membranes can be easily isolated and their absorption spectrum does resemble the vertebrate cone pigment, iodopsin (Figure 13.7), for it has its maximum absorption in the visible at 560 nm and at 280 nm in the ultraviolet. The ratio of the extinction at 280 and 560 nm is 2:1, and the complex is photosensitive. The molecular weight of the complex was found to be 26,000 in a molar ratio of 1:1 pigment to protein. The pigment was named *bacteriorhodopsin* because it shows similar properties to rhodopsin. The chromophore was identified as retinal$_1$. The finding of a visual pigment in a membrane of this bacterium raises many interesting questions about retinal and the evolution of visual pigments.

To identify whether retinal was specifically the chromophore, the bacteria were extracted and retinal was isolated by chromotography (Wolken and Nakagawa, 1973). The absorption spectrum of the retinal spot eluted from the chromatograms with chloroform showed a peak at 376 nm (Figure 13.8a). The absorption peak at 376 nm is a few nanometers blue-shifted when com-

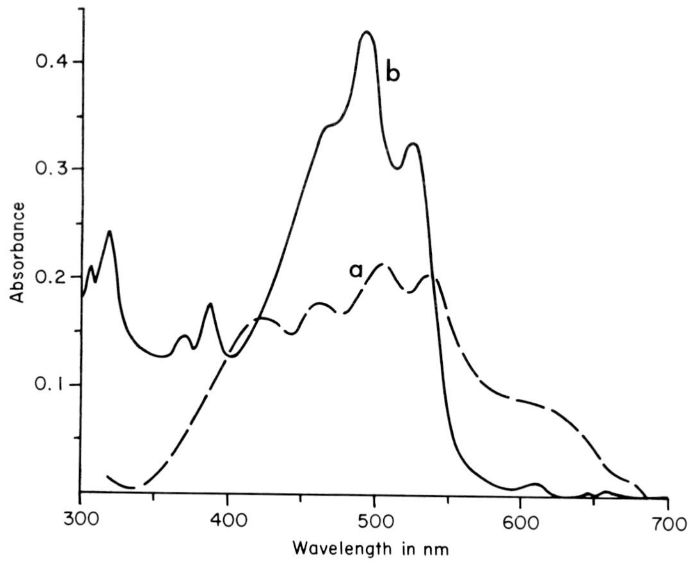


FIGURE 13.6 *Halobacterium halobium.* (a) Absorption spectrum obtained by scanning the cell with microspectrophotometer, (b) absorption spectrum of extracted bacterioruberin $C_{50}$-carotenoid.

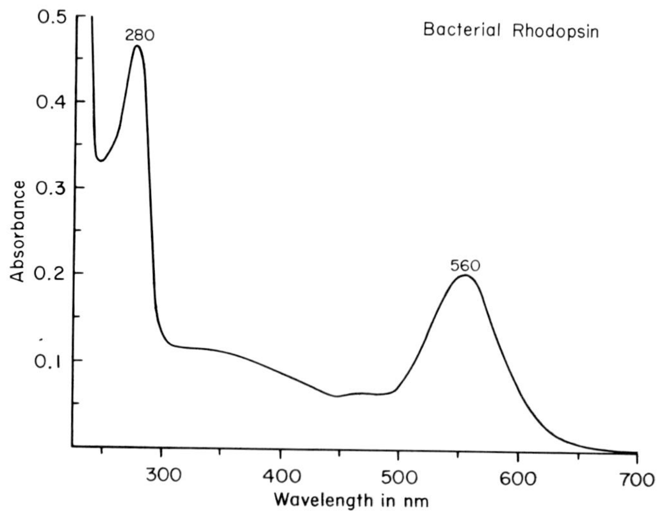


FIGURE 13.7 Absorption spectrum of isolated "purple membrane" bacterial rhodopsin (compare to iodopsin Figures 11.1 and rhodopsin 11.4).

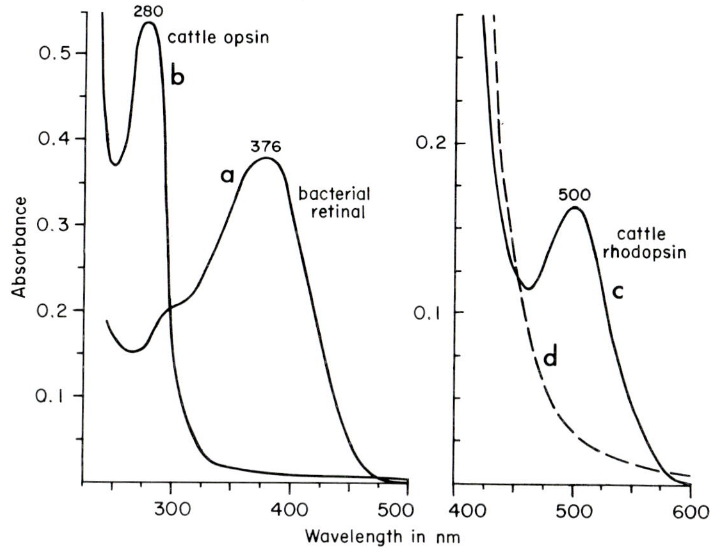


FIGURE 13.8 Formation of cattle rhodopsin from *Halobacterium halobium* retinal and cattle opsin. (a) Isolated retinal from *Halobacterium halobium,* absorption spectrum in chloroform, (b) isolated opsin fraction of cattle retinas in 1.8% digitonin, (c) absorption spectrum of cattle rhodopsin formed upon incubation in the dark of bacterial retinal with cattle opsin, and (d) light bleached spectrum of (c).

pared to the absorption spectrum of all-*trans*-retinal$_1$ (Eastman, 903720) in chloroform, which is at 380 nm. This could indicate the presence of the *cis* isomer rather than all-*trans*-retinal. To test whether in fact it was retinal$_1$, it was then reacted with the Carr–Price reagent. This reaction produced a characteristic blue color with an absorption peak at 664 nm, indicating that we had isolated the 11-*cis* or another *cis* isomer of retinal$_1$, now identified as the 13-*cis* isomer in the bacterial rhodopsin. From its absorbance, the concentration of retinal was calculated to be 0.64 $\mu$g/gm of wet weight bacteria.

We attempted to determine whether this bacterial synthesized retinal would form rhodopsin when combined with opsin. The visual protein, opsin, was prepared from cattle retinas, chosen since bovine opsin is one of the most stable opsins. The procedure followed was that of Hubbard *et al.* (1971). Opsin was finally extracted with saturated digitonin (1.8%) solution in M/15 phosphate buffer at pH 6.3. The cattle opsin absorption spectrum showed that it was a typical opsin with its major absorption peak at 280 nm (Figure 13.8b).

Since it is known that the 11-*cis* isomer of retinal will complex with opsin to form rhodopsin, we took the bacterial retinal in the 1.8% digitonin solution buffered to pH 6.3 M/15 phosphate buffer and incubated it with an excess of the cattle opsin for 3 hours in the dark at room temperature. The absorption spectrum showed that cattle rhodopsin was in fact formed (Figure 13.8c) and that the absorption peak was shifted to near 500 nm. The rhodopsin formed was photosensitive, and the visible peak at 500 nm bleached when irradiated (Figure 13.8d).

The fact that retinal can be isolated from a very halophilic bacterium like *H. halobium* and that it's retinal will complex with vertebrate opsin to form the visual pigment, cattle rhodopsin, suggests the universality of the retinal molecule.

Why retinal is synthesized by organisms that do not possess eyes, and whether it participates as a photoreceptor molecule in these organisms or has some other function awaits further study. S. H. Hutner (private communication, 1973) suggested that the function of vitamin A and retinal in the cell is associated with radiation resistance and that one of its original functions was for protection of DNA against radiation destruction and for DNA repair. We have already indicated that vitamin A is necessary for growth and life processes other than vision, so its role should be far more universal than was previously thought.

Returning to the bacteriorhodopsin in the purple membrane fragments, some similarities can be drawn to the photobleaching of cephalopod rhodopsin (Figure 12.1). Bacteriorhodopsin does not bleach in the light, but in the presence of sodium borohydride ($NaBH_4$), the 560 nm peak is bleached by light and a new peak at 330 nm appears in the spectrum. The 560 nm peak can also be bleached in the presence of hydroxylamine ($NH_2OH$) to form the retinaldehyde oxime.

In the photobleaching process of bacteriorhodopsin, Oesterhelt and Stoeckenius (1973) suggested that an intermediate of a few milliseconds lifetime occurs with an absorption peak at 415 nm. Scheme 13.1 illustrates and summarizes that photobleaching of bacteriorhodopsin.

$$\underset{(560\ \text{nm})}{\underset{C_{19}H_{27}CH{=}N\text{-protein}}{\text{Bacteriorhodopsin}}} \underset{\text{millisecond}}{\overset{h\nu}{\rightleftharpoons}} \underset{(415\ \text{nm})}{\text{bleached form}} \xrightarrow{NaBH_4} \underset{N\text{-retinylprotein}}{C_{19}H_{27}CH_2NH\text{-protein}}$$

$$\text{bleached form} \xrightarrow{NH_2OH} R{-}CH{=}NOH + NH_2\text{-protein}$$

Scheme 13.1

Therefore, continuous light irradiation does not bleach the pigment completely but a very fast recovery reaction to the original state takes place. However, all the experimental studies indicate that bacteriorhodopsin has similar properties to the visual pigment rhodopsin, being a retinal linked to the protein opsin through a Schiff-base formation with the amino group of lysine.

What function the rhodopsin purple membrane performs in these bacteria is still questionable. Danon and Stoeckenius (1974) have found that under anaerobic conditions in the dark, the ATP of the bacterial cells decreases significantly and light is required to restore the original ATP concentration, but in the absence of light, oxygen is required. Thus the halobacteria have adapted to their environment by incorporating rhodopsin in a cell membrane that converts light energy to chemical energy, like the chloroplast membrane, and provides an alternative to oxidative mechanisms for energy production (Mitchell, 1961, 1966).

Several important findings that stand out in these comparative studies are (1) all visual photoreceptors are associated with membranes that form a lattice structure very much like a crystal, (2) they contain a common pigment molecule with retinal as its chromophore attached to opsin, its protein, by an amino group of lysine through a Schiff-base linkage, and (3) retinal is a more universal molecule for photoprocesses than was previously thought, for it is found in primitive organisms that do not possess eyes.

# XIV
# Photoperiodic Phenomena and Memory

All living organisms are affected directly or indirectly by light. We have already discussed in some detail phototropism and phototaxis and have shown how these reactions to light in primitive organisms led to the development of photoreceptors and specific photopigments which were utilized in photosynthesis by plants and in the visual system of animals.

## Photoperiodism

A phenomenon which involves light and has evolutionary significance is the rhythmic behavior exhibited by living systems from microorganisms to man. It has been observed since ancient times that the leaves of plants fold at night and rise in the morning. Many organisms show such rhythmic behavior patterns set by the periods of time in the light and in the dark. This phenomenon is called photoperiodism. Although related in some way to photosynthesis in plants and vision in animals, photoperiodism is independent from them in that it is controlled by a different photoreceptor system.

Pioneering and extensive studies of photoperiodism were made by Bünning (1958, 1973). Bünning referred to the period in light as *photopic* and in the dark as *scotopic,* terms which were already used for the visual photoreceptors, the retinal cones which function at high light levels and the rods that function at low light levels, implying that there were equivalent photoreceptors active for photoperiodic phenomena.

It had long been supposed that the alternating periods of day and night impose the diurnal rhythm. But, if rhythm is related to environmental periodicity, it should disappear when the organism is placed in continuous darkness. Experiments show that the rhythm is found to persist in continuous darkness (or in very dim light) indicating that rhythmicity is not under direct environmental control. Rather it is endogenous, that is, it is controlled from within the organism itself.

Endogenous clocks are called *circadian* when the natural period is approximately 24 hours (usually between 22 and 26 hours). A change in the pattern of external stimuli results in a shift in phase of the endogenous rhythm. When the plant or animal is raised under constant conditions, a single stimulus, such as transition from continuous darkness to continuous light or a short period of light interrupting continuous darkness, is necessary to evoke the periodicity. A signal from the outside, usually the onset of dawn or dusk, keeps the clock in step with the natural day–night cycle. An endogenous rhythm within cells suggests some clocklike mechanism and is commonly referred to as the *biological clock.* "Interval timers" less than 24 hours and longer timers, "seasonal clocks," also exist in organisms.

In a wide diversity of organisms behavioral and physiological activities exhibit a 24-hour periodism even in the absence of external cues. Although these rhythms are circadian when isolated from light or temperature cycles, they may synchronize to the rhythm of the earth's rotation. However, in many organisms circadian rhythms decay in continuous light. The question here is whether biological clocks are *photoperiodic,* or are controlled by some other mechanism (Winfree, 1975).

Let me now cite two examples of diurnal periodicity in photosynthetic flagellates. Pittendrigh (1954) and Bruce and Pittendrigh (1956, 1957) studied the phototactic response of *Euglena.* In their experimental procedure a narrow beam of light was passed through a suspension of *Euglena,* acting as a "light trap" to attract the *Euglena.* The light beam also served as a sensing system to measure the number of organisms that had been attracted to the light and their response time. The test beam was turned on for 30 minutes at 2-hour intervals. This produced an endogenous diurnal rhythm in the phototactic response that was always very close to 24 hours. In the unicellular photosynthetic dinoflagellate *Gonyaulax polyhedra,* Hastings and Sweeney (1957) demonstrated a diurnal rhythm of luminescence which persists under conditions of constant low light and constant temperature.

In the way of explanation, it was postulated that two systems exist which are coupled to one another. One system is light-sensitive and independent of temperature, the other is temperature-sensitive but independent of light. If organisms are independent of their external environment for timing information, they are thought to possess autonomous biochemical clocks capable of measuring out absolute time. But there is evidence to indicate that there is also an external timing system that is set by the period of light and by geophysical forces (Brown *et al.*, 1970).

We can now return to photoperiodism in higher plants, for here information is available to relate photoperiodic behavior with a particular molecule.

## Photoperiodism and the Flowering of Plants

In plants, photoperiodism, the response to variations in wavelength of light and in the length of day and night, controls growth and flowering. For example, continuous red light at 660 nm was found to be effective in altering the plants' response, that is, in preventing flower formation. But it was also found that if a flash of red light was followed immediately by a short interval of far-red light with a maximum of 730 nm, the effect of red light was cancelled out. Then the plants acted as if their night time had never been interrupted, that is, they did flower. There appear, then, to be two light reactions, one with a maximum in the red and one with a maximum in the far-red.

### Phytochrome

The pigment responsible for this red and far-red effect is the chromoprotein *phytochrome* (Siegelman and Firer, 1964; Hendricks, 1968). Phytochrome has two distinct states, one with a maximum absorption in the red near 660 nm and another with a maximum absorption in the far-red near 730 nm (Figure 14.1).

$$P_r \underset{}{\overset{660\text{ nm}}{\rightleftharpoons}} P_{fr} \xrightarrow{\text{darkness}} P_r$$

The structure of phytochrome as worked out by Rüdiger (1969) is illustrated in Figure 14.2. The chromophore of phytochrome is a linear tetrapyrrole (see porphyrin structure, Figure 3.1), and is structurally related to the phycocyanin pigment of the blue–green algae (Figure 3.10). The absorption spectra of phytochrome and phycocyanin also appear to be similar. Phytochrome with its system of conjugated double bonds resembles a carotenoid molecule (Figure 3.6). The molecular weight of phytochrome isolated from various plants has been estimated to be from 50,000 to 150,000. The average molecular weight of phytochrome would be about 60,000, computed on the assumption that there is one chromophore molecule associated with one protein molecule.

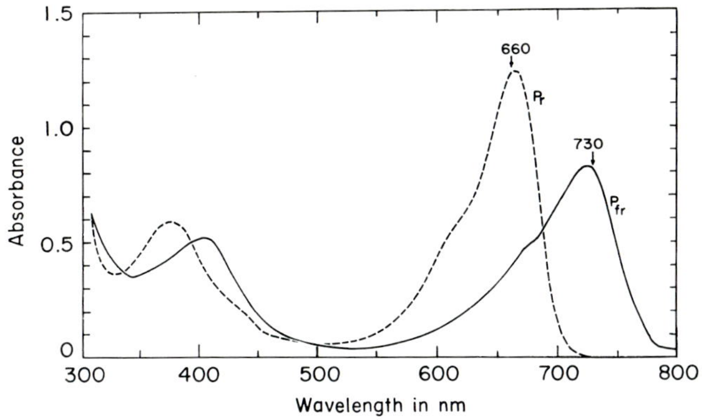


FIGURE 14.1 Phytochrome, absorption spectrum of the red $P_r$ form (------) and the far-red $P_{fr}$ form (——), isolated from oats. From Hartmann (1966, p. 352).

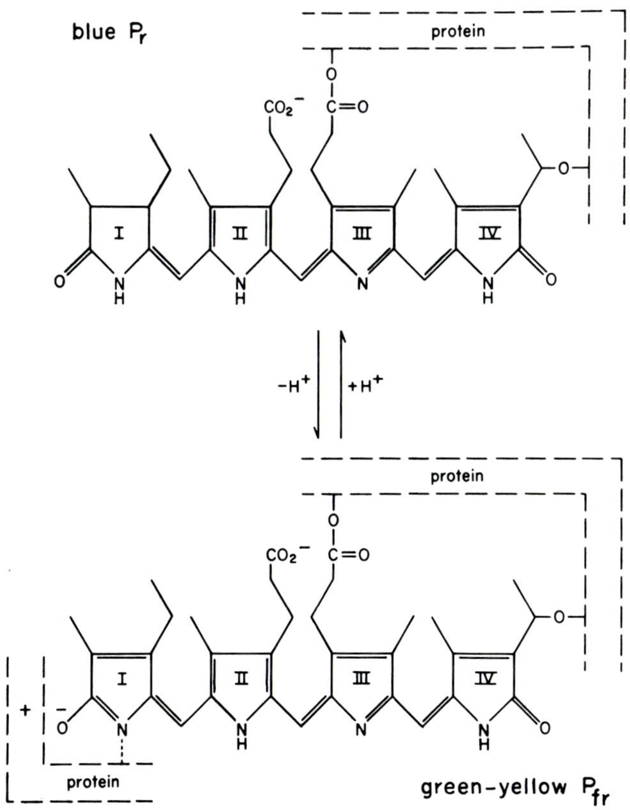


FIGURE 14.2 Phytochrome. Molecular structure, proposed by Rüdiger (1969).

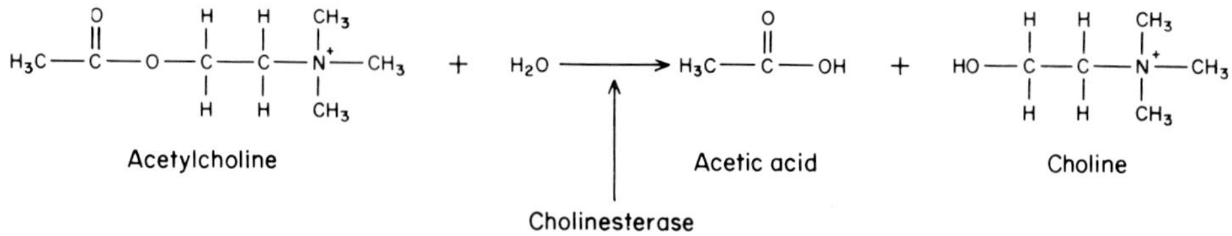


FIGURE 14.3 Acetylcholine and degradation by acetylcholine esterase to acetic acid and choline.

In the photochemistry of phytochrome, Butler *et al.* (1964) suggested that there are similarities to the behavior of the visual pigment rhodopsin. Hendricks and Siegleman (1967) proposed two possible mechanisms for the phototransformation of phytochrome based on the *cis* to *trans* isomerization similar to the mechanism of retinal in rhodopsin (Figure 11.6). The phytochrome transformation from the red absorbing to the far-red absorbing form involves a change in protein conformation, that is, a change in molecular shape which may alter the receptor membrane properties.

Jaffe (1970) found that acetylcholine and hence acetylcholine esterase play a role in certain phytochrome mediated phenomena in plants, and that acetylcholine is related to the red light effect. This finding that plants contain acetylcholine is of considerable importance from an evolutionary point of view, for acetylcholine is a neurotransmitter molecule associated with muscle and nerve excitation in animals (Figure 14.3). This brings us back to the point we made earlier that plants possess a photoneuromuscular system, as we have described for *Phycomyces* and *Euglena.*

There is no experimental evidence so far that phytochrome is present in animals, although researchers have looked for it. However, in some animals the photoperiodic effect may be due to a similarly structured molecule. For a more comprehensive review of phytochrome, its structure and photochemical properties in plant photoperiodism, reference should be made to the reviews by Briggs and Rice (1972) and Shropshire (1972).

## Marine Organisms and Tidal Rhythm

Great numbers of lower organisms living in the seas display rhythmic behavior stimulated by the timing of the tides, including semilunar breeding cyles. When these organisms are removed to the laboratory they continue to rise and fall in synchrony with the timing of the tides. One of the earliest examples of this behavior to be studied was the planarian *Convoluta roscoffensis,* which is peculiar to the coast of Brittany, France (Keeble, 1910). *Convoluta* is green for it lives symbiotically with a *Chlamydomonas*-like alga whose flagella project through the skin of *Convoluta* and propel the organism through the water. *Convoluta* emerge onto the surface of the sand at low tide and burrow into the sand at high tide. The reason that *Convoluta* come to the surface at low tide is to allow the symbiotic alga to photosynthesize, for there is a direct relationship between photosynthesis—a necessary energy source—and the metabolism of *Convoluta.*

It was then found that numerous marine organisms exhibit rhythmic behavior coinciding with the cycles of high and low tides (Fingerman, 1960; Palmer, 1974). The behavioral cycles observed in marine animals set by the tidal rhythms is associated with color and pattern changes in the skins of these animals. It is known that crustacea and molluscs as well as cold-blooded vertebrates possess *chromatophores* which contain yellow, brown, and red pigments which are melanins. The black pigmented chromatophores are called melanophores. Collectively the chromatophores act to produce color and shade changes in the skin of the animal by means of their concentration or dispersal. For many of these animals the eye is the receptor organ which through hormonal action initiates the expansion and contraction of the chromatophores. A variation of this pigment effector system occurs in certain lower vertebrates, e.g., amphibians and lizards in which the pineal organ in the brain participates in the control of adaptive pigmentation.

Before we turn to the pineal and the brain, it may be remembered that we have previously indicated that the general body surface of many eyeless and blinded animals is remarkably light-sensitive. This diffuse photosensitivity over the whole or part of the animal's body was defined as the *dermal light sense.* But even deeper tissue cells can be photosensitive, as in certain marine animals which exhibit responses if their nerve or ganglion cells are exposed to light. For example, Chalazonitis (1964) showed photosensitivity in the neurons of *Aplysia* which are deeply situated in the animal. The photoreceptor molecule believed to be responsible for the photoexcitation has absorption maxima at 418, 542, and 579 nm, but another pigment was found with absorption maxima at 463 and 490 nm. From these spectra it was suggested that the former is a heme protein (Figure 3.3) while the latter is a carotenoid (Figure 3.15); but the photoreceptor pigment has not been specifically identified. Why deeper tissue cells, such as nerve cells, function as photoreceptors in these animals is little understood at present.

## The Pineal

An organ in the brain which functions as a sensitive "biological clock" is the pineal. It has a long history for it was believed to be the vestigial organ referred to as the "third eye" of our reptilian ancestry. René Descartes (1637) thought that the pineal was the seat of the soul. He visualized the events of the world to be perceived through the eye and transmitted through a series of fibers to the pineal organ in the brain.

In living vertebrates as well as extinct vertebrates the pineal appears to have evolved active photoreceptor structures, and the level of function in the present-day mammalian pineal continues to be regulated by light. A striking finding is the similarity between the structure of the photoreceptor cells of the pineal in lizards and amphibians and that of the retinal cones of the vertebrate eye (Eakin, 1965, 1973; Kelly, 1965, 1971). The action spectrum for the frog pineal indicates a maximum near 560 nm which suggests that the photoreceptor pigment molecule is comparable to the cone visual pigment, iodopsin (Figure 11.1). Light stimulation of these photoreceptor pineal cells causes the pineal to generate action potentials. A large body of information has now become available about the physiological factors which control pineal activity (Wurtman *et al.*, 1968).

Let me now cite some properties of the pineal which bear on photoperiodicity and behavior. It was observed that if crushed pineal glands were introduced into water in which tadpoles were swimming, the tadpole skin color bleached. Lerner *et al.* (1958) were the first to see a relationship between the extract of the pineal glands and skin coloration. He isolated a substance from bovine pineal glands and called it melatonin because it caused the contraction of the melanin granules. The action of melatonin on the skin indicates that the pineal organ of fish and amphibians causes changes in skin pigmentation in response to light. Thus, it was proven that melatonin was a hormone, that is, an active substance produced by the pineal, and not simply a vestigial organ. The pineal serves as a neuroendocrine-chemical transducer, which is activated by light (Axelrod, 1974). Melatonin, the molecule responsible for this activation, was found to be the indole, 5-methoxy-*N*-acetyltryptamine (Figure 14.4).

The precursor chemical necessary for melatonin synthesis was then found to be serotonin. Serotonin is relatively widespread for it is found in the cephalopod molluscs, the amphibians, and in the pineal of all vertebrate species; it is also found in numerous plants such as bananas, figs, and plums.

The synthesis of melatonin begins with the amino acid, 5-hydroxytryptophan. Enzymatic action removes the carboxyl group (—COOH). The product of this reaction is serotonin. Another enzymatic reaction acetylates the molecule to form *N*-acetylserotonin which is then methylated to yield melatonin (Figure 14.4). The methylating enzyme, hydroxyindole-*O*-methyltransferase is found only in the pineal of mammals. The activation of this pathway is controlled through a regular oscillating circadian rhythm.

In some birds the photoreceptive regions for circadian and photoperiodic control are located mainly in parts of the brain near the hypothalamus. Re-

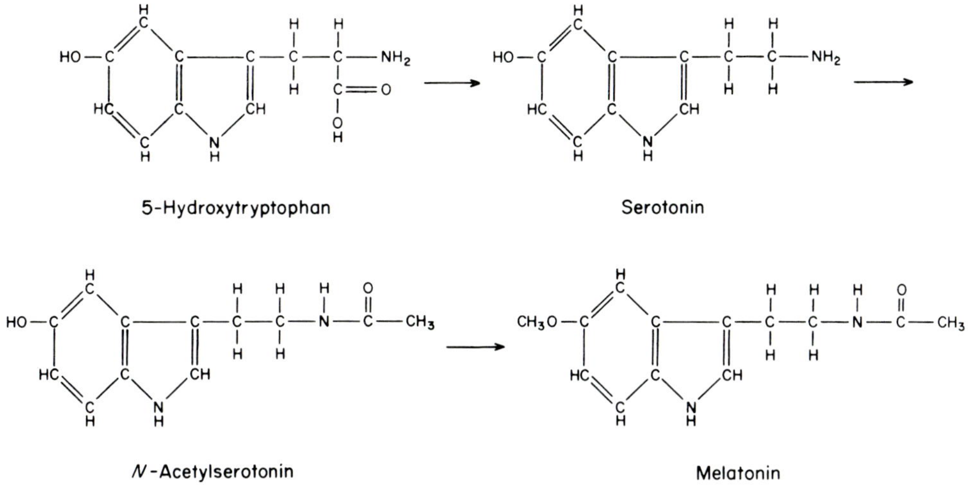


FIGURE 14.4 Synthesis of melatonin (according to Axelrod, 1974).

moving the eyes or cutting the optic nerves does not reduce the responsiveness to light, but painting the top of the skull black with india ink does interfere. In the sparrow the pineal body, although involved in a circadian rhythm of activity, does not act as a photoreceptor. Removal of the pineal causes loss of the endogenous rhythm under constant illumination, but light–dark related cycles continue.

The ability of melatonin to modify gonadal function suggests that its secretion may also have something to do with the timing of the estrous and menstrual cycles. Probably in most vertebrates photoperiodic control involves a sequence of neural and hormonal events.

We can mention another interesting finding and that is that serotonin production and various mental states may be related. The structures of serotonin and hallucinogenic (or psychotomimetic) agents are similar. For example mescaline, extracted from the peyote cactus, is structurally similar to noradrenaline (Figures 14.5 and 14.6). Serotonin is known to have the same type of physiological effect as adrenaline. It is interesting to find that certain structural features of serotonin bear a relationship to lysergic acid diethylamide (LSD). We can speculate then that these drugs exert their psychical effects by competing with the enzymes and receptor membranes of the pineal which are involved in the metabolism and activity of serotonin.

It is interesting to note that humans possess the same neurotransmitters (acetylcholine and serotonin) that are involved in periodic phenomena.

The pituitary gland in vertebrates regulates the output of other glands, and through them it exercises control over a wide variety of metabolic and

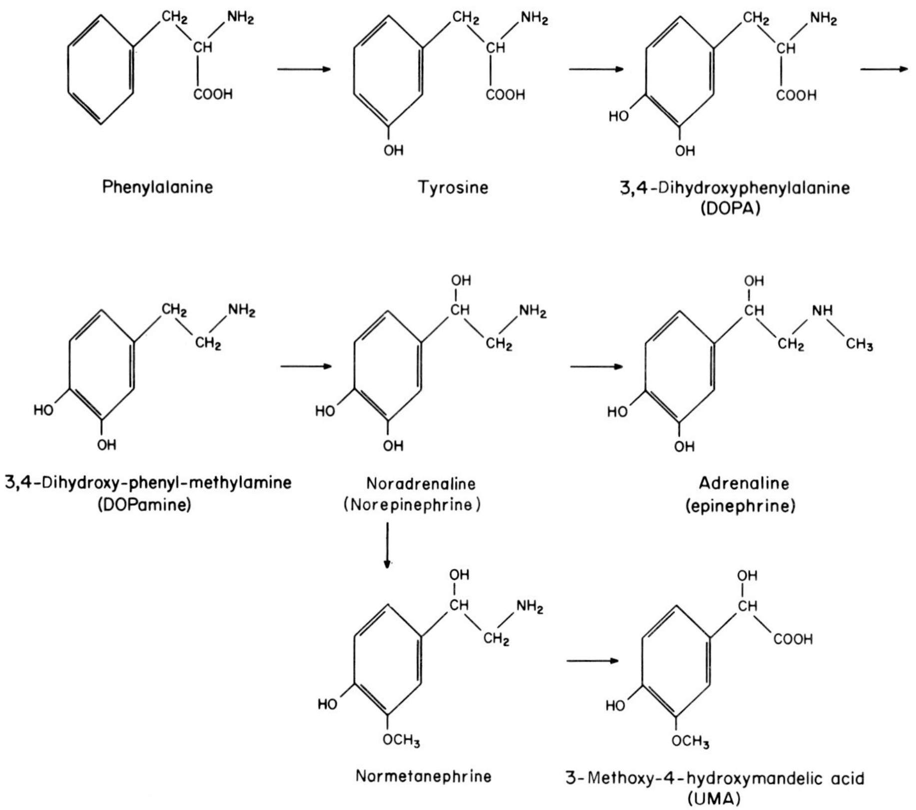


FIGURE 14.5 Synthesis of noradrenaline and related compounds from phenylalanine and tyrosine.

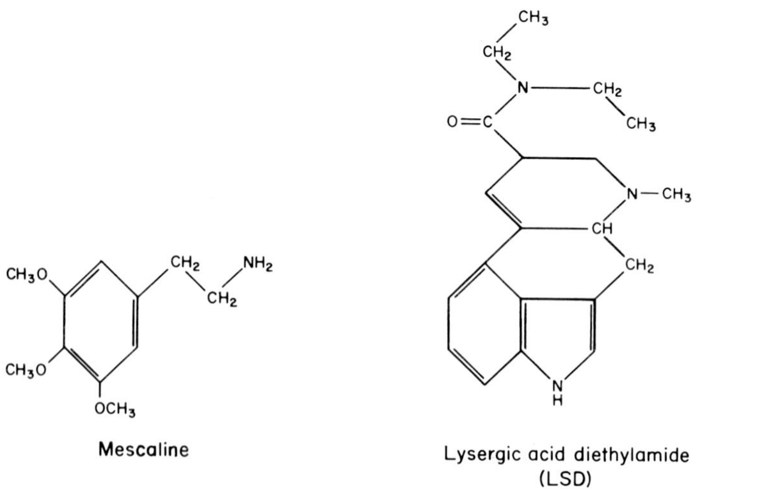


FIGURE 14.6 Structural relationships of mescaline and LSD to DOPA and noradrenaline to the catecholamines. (Refer to Figures 14.4 and 14.5.)

sexual processes. It also plays a major part in daily and seasonal cycles. In one species of lizard the skin is green at night and brown by day. This circadian alternation continues under constant illumination unless the pituitary has been removed.

All of these complex photoneuroprocesses, clocking mechanisms as diverse as the control of flowering of plants, the feeding and diapause of insects, the pigmentation of animals, the gonad development and reproductive cycles in animals, and even the migration of birds must have been—and are continually being—explored in evolution.

## Memory

> Memory is an organ to apprehend the Impression that is made in Time [By analogy] That the record in memory is like—Impressions of light impinged on matter—retain these impressions—though for no long time.
>
> Robert Hooke, 1705, "Posthumous Works"

Photoperiodism and biological clockwork imply some kind of memory system which functions at the molecular level in plants and animals. The question which we can now ask is where and how is information stored and how is it recalled in living systems?

Most living organisms receive information from their environment, and the handling of this information became, in animals, the essential function of the nervous system. The capacity to acquire impressions from the environment is particularly well marked within the nervous system of vertebrates.

Three types of information are important to living systems. They are (1) the genetic information which does not get feedback from the organism, but is passed on from generation to generation, (2) sensory information, which has considerable feedback into the storage system, but is not passed on from generation to generation, and (3) communicated information, which does have feedback and is passed to the next generation. It is the sensory and communicated information, or memory, which concerns us here.

Learning and memory are intimately related and are two of the most complex aspects of the brain. If we define learning as the acquisition of a response to a stimulus and memory as the storage of this response we find that learning (a form of memory) is not limited to animals with highly developed nervous systems. The ability to learn in terms of acquiring a response to a stimulus goes back to primitive cells. For example, *Phycomyces* and the protozoans (amoeba, paramecia, *Euglena,* etc.), demonstrate "simple association" in response to stimuli and remember this response for a limited time, a form of "short-term memory." Such limited learning–memory processes no doubt serve a survival function for these organisms. On the other hand, complex learning processes associated with long-term memory are obtained within the more highly developed nervous systems of animals. However, these complex systems must have evolved from far simpler systems and as we shall find, the study of "memory" in such primitive organisms may give us some insights into the memory processes in more complex nervous systems.

To discover the physical–chemical basis of memory two avenues of research are being explored. One is an electrophysiological basis for memory, or the search for the electrical engram. Since it is beyond the scope of this book to discuss in detail these electrophysiological researches the interested reader will find them in Pribam (1971) and Schmitt and Worden (1974). The other is a chemical basis for memory.

The research on the chemical basis for memory is more relevant to our interests. Let us briefly review some recent research to discover whether there is a chemical basis for information processing and memory. One of the earliest experiments in chemical transfer of memory was performed by McConnell *et al.* (1959) using trained planarians, organisms which have the ability to regenerate. McConnell claimed that if a trained planarian was cut in half transversely and both halves were allowed to regenerate into complete organisms the regenerated heads and tails both showed siginificant retention of the conditioned response learned. He concluded that the "memory molecule" could migrate and was not specifically located in the head. Additional work indicated that retention of some part of the original learning could be found in regenerated planarians with no intact structure from the trained parent. When trained planarians were fed to untrained cannibalistic planarians, some of the information acquired by the trained planarians seemed to "transfer" to the cannibals by ingestion (McConnell, 1962, 1966; McConnell and Shelby, 1970). These experiments suggested to McConnell that some of the memory had been coded chemically in the RNA molecule. The possible relation between memory and RNA led to the hope that learning could be transferred through "memory molecules."

The interest in RNA synthesis accompanying memory imprinting was generally related to the belief that the ultimate step in the memory process is the synthesis of specific proteins. Proteins or their aggregates were regarded as the long-lived storage records which retain imprints of experience throughout life. Hydén (1967) had also implicated RNA in the relation of learning to memory. Later Hydén and Lange (1968) determined that a specific protein, S-100, with a molecular weight of 21,000 increases with learning.

George Ungar (1970), a proponent of the chemical basis for memory, was concerned with the role of amino acid sequence in small polypeptides and thought that some intact peptide was necessary in the processing of acquired information. He trained rats to fear the dark, and from these rat brains he obtained an extract, which when injected into untrained rats caused them also to fear the dark. Ungar also claimed to have isolated "memory molecules" from goldfish trained to avoid certain colors. The fraction isolated was a polypeptide containing 8 to 12 amino acid residues which he termed "scotophobin." The amino acid residues identified in the polypeptide were aspartic acid, glutamic acid, glycine, lysine, serine, and tyrosine. Ungar's experiments ruled out RNA as the memory molecule. However, whether the possibility that the active peptide is combined with RNA in some complex was not eliminated.

From these experimental results, the evidence indicates that the first step of the recording process is not a chemical synthesis of either RNA or a protein. An alternative explanation is that it produces a physical change in existing molecules within their receptor membranes. Only at a later stage, when consolidation and fixation occur, do chemical processes become involved. This physical change most likely comprises conformational changes of the macromolecules, induced by either direct contact with the stimuli (e.g., retinal of rhodopsin, in visual excitation) or by concentration gradients of ions or small molecules. Such changes could alter the permeability of membranes and release neurotransmitter molecules (e.g., acetylcholine and serotonin).

In order to further explore this idea let us turn again to the photoreceptors which have the capacity to exhibit a form of short-term memory.

## Photochemical Memory

A memory is characterized by the following features. Information is introduced into the "memory molecule" by a certain signal, for instance by a flash of light at a particular wavelength, and this information is stored, or "memorized," until a signal of another wavelength of light erases the information and restores the "memory" to its original state. This photochemical cycle must be reversible and reproducible. A simple picture of the process is a reversible system that is driven in one direction by light and reversed in the dark:

$$A \underset{}{\overset{\text{light}}{\rightleftarrows}} B$$

Photoreceptor molecules are light sensitive and therefore can be considered model systems for photochemical memory, for they receive and process light information to the cell. They are chromophore–proteins, enzymes that can be photosensitized by the light-absorbing chromophore. Photoreceptor molecules thus have the potential for participating in photoregulated events. An excellent example of a photoreceptor memory molecule by our definition is phytochrome (Figure 14.2). Another example is protochlorophyll, the precursor molecule for chlorophyll, which is practically colorless, but in the light turns green and becomes the chlorophyll holochrome. The cytochromes can also function as memory molecules. Phytochrome, protochlorophyll, and cytochrome are all porphyrin structures. These molecules when irradiated by light, change color and their absorption peaks shift and the process is reversible in the dark. This is true of all the photoreceptor pigments.

Thus chromophores are information-receiving molecules that are complexed to proteins or polypeptides with a specific amino acid in the protein sequence. For example, the visual pigment rhodopsin in the retinal rods is a retinal–opsin protein (Figure 11.14). Retinal is attached to an amino group, lysine, of opsin through a Schiff-base linkage. Retinal is yellow, but when complexed to its opsin turns pink or red. Upon light excitation, rhodopsin bleaches and the retinal isomerizes to the all-*trans,* and in the dark retinal transforms to the *cis* isomer (Figure 11.6). The *cis–trans* isomerization brings about conformational changes in the rhodopsin molecule with shifts in the absorption peaks (Figure 11.2).

Previously we described a "memory" system in *Phycomyces.* If the sporangiophore receives a flash of light at a critical time in its development and is then grown in the dark, it "remembers" where the flash of light came from and will continue to search for it. In addition, when the *Phycomyces* sporangiophore is cooled ($<5°C$) the sporangiophore stops growing, but a light signal can be perceived and stored so that when *Phycomyces* is returned to room temperature (22°C) the light-growth response takes place. *Euglena* also demonstrates a phototactic memory system (Diehn and Tollin, 1966a,b). The photoreceptor molecule in both of these primitive organisms appears to be a flavoprotein which becomes photoreduced to a flavin semiquinone in the light and is regenerated in the dark. Their spectral peaks, in light and darkness, mimic to some extent that of the rhodopsin molecules (Figures 3.14 and 11.8).

There are other naturally occuring pigments, such as the chromones (benzo-γ-pyrone), pyrans and flavones (2-phenylchromone) which are particularly widespread in many plant leaves and flowers. Many such molecules are found in animal tissue. One such molecule is a quinhydrone, found in the hypobranchial gland of molluscs, which is colorless but when irradiated with light turns purple (Baker, 1974). Photochemical changes in these compounds can be brought about by ultraviolet and visible light and therefore may also be implicated in photochemical memory storage.

## Photochemical Memory and Evolution

If man has evolved the most highly developed of the sensory nervous systems, and these refinements are due to evolution, then its history must go back to relatively simple photosensory systems, as found in *Euglena* and *Phycomyces*.

The information that an organism receives from the environment is responsible for initiating its action. The amounts of information involved may be very large. Thus a single bacterial cell has been estimated, from its RNA, to contain $10^{12}$ bits of bound information (Lehninger, 1965). A typical growth of one cell division every 20 minutes would require a developmental rate of $10^9$ bits/second.

The sum of the sensory inflow from all sensory organs to the nervous system in man has been estimated at $10^7$ bits/second; in a lifetime of $10^9$ second, this would amount to $10^{16}$ bits. The maximum likely storage capacity of the human brain, which has approximately $10^{10}$ nerve cells in the cerebral cortex has been estimated at around $10^{10}$ to $10^{11}$ bits (Young, 1971). Photosensitive pigments spread as monolayers on the lamellae of the photoreceptors (Figures 5.7 and 11.14) are particularly well suited for processing considerable information. The assumption is that each photoreceptor molecule can receive a single bit, one photon, and there are from $10^6$ to $10^9$ photoreceptor molecules on the photoreceptor lamellae. According to the calculation of Kühn (1968) it should be possible to store a single bit on an area of 100 Å$^2$.

Therefore we can conceive of a very simple model for the photochemical system (Figure 14.7). In this model information is introduced into the photoreceptor molecule by a flash of light at a particular wavelength. This information is "memorized" and stored in the dark, but can be recalled by the same wavelength of light. In addition, a signal of light of another wavelength of light can erase the information from the first signal and restore the molecule to its original state. In a sense it would be like a computer system which functions through photochemical changes rather than electronic.

Let us now consider such a molecule in its nonsensory state A. It receives a signal, undergoes a conformational change, and enters the memory state B. This photochemical cycle must be reversible, for after a finite time another sig-

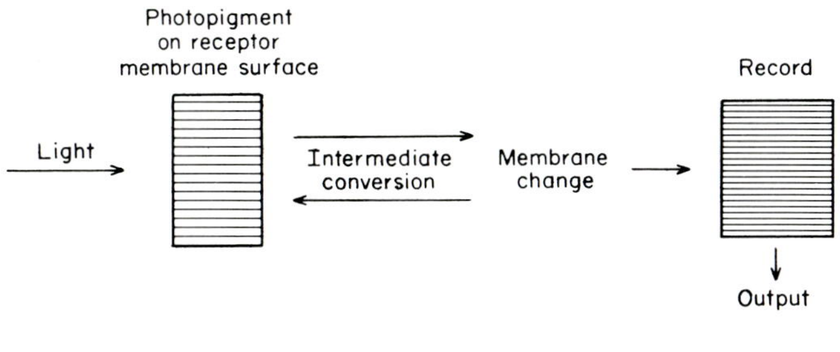


FIGURE 14.7 Schematic for conformational change of photopigment in the photoreceptor membranes upon photoexcitation.

nal returns it to state A. But what happened to the "memory" in state B? This will depend on where the molecule is. The molecules reside on the surfaces of photoreceptor membranes and upon light excitation there is a conformational change, which may be very fast, of the order of nanoseconds, or slow, and have a lifetime of the order of hours. This conformational change in the molecule induces changes in the membrane which lead to more permanent changes in the cell. Thus, the photoreceptor structure can function as a short-term memory system.

## Photoprocesses and Lifetimes

We have discussed how light provides a signal to the cell, which clocks its time. But does it also signal the end of the organism's time? Organisms from bacteria to man have fixed lifetimes; for example, the lifetime of a bacterium is the order of minutes, protozoa and animal cells hours, the mayfly one day, the frog six years, the dog 15 years, and man near 100 years. Although we may ponder the question—for there may be no answer—we can try to indicate experimental directions worth considering and to speculate whether light is related to lifetime clocks and aging.

If we put aside the genetic basis for lifetimes associated with the transcription and translocation processes through messenger and transfer RNA molecules, what other factors can be considered?

One suggestion is that the hormonal processes associated with the brain and nervous tissue break down. In animals this points to the brain which regulates the release of neurotransmitter molecules, for example, acetylcholine and serotonin (Figures 14.3 and 14.4) which we have discussed in the control of photoperiodic phenomena.

Another direction was indicated by Strehler (1962) which points to the fact that there is an accumulation of yellow pigments with aging and senescence. These yellow pigments are found to be associated with fluorescent granules, referred to as cellular microbodies. The microbodies appear to increase in nerve, brain, muscle, and sex glands with age. In many cases these cell inclusions appear to be crystalline. Their real function is unknown. Some of these microbodies indicated the presence of alkaline phosphatase and esterase activity and were identified with lysosomes, as characterized by De Duve *et al.* (1955).

But attempts to isolate and extract the pigment from these microbodies have not been successful. However, from extractions made from the nerve cells an absorption spectrum was obtained. The absorption spectrum has a maximum at 370 nm and its fluorescence emission spectrum was maximal in the 440 to 460 nm and 530 to 560 nm range. This could be a pteridine derivative, but it more closely resembles a flavin complex. At that time it was loosely termed an "aging" pigment. Also a similar or a different pigment found in the secretory granules was referred to as a "senility" pigment (Gatenby and Leslie-Ellis, 1951). Further extraction indicated it was a complex which could be partially solubilized in chloroform–methanol and the soluble fraction contained cephalin and unsaturated lipids similar to the lipids found in membranes and nervous tissue. The insoluble residue was an unidentified brown pigment, which is sometimes referred to as the yellow aging pigment lipofuscin. It was hypothesized that lipofuscin accumulates with age and is associated with the progressive decreased rate of oxygen diffusion into tissue cells. More recent analysis (Karnaukhov, 1973) shows that lipofuscin contains carotenoids, myoglobin, and flavoproteins.

This brings us back to the flavin complexes. We have already encountered yellow pigments and this suggests to us that the flavins could be involved, for the flavins are yellow, fluoresce and are photosensitive. We have already shown how a flavoprotein could function in photoprocesses for phototropism and phototaxis. A further finding was that the enzyme D-amino acid oxidase (FAD), whose prosthetic group is a flavin, changes with age and is found in high concentration in nervous tissue. In addition the enzyme is found in the liver and kidney of almost all vertebrates, including birds, reptiles, and fish. It is also found in the invertebrates, the cephalopod molluscs, and the arthropods, as well as in plants, fungi and various strains of bacteria. So it is a generally universal enzyme.

What is curious though, is that a yellow pigment, neurotransmitter molecules, crystalline microbodies, and membranes seem to be somehow involved and are found to accumulate in aged animals. Light also plays some role, but since we know so little of the processes they remain for the present in the realm of speculation. However, these findings do suggest directions for future investigations.

# XV
## Bioluminescence

> The ability to respond to stimuli is one of the characteristics of living things which appeared early in the course of evolution, almost as soon as the aggregation of large molecules to form a cell. The evolution of *stimulus reactions* has not been sufficiently discussed, so intent has the evolutionist been on morphological characteristics, but it is certainly true that response to stimuli by luminescence is found as far down in the evolutionary scale as response to stimuli by movement of one kind or another.
>
> E. Newton Harvey, "Bioluminescence," 1960

The photobiological phenomena that we have discussed so far are those that are dependent on light absorption in the visible part of the electromagnetic spectrum of energy. But it has been known since ancient times that organisms emit visible light. Aristotle is supposed to have noted that rotting meat phosphoresces. However, Robert Boyle (1672) is credited as being the first to demonstrate that rotting wood and decaying fish and meat require air to phosphoresce.

The emission of light as a result of chemical reaction is chemiluminescence and the emission of light by living organisms is bioluminescence. Therefore bioluminescence is a chemiluminescence. Molecules which fluoresce by absorbing the energy of radiation are most likely to chemiluminesce by absorbing the energy of a chemical reaction. That is, the energy of a chemical reaction goes to a molecule whose electron is thereby raised to a higher energy level. A quantum of light is emitted on return of the electron to its original ground level (Figure 2.6). By studying the lifetime and the spectrum of emitted light, one can learn something about the nature and fate of molecules in their excited states. Such analysis of emitted light is an important tool, as are absorption spectra, for photochemists and photobiologists (refer to Seliger and McElroy, 1965; Porter, 1969; Cundall and Gilbert, 1970).

Bioluminescence is distributed widely but very erratically throughout all the phyla. Among the plants, certain bacteria, fungi, and photosynthetic marine flagellates possess it. Bioluminescent animals exist in almost half of the 27 phyla: protozoa, hydroids and radiolarians, sponges, marine worms, jellyfish, insects, crustaceans, and molluscs. Among the vertebrates, some fish are bioluminescent but no amphibians, reptiles, birds, or mammals.

Bioluminescence is more prevalent in marine than in terrestrial animals. In the oceans, especially at great depths, there is large-scale bioluminescence. For reasons that still are not known, most marine bioluminescence is blue, but the color of the emitted light is a function of the system of the organism. The relationship among depth in the sea, bioluminescence, and the bioluminescent organisms has been extensively studied by Clarke (Clarke and Backus, 1956; Clarke and Denton, 1962), Hardy (1956), Denton *et al.* (1970, 1972), and Dilly (1973). Among freshwater organisms bioluminescence is almost unknown, with the exception of a limpet, *Latia neritoides,* found only in New Zealand.

Light is produced by animals with and without eyes, by sessile as well as motile animals. Luminescent bacteria are sometimes located in or near the eye of marine animals. Certain marine organisms, including fish, have luminescent organs whose light is sufficient to excite the eye of these animals. An important difference between animal bioluminescence and that of bacteria and fungi is that in these organisms the bioluminescence is continuous and steady, but is affected by temperature changes. Animal bioluminescence occurs only upon stimulation and the excitation appears to be associated with a nerve impulse. Much of our knowledge of bioluminescent organisms and their distribution throughout the plant and animal phyla is due to the extensive investigations of E. Newton Harvey (1952, 1960).

Our purpose here is not so much to elucidate the biochemical mechanisms of bioluminescence, but to see if we can find evolutionary clues related to receptor systems. For example, are there similarities in the chemistry between light reception and light emission, and is the phenomenon an early evolutionary relic left over from the time when there was a sudden appearance of oxygen in the atmosphere?

Dubois (1885) was one of the first to try to arrive at some chemical basis for bioluminescence. He extracted from the click beetle, *Pyrophorus,* a closely related species to the firefly, two substances, one an oxidizable substrate molecule which he later named *luciferin* and another molecule, an enzyme, named *luciferase.* On mixing the two together in the presence of oxygen, light was generated and he noted that no perceptible heat accompanied the light emission. This demonstration of bioluminescence is now a classic reaction:

$$\text{Luciferin } (LH_2) + \text{luciferase } (E) \xrightarrow{O_2} P^* \underset{\substack{\text{electronically} \\ \text{excited product}}}{\xrightarrow{\hspace{3cm}}} P + \text{light}$$

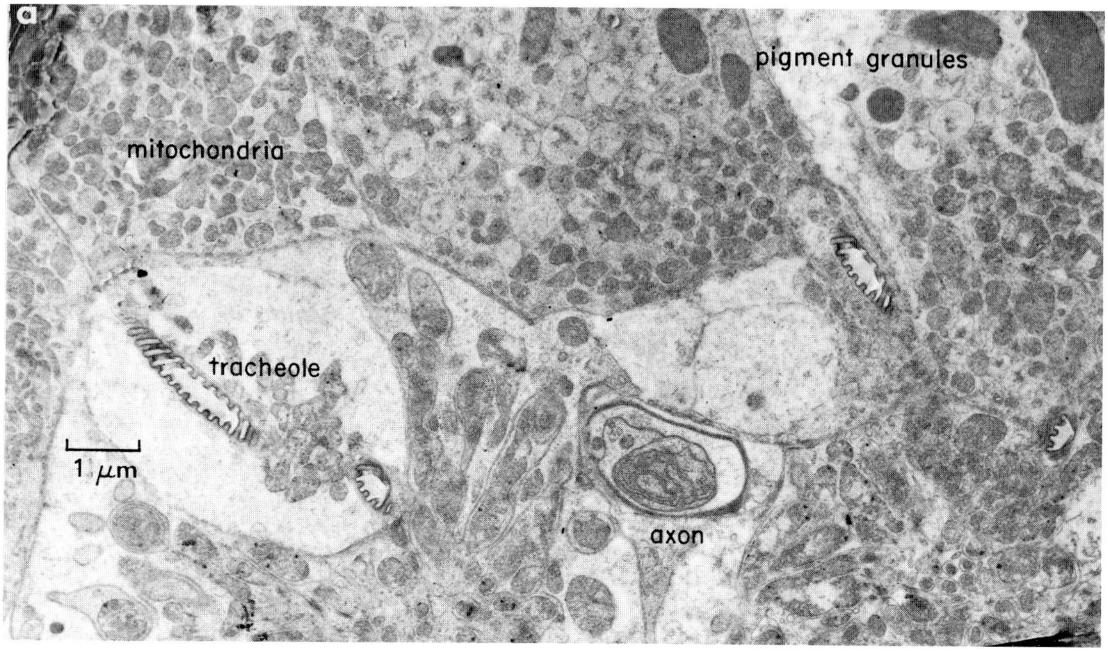


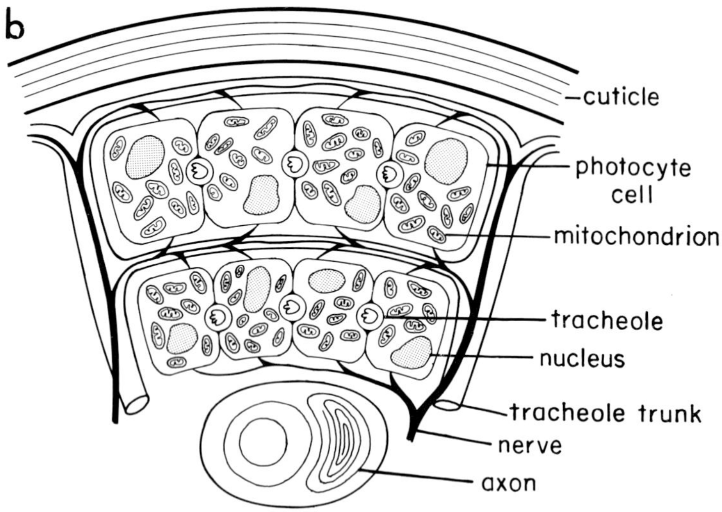


FIGURE 15.1 (a) Firefly lantern, *Photuris pennsylvanica,* photophore, electron micrograph of photocyte section of cells; (b) schematic of the photocyte cell.

Since these early observations, numerous investigations have been pursued to elucidate the phenomena. Much of our present knowledge of the biochemistry of bioluminescent organisms comes mainly from the more recent studies of Seliger and McElroy (1965), Johnson (1967), Hastings (1968), McElroy *et al.* (1969), Cormier (1973), and McCapra (1973).

We will mention here only a few bioluminescent organisms that have been studied in some detail and indicate briefly the findings of the chemistry of the process.

The best known example of bioluminescence is the firefly. The two most common species studied are *Photuris pennsylvanica,* whose lantern emits a greenish light, and *Photuris pyralis,* whose light is yellow (Figure 15.1). The most dramatic displays of fireflies to be found in the world occur along certain riverbanks in the Orient. In most cases, the female of the species is the luminous one. Such displays, some believe, represent communal efforts to attract prey or divert predators rather than serving as individual mating signals.

In different species of firefly, bioluminescent emission varies from about 550 to 595 nm. These variations have been associated with differences in the luciferase rather than the luciferin. Firefly luciferase has been isolated and crystallized and the structure of luciferin is known (Figure 15.2). The luminescent reaction requires the enzyme luciferase to catalyze the reaction between luciferin and ATP, forming a luciferin–AMP compound and releasing pyrophosphate. The overall reaction would be, according to McElroy and Seliger (1962):

$$LH_2 + ATP + O_2 \xrightarrow{E} L{=}O + AMP + PP + H_2O + \text{light}$$

At present, investigators are inclined to the theory that through the eye the individual firefly's central nervous system judges the timing of the discharges. We have already described the structure of the firefly compound eye (Figures 8.8–8.10) and indicated that the lantern emits light that can excite the eye. Here we may find some relationship between the light emitter and the visual photoreceptor. The important thing to note here is that the light emission is under nervous control and the chemistry of the luminescent reactions may be associated with a nerve impulse through the neurotransmitter

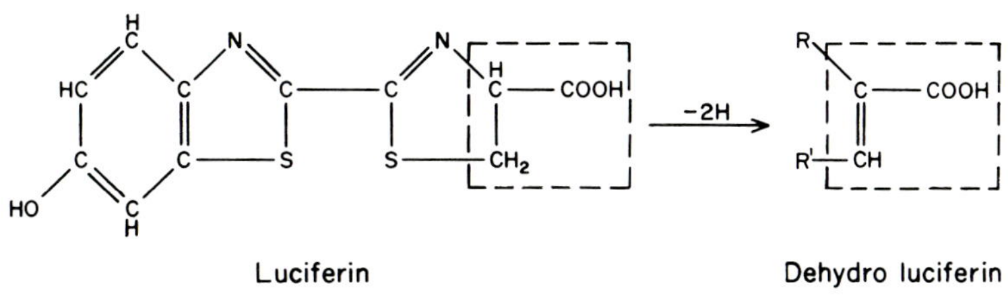


FIGURE 15.2 Firefly luciferin structure.

molecule acetylcholine, which we have already discussed in relation to photoperiodic phenomena.

Another organism that has been actively studied is the crustacean *Cypridina hilgendorfii*. This small crustacean is about 2 × 3 mm and is found mainly in the coastal waters of Japan and also off Jamaica. It possesses a gland near the mouth which, when it is disturbed, squirts luciferin and luciferase into the sea, where they mix and react with $O_2$ to give a flash of light. The reaction of the luciferin with $O_2$ consumes one $O_2$ molecule per luciferin molecule.

The purified luciferase was found to have a molecular weight of about 50,000. It lacks a prosthetic group or a metal that could serve as a light-emitting chromophore. Luciferase has no appreciable absorption in the visible and the luminescence optimum pH is 7.5. The *Cypridina* luciferin structure is also known (Figure 15.3). Its absorption spectrum in the ultraviolet is near 220, 270, and 312 nm. In the visible its absorption peak is around 425 nm and it fluoresces yellow at 514 nm. The bioluminescence is in the blue around 465 nm.

$$2LH_2 + O_2 \xrightarrow{E} 2L + 2H_2O + \text{light}$$

In the coelenterates, the jellyfish *Aequorea aequorea* possesses blue luminescent photophores around the edges of the mantle. Although it contains a luciferin–luciferase system, oxygen is not necessary. The system does require the addition of a calcium ion for full luminescence. The other well-investigated coelenterate is the sea pansy *Renilla reniformis*. *Renilla* is a pinkish violet colonial coelenterate which lives on the sea floor at depths of 6 to 15 meters. When a sea pansy is stimulated mechanically or electrically, it sends a concentric wave of greenish light from the point of stimulation across its body surface. The wave of luminescence is controlled by some kind of nervous net-

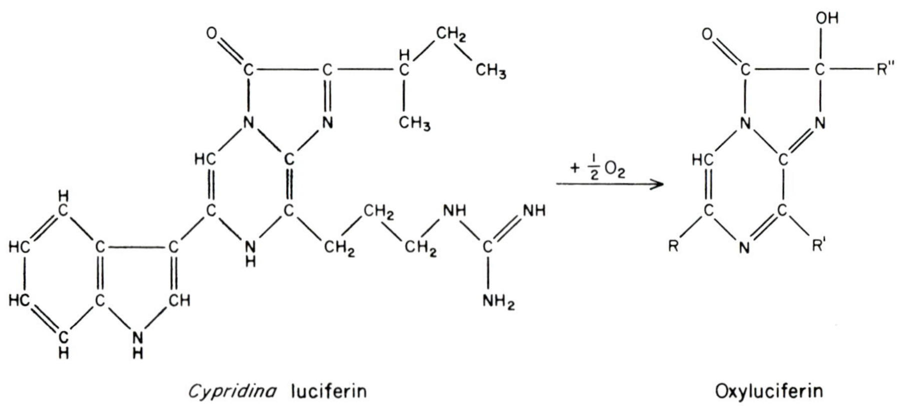


FIGURE 15.3 *Cypridina* luciferin structure (according to Hastings, 1968, p. 601).

work and also requires calcium for luminescence. Its luciferin closely resembles *Cypridina* luciferin in structure and behavior (Figure 15.3).

Meat and dead fish have been noted to be luminescent. In many cases this is due to the tissue being infected with bioluminescent bacteria. A very large number of bacteria are luminescent and are either parasitic, saprophytic, or symbiotic with certain animals. It is difficult to determine whether the bacteria are parasitic on or symbiotic with the animals with which they are associated.

Most of the bacterial research has been done by Hastings (1968) and his collaborators using *Photobacterium fischerii* and *Photobacterium phosphoreum*. Attempts to isolate a luciferin–luciferase reaction in bacteria proved to be difficult, but it is now possible to isolate luciferase in a pure form. However, there is no luciferase as in the classical reaction. There are two cofactors, though, dihydroflavin mononucleotide ($FMNH_2$) and a straight-chain aldehyde.

Two possibilities for a mechanism remain: the aldehyde on oxidation by an as yet undetermined mechanism transfers energy to a modified FMN molecule, which then fluoresces; or a chemical combination of aldehyde and $FMNH_2$ occurs, resulting in a "luciferin-like" molecule (Figure 15.4), which is oxidized with light emission. In the second possibility some means of recycling FMN must operate. A working hypothesis which satisfies most of the evidence is shown in Scheme 15.1.

In most living bacteria the emitted light is in the blue–green around 490 nm, but in some bacteria it ranges from 475 to near 520 nm. The fluorescence of FMN has a maximum around 530 nm which differs from the light emission of any of the bacteria. Therefore, it is thought that FMN is not the emitting molecule. However, the role of reduced FMN in the light emission of bacteria is still not completely understood.

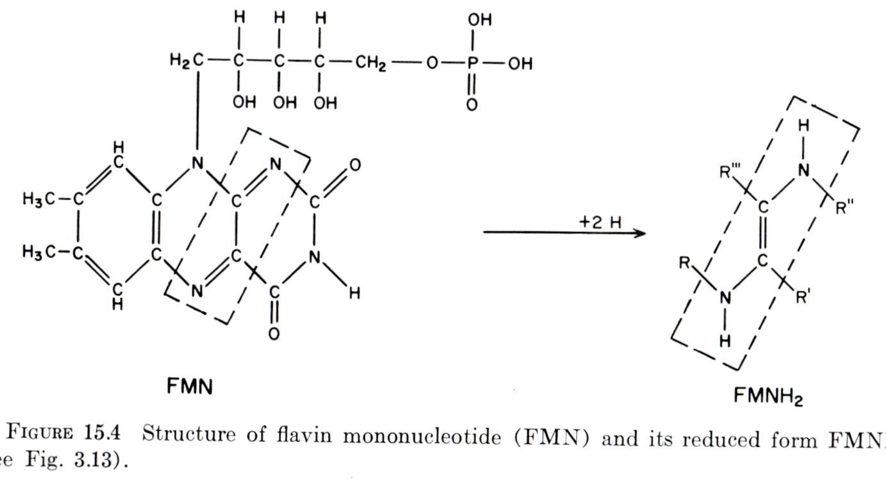


FIGURE 15.4 Structure of flavin mononucleotide (FMN) and its reduced form $FMNH_2$ (see Fig. 3.13).

$$\text{Luciferase} + \text{FMNH}_2 \rightleftharpoons \text{Luciferase}\cdot\text{FMNH}_2$$

$$\text{FMNH}_2 \xrightarrow{O_2} \text{FMN} + H_2O$$

$$\text{Luciferase}\cdot\text{FMNH}_2 + \text{R}\cdot\text{CHO} \rightleftharpoons \text{Reactive complex}$$

$$\text{Complex} + O_2 \longrightarrow \text{Luciferase} + \text{FMN} + \text{R}\cdot\text{CO}_2\text{H}$$

SCHEME 15.1

Although much remains to be worked out, a more precise knowledge of the chemistry of bioluminescent bacteria will lead to a more complete understanding of the process in general.

We can now compare bacterial bioluminescence with the fungi. The luminescence of rotting wood and plants is usually the result of these luminous fungi. The luminous fungi are confined to the basidiomycetes.

In some reactions of fungal luminescence the luciferin is reduced adenine dinucleotide phosphate before the active light reaction. It also requires an electron acceptor, a soluble dehydrogenase, molecular oxygen and luciferase. These reactions are indicated as follows:

$$\text{L} + \text{NADH} + \text{H}^+ \rightarrow \text{LH}_2 + \text{NAD}^+$$

$$\text{LH}_2 + \text{O}_2 \xrightarrow{\text{E}} \text{light} + \text{H}_2\text{O}$$

Here too, much of the chemistry of luminous fungi requires further elucidation of the reaction mechanisms.

## Photophores and Photoreceptors

The special organs for bioluminescence are named photophores. The photophores vary from relatively simple to very complex structures. The simplest of such structures would be a crystal, and Hastings (1968) has isolated from *Gonyaulax* particles, *scintillons,* which are rhombohedral crystals. These crystals luminesce in the presence of $O_2$ and at pH 5.7. Crystals have been observed in the photophores of other organisms and we have already discussed how crystals can function as photoreceptor structures as well.

Biologists have long observed similarities in structure between the eye and their visual photoreceptors to the bioluminescent photophores. The firefly, *Photuris pennsylvanica,* bioluminescent lantern structure was described by Smith (1963) and studied by us (Figures 15.1 and 15.5) in search for relationships between the firefly eye photoreceptor and its photocyte. There are of course many different types of photophores among the various luminescent species.

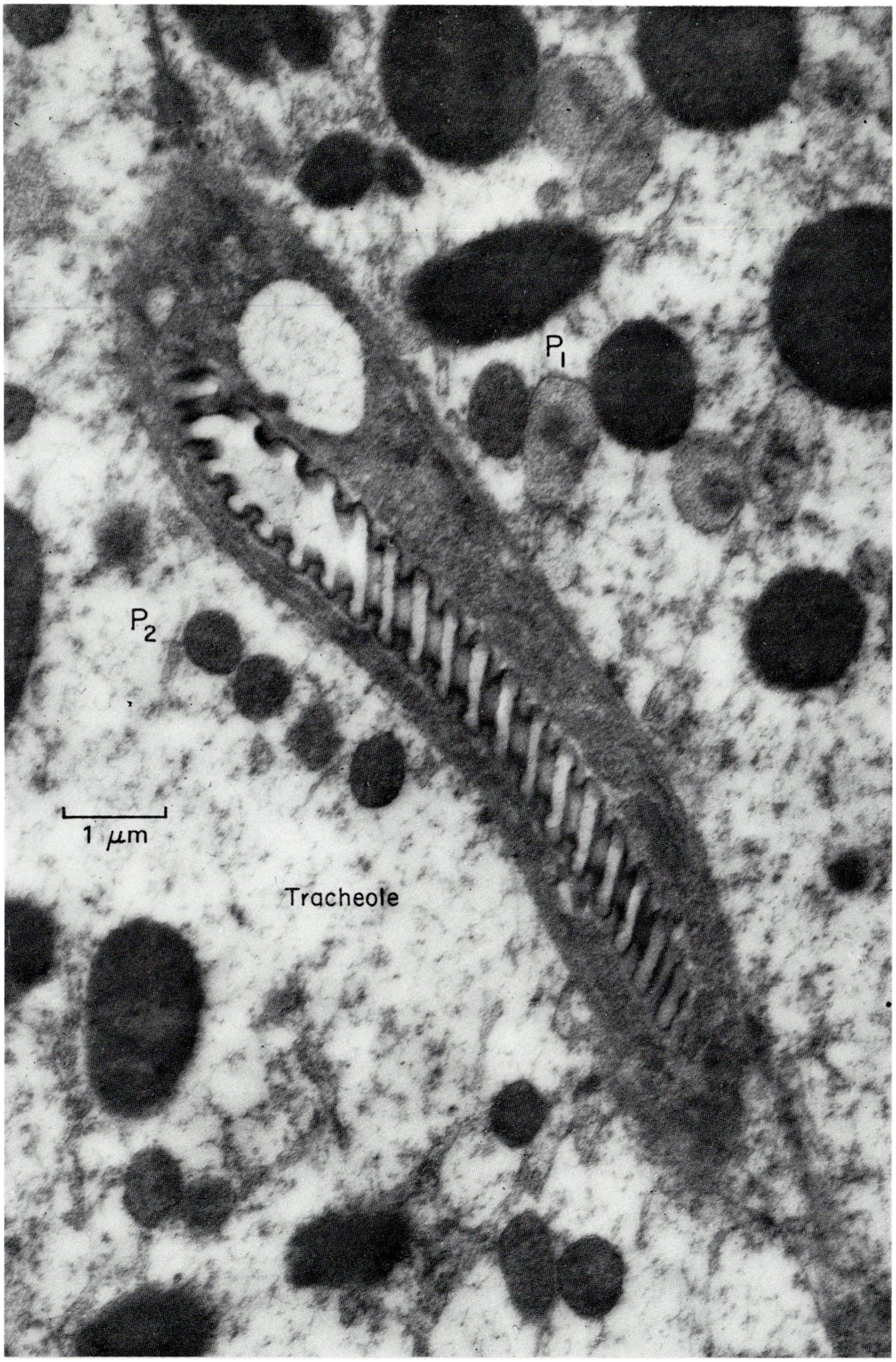


FIGURE 15.5 Enlargement of tracheole, longitudinal section, situated between two photocyte cells, $P_1$ and $P_2$ (see also Fig. 15.1).

One type consists of a luminous layer surrounded by a cup of pigment cells, much like an ocellus (Figure 8.2) in deep sea fish, squid, and shrimp. Other photophores contain a corneal-like structure or lens and a layer of the light-producing cells, like the retina with its photoreceptor cells and their nerve connections.

Another relationship between the photophores and the retinal photoreceptors is that the luciferin–luciferase system, like the rhodopsin system, is associated with specific membranes. In certain fish and marine worms electron microscopy showed that photophores have lamellae that compartmentalize the luciferin–luciferase biochemical system into separate membranes (Bassot, 1966). In the Japanese squid, *Watasenia*, Okada (1966) observed membranes, microvilli, which have a structure similar to that of rhabdomeres (see Figures 8.10, 8.12, and 8.15). At the level of the membrane there could be another relationship between the retinal rod and the photophore. That is, in light absorption by the retinal rod, calcium is released and this is believed to be associated with the generation of a nerve impulse. Similarly, during light emission in the sea pansy, *Renilla*, the nerve impulse releases calcium and light is generated.

It would be extremely interesting from an evolutionary point of view if the development of the photophore and the rhabdomere have similar origins. But whether or not the photophores and the photoreceptors have similar origins, the relationships between light production and light reception were evolutionary developments which served to increase the efficiency of these processes.

## Bioluminescence and Evolution

The widespread occurrence of luminous organisms may be an important key to uncovering its origin and evolution. But there appears to be no discernable evolutionary pattern of bioluminescence among the various organisms. But bioluminescence must have had some evolutionary significance and performed a functional role which is seen in the organisms that still possess it.

According to Seliger and McElroy (1965), bioluminescence originated as a biochemical means for the removal of oxygen and hence the reactions were detoxifying processes which were necessary for the survival of early anaerobic organisms. They proposed that the use of organic reducing substances to remove oxygen by direct reduction led to the formation of an excited state which could emit light—and so was the basis for the origin of bioluminescence in these organisms. However, Seliger no longer holds to this view.

Bioluminescence probably developed in a sporadic fashion throughout the course of evolution. Evolution at the molecular level proceeds by small changes in preexisting molecules. It may well be that luciferins and luciferases evolved from molecules that once had entirely different functions.

Although some closely related phyla appear to use similar biochemical systems and luciferin and luciferase as the basic reactants, these are generic names and they can be different molecules isolated from different organisms. For no one chemical system serves all types of luminescent species.

In bacteria bioluminescence may have an important functional biochemical role. This hypothesis was suggested by Hastings (1968) and is in contrast to that of Seliger and McElroy (1965); for present-day anaerobes, which are still subject to oxygen toxicity in the environment, have not been found to utilize bioluminescence as a detoxification mechanism. However, Hastings (1968) thinks that the bioluminescent reactions result in the formation of an active oxygen molecule, which then acts in the cell as a high energy oxidant. Another thought in this regard is that the depths of the ocean probably represent an ancient and constant environment, where the very low external light level may be an environmental factor with regard to the adaptation of a special mechanism by an organism to produce active oxygen. The active oxygen then could provide a chemical mechanism to allow cells to metabolize compounds not readily attacked by its enzymes.

There are several other points worth mentioning here. One is that flavins which we have indicated are involved in photoprocesses, are associated in luminescent reaction of the firefly and in the luminous exudate of earthworms. The second is that the reactions which result in bioluminescence could well be very close to the reversal of photosynthesis and that these two phenomena can be found in the same organism, *Gonyaulax,* which is photosynthetic and exhibits bioluminescence. The third, which we have already alluded to, is that there is a relationship to the visual photoreceptor system.

The relationship of bioluminescence to photoreception, the photoreceptors, and to the cellular energetics in evolutionary development is far from understood and remains to be discovered.

# XVI

# Summary

Light from the sun has been associated with life since the very origin of life itself. Such beliefs are born out by the fact that all organisms, from bacteria to man, exhibit some form of photosensitivity to solar radiation. This photosensitivity goes all the way back to the earliest known forms of life that directly utilized solar radiation.

As plants and animals evolved they adapted to a very narrow band in the electromagnetic spectrum of energy. Ultraviolet radiation from 200 to 300 nm is absorbed by proteins and nucleic acids, and such absorption produces damaging effects on cells, mutations, and even death. Evolving organisms were able to repair these effects through a phenomenon known as photoreactivation. Photoreactivation continues to be effective in bacteria, fungi, plants, and animals although ozone in the upper atmosphere filters out most of the damaging ultraviolet radiation. Infrared radiation beyond 900 nm is mostly absorbed by atmospheric water vapor and by the water that surrounds living cells. Therefore the limits of radiation effective for photobiological phenomena lie between 300 to 950 nm. This corresponds to the spectrum of solar radiation that reaches the surface of the earth with a maximum near 500 nm (Figures 2.4 and 3.15). To effectively utilize this energy photoreceptor systems developed which made life on earth possible.

Photobiological phenomena require specific molecules, a pigment or pigment system, to absorb the light energy. Thus, we discussed the structure of porphyrins, chlorophylls, carotenoids, flavins, and other pigment molecules that via light absorption initiate these phenomena. Therefore what we were searching for were molecular structural relationships common to all photoreceptor processes. To establish the generality of our conclusions the photoreceptor structures of a variety of organisms were examined along phylogenetic lines. In doing so we tried to see whether there were evolutionary trends that could be discerned in the photoreceptor systems at the molecular level.

Photosynthesis is the direct utilization of the visible part of the electromagnetic spectrum of energy. This evolved in ancestral bacteria, making it possible to convert light energy to chemical energy for cellular metabolism. These bacteria were able to utilize inorganic compounds such as hydrogen sulfide and hydrogen gas as well as organic compounds in their environment as a source of hydrogen required in photosynthesis. A second type of photosynthesis then evolved in the blue–green algae in which the utilization of hydrogen was accomplished by the photolysis of water. As a result, free oxygen entered the atmosphere, thus becoming available for further evolution and necessary for all animal life.

The bacteria and blue–green algae are prokaryotes. They do not have membrane-bound organelles, such as a nucleus, mitochondria, and chloroplasts which are associated with more highly developed eukaryotic cells. It is interesting to note that sterols and polyunsaturated fatty acids, the major constituents of eukaryote organelle membranes, require molecular oxygen for their synthesis. These compounds were believed to have been synthesized only after the evolution of photosynthesis. But biochemists have found that these compounds and the nicotinamides, NAD and NADP, which are synthesized by eukaryotes using molecular oxygen, are also synthesized by bacteria using different anaerobic pathways. Thus, biochemists have raised some doubts regarding the concept of a single line of biochemical and biological descent from the prokaryote to the eukaryote type cell. This dilemma is presently reconciled with an old hypothesis that ancestral bacteria and blue–green algae established a symbiotic relationship from which further evolution took place. Therefore, modern eukaryotes may not be products of direct evolutionary descent from a single ancestral prokaryotic cell, but may have shared the genetic and metabolic systems of diverse evolutionary pathways (Margulis, 1970; Cohen, 1970, 1973; Bogorad, 1975).

To explore organelle evolution, particularly that of the chloroplast, we chose the protozoan algal flagellate, *Euglena gracilis*, a eukaryote. In the light *Euglena* synthesizes chlorophyll a and b (as do higher plants), develops chloroplasts, and carries on photosynthesis. But when grown in the dark, *Euglena* no longer synthesizes chlorophyll or retains chloroplast structures; it obtains its energy by chemosynthesis, a process typical of all animal cells. This is a reversible system in *Euglena* providing mutations do not occur. Mutations can be brought about by physical means, e.g., elevated temperature, or by chemical means. These *Euglena* mutants do not synthesize chlorophyll or develop chloroplasts even in the light. In Euglena, then, the relation of the chloroplast to the organism is dynamic—and the organism can survive with or without chloroplasts.

From the point of view of evolution, an early stem organism, like *Euglena,* with a photosynthetic apparatus may at one point have been the progenitor of both plants and animals—an organism having the advantage over other simple living organisms in an environment where organic food was scarce. Later, when life was well established and organic food plentiful, this type of organism may have undergone a similar transformation to that seen in *Euglena,* losing its chloroplasts and becoming an animal cell. However, *Euglena* may be just one relic member of a large group of organisms from which plant and animal cells evolved by separate routes.

In photosynthesis, the light reaction is the photolysis of water and the dark reaction is the reduction of carbon dioxide to carbohydrates. The initial photophysical event is the absorption of light by chlorophyll a, in the chloroplast, to produce an excited chlorophyll molecule in which an electron is raised from its normal energy level to a higher energy level. Such excited electrons are transferred from chlorophyll to ferredoxin and to cytochromes via flavins and quinones. During this cyclic flow of electrons, the energy which the electrons initially acquired is transferred through oxidation–reduction reactions. The oxidation–reduction reactions are driven by two photochemical pigment systems, photosystem I and photosystem II. The reduced electron acceptor of photosystem I reduces $NADP^+$ via ferredoxin; the oxidized electron donors of photosystem II produce oxygen by the oxidation of water. These two photopigment systems provide the high energy phosphates (reduced NADP and ATP) needed for the synthesis of carbohydrates and proteins from $CO_2$ and water (Figure 5.2).

The chloroplast is the photoreceptor organelle for photosynthesis. So we examined the chloroplast to see if there is a structural basis for the two photochemical pigment systems in photosynthesis. The chloroplast is a highly ordered structure of tightly packed lamellae, with spacings of the order of 200 Å (Figures 5.5 and 5.6). This structure has similarities to the lattice structure of crystals (Figure 6.13). It was further shown that the number of chlorophyll molecules per chloroplast, from photosynthetic algae to higher plants, is of the order of $10^9$ molecules. The number of chlorophyll molecules is directly related to the number of lamellae and suggests a mode of growth regulation on the molecular level for development. The chlorophyll molecules are spread as monolayers on the surfaces of the lamellae as depicted in our molecular model (Figure 5.7). This maximizes the surface area of each chlorophyll molecule for light absorption and for energy transfer.

In animals various kinds of photosensory cells evolved, ultimately giving rise to eyes and vision (Figure 13.1). As the invertebrates developed, the photoreceptors became restricted to certain regions of the epidermis which later invaginated, forming simple eyes or ocelli. In examining these eyes there seems to be no smooth phylogenetic development from ocelli to compound and to refracting eyes, for the type of eye has been found to differ widely even among species of the same phylum. In the vertebrates, however, there appears to be no such variation; by the time vertebrates emerged they had already evolved a functional refracting eye.

It must be said that all visual photoreceptors, whether the rhabdomeres of invertebrates or the retinal rods and cones of vertebrate eyes, are formed of membranous processes of the receptor cell. These membranes appear as microtubules for the invertebrates and as lamellae for the vertebrates (Figure 13.2) and show a striking similarity to the chloroplasts (Figure 5.5).

At the molecular level, not only do we find similarities in the molecular structure for all photoreceptors, but we see that all visual photoreceptors depend for their function on a single molecular group of pigments, the carotenoids and their derivatives. Evidently an important change occurred during the evolutionary development of animals which made them dependent on the ingestion of plants for their source of carotenoids. It was not the ingested plant carotenoids themselves but their degraded derivative, vitamin A, that became necessary for animal life. The carotenoids, then, are seen to play a central role in the biochemical evolution from the plant $C_{40}$ ($\beta$-carotene) $\rightarrow$ animal $C_{20}$ (vitamin A) $\rightarrow$ eye, retinal (vitamin A aldehyde).

All visual pigments thus far isolated from invertebrate and vertebrate eyes are rhodopsins which contain retinal$_1$ or retinal$_2$ (derivable from vitamin $A_1$ or vitamin $A_2$) as the chromophore (Figure 13.3). Rhodopsins in the retinal rods and cones are retinal–protein complexes. The visual protein, opsin, is species specific and determines the absorption maxima and molecular weight of each rhodopsin.

It is only the 11-*cis* geometric isomer of retinal that is the functional molecule complexing with opsin to form rhodopsin. The action of light is to release 11-*cis* retinal to the all-*trans* form. Therefore, in vision light plays an entirely different role; not for chemical work as in photosynthesis, but through photoexcitation of a visual pigment in the retinal photoreceptors it triggers nervous impulses.

It was thought for a long time that retinal is found only in animals with eyes. But recent findings that the very halophilic bacteria synthesize retinal which is complexed to a protein to form rhodopsin in the cell "purple" membrane (Figure 13.7) raise questions about the evolution of visual pigments.

These bacteria live in high salt concentrations, high temperatures, and intense sunlight but are not very phototactic. Therefore the rhodopsin in their membrane could be a byproduct of their metabolism and have no function. On the other hand, it could have a vital function, for example in preventing photodestruction of the cell or actively participating in the cell membrane in metabolic processes such as photophosphorylation. The fact that retinal is found in organisms without eyes suggests, as with vitamin A, a more generalized biological function than simply for vision. This phenomenon can be viewed as resulting from the evolution of a type of molecule already found in nature. Therefore, retinal, once it was synthesized and became incorporated into the visual system, was then used for all visual systems that evolved. This universality demonstrates natural selection at the molecular level (Wald, 1957).

Light and dark periods control the organism's life rhythm. They can set the rhythmicity of the organism's behavior—its "biological clock." Photoperiodism is found to occur in plants, from leaf movements to the control of flowering cycles, and in animals, from movement and navigation to reproduction cycles (Ward, 1971).

The photoperiodic behavior of higher plants is due to the phytochrome molecule. The structure of phytochrome (Figure 14.2) resembles an open ring porphyrin (compare Figures 3.2 and 3.10 to Figure 14.2), and with its system of conjugated double bonds resembles a carotenoid molecule (Figure 3.7).

In the photochemistry of phytochrome there are similarities to the visual pigment rhodopsin; Hendricks and Siegelman (1967) proposed a mechanism for the phototransformation of phytochrome based on a *cis* to *trans* isomerization, similar to that of retinal in rhodopsin (Figure 11.14). The phytochrome transformation from the red-absorbing to the far-red-absorbing form, as in other photoreceptor molecules, involves a change in protein conformation, that is, a change in molecular shape that alters the receptor membrane properties to which it is bound.

Acetylcholine is a neurotransmitter molecule associated with muscle and nerve excitation in animals. But acetylcholine is found and participates in certain phytochrome-mediated phenomena in plants (Jaffe, 1970). This finding suggests the intriguing possibility that the origin of photoperiodism in animals may be related to the presence of a neurotransmitter molecule associated with a molecule that is similar in structure to phytochrome.

Eyes are not the sole means of photoreception, for the general body surface of many eyeless and blinded animals is remarkably light-sensitive. This diffuse photosensitivity over the whole or part of the animal's body is defined as the *dermal light sense*. But even deeper tissue cells, e.g., nerve or ganglion cells, are also photosensitive. Why these cells in animals should be light-sensitive and function as photoreceptors is still unclear.

Crustacea and molluscs, as well as cold-blooded vertebrates, possess chromatophores. The black pigmented chromatophores are called melanophores because they contain the pigment, melanin. Collectively, the chromatophores act to produce color and shade changes in the skin of the animal. For many of these animals the eye is the receptor organ which, through hormonal action, initiates the expansion and contraction of the chromatophores that bring about these color changes. A variation of this pigment effector system occurs in certain lower vertebrates, for example, amphibians and lizards, in which the pineal organ of the brain is photosensitive and participates in the control of adaptive pigmentation.

The main line of pineal evolution appears to have evolved active photoreceptors, for there is a striking similarity in structure between the photoreceptor cells of the pineal and those of retinal rods and cones (Figures 10.5–10.8). The level of function in the present-day mammalian pineal continues to be regulated by light. Action spectra indicate a maximum near 560 nm for the frog pineal response, which immediately calls to mind the visual cone pigments.

The pineal serves as a neurochemical transducer which is activated by light. We look at the pineal, then, as a sensitive neuroendrocrine transducer which functions as a "biological clock" (Axelrod, 1974). The molecule responsible for this activation is melatonin (Figure 14.4). It is of interest to note that the neurotransmitter molecules acetylcholine and serotonin (the precursor molecule of melatonin) appear to play a role in photoperiodic phenomena. Neurotransmitter molecules are present in primitive organisms (Table 5.2). For example, acetylcholine, epinephrine, and norepinephrine occur in protozoa and in invertebrate nervous systems (Lentz, 1968). The apparent universality of these molecules in nervous systems tempts us to suggest that a simple modification within the cell allowed for the coupling of excitatory and conductile properties, permitting transmission to occur after stimulation.

Because photoreceptors receive and process information for the cell, we consider them rudimentary "memory" systems. For example, information can be introduced into a chromophore, the "memory molecule," by a light signal at a particular wavelength. This information is stored or "memorized" until a signal of another wavelength of light erases the information and restores the molecule to its original "memory" state. This cycle is reversible and reproducible. Such a memory system does in fact function in the photoperiodism of plants embodied in the red and far-red absorbing forms of phytochrome and in vision with the *cis–trans* isomerization of retinal.

Primitive organisms also possess a kind of memory system. In *Phycomyces* phototropism we showed that if the organism receives a light signal at a critical time in its development it remembers where the light signal came from and continues to search for it, even in the absence of light. The phototactic behavior of *Euglena* also exhibits a kind of primitive memory system, that of simple association. The photoreceptor molecule in both these primitive organisms appears to be a flavoprotein complex. The flavin becomes photoreduced to a flavin semiquinone in the light and is regenerated in the dark. These light ↔ dark adaptations, then, bring about changes in the chromophore and result in conformational changes in the protein, which in turn permit excitation to occur. The system is then restored to its original state in the dark.

How then did the photoreceptor originate and evolve? We suggested that the photoreceptor arose from a crystalline body, as in the chromatophores of bacteria, or the crystals in the phototropic growth zone of the fungus *Phycomyces* (Figure 6.13), and in the paraflagellar body of the protozoan flagellate *Euglena* (Figure 7.1). The crystalline lattice eventually metamorphosed into a lamellar structure as seen in chloroplasts (Figures 5.5 and 5.6) and the visual photoreceptors (Figures 10.7 and 10.8).

Biological membranes depend on the physicochemical properties of lipids and proteins. The lipids, the sterols, and the phospholipids are found in concentrations of greater than 30% in a typical membrane. The lipids have the unique property of forming mono- or bimolecular layers. This is due to the presence of hydrophobic water-soluble groups at one end of the molecule and hydrophilic lipid soluble groups at the other end of the molecule.

Experimental studies indicate that the photoreceptor lamellar structure is related to its molecular state; for example, its membranes exhibit properties of a "liquid-crystal" which undergoes phase transition, depending on the internal environment of the cell and the external environment, particularly light and temperature.

A phospholipid liquid crystal system will show different structures when the ratio of water to the phospholipid changes (Brown and Doane, 1974). The first structure formed on addition of water to the crystal is a lamellar structure. With increased water a hexagonal structure is formed, and as additional water is added a micelle is formed. And the process is reversible by removal of water. One is tempted to speculate that the addition of water to a primal crystalline photoreceptor was the critical factor in the evolution of the lamellar structure.

The photoreceptors of animal cells may have resulted from the differentiation of the flagellum to which a photosensitive pigment became absorbed. With time the pigment became an intimate part of the flagellar structure linking the light sensitivity of the pigment with the motor action of the flagellum. Electron microscope studies of various photoreceptors show flagellar processes associated with all sensory structures, from the flagellum of the protozoan *Euglena* to the flagellum found between the outer and inner segments of vertebrate retinal rods (Figure 13.1).

The photoreceptor's true origin may go back to the evolution of an invagination in the cell membrane itself. Eakin (1972) classified the photoreceptors on this basis, and termed those that evolved from cilia or flagella as *ciliary*

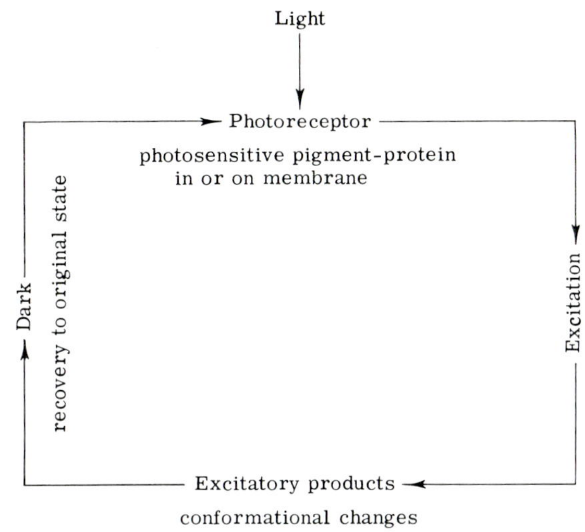


FIGURE 16.1 Photoreception process.

and of the vertebrate line, and those evolving from the cell membrane as *rhabdomeric* and of the invertebrate line. Whatever the mechanisms used in the origin and development of the photoreceptor, we know this to be true: that a pigment complexed with a protein to form a photosensitive molecule. This pigment–protein complex became ordered within a membrane, and the result was a functional photoreceptor structure.

The process of photoexcitation exhibits a pattern of activity which is fundamentally the same over a wide range of photobiological phenomena, and the photoreception process may be generally formulated as shown in Figure 16.1.

Light and life are intimately tied through their photoreceptor systems. For, once the cell could efficiently trap the visible bands of energy to perform specific functions—the energetics—life and evolution were and continue to be possible on earth.

To summarize, we began with the synthesis of pigments that absorb in the visible part of the electromagnetic spectrum which gave rise to photoreceptors and photoreceptor systems. To develop this thesis a wide range of photobiological phenomena was investigated. We searched along phylogenetic lines for similarities in the molecules and molecular structure of photoreceptors. The phylogenetic basis for the distinctions among plants and animals is purely morphological. But in examining the molecular basis for such phylogeny of the photoreceptor systems, we begin to see that these morphological differences become less distinct and that a common molecular basis for all photoreceptor systems emerge despite their myriad variations. (Figures 13.1 and 13.3).

# Bibliography

Aaronson, S., and Baker, H. (1961). Lipid and sterol content of some protozoa. *J. Protozool.* **8,** 274.

Abelson, P. H. (1966). Chemical events on the primitive earth. *Proc. Nat. Acad. Sci. U.S.* **55,** 1365.

Abrahamson, E. W., and Fager, R. S. (1973). The chemistry of vertebrate and invertebrate visual photoreceptors. *Curr. Top. Bioenerg.* 125–200.

Abrahamson, E. W., and Ostroy, S. E. (1967). The photochemical and macromolecular aspects of vision. *Progr. Biophys. Mol. Biol.* **17,** 179–215.

Agassiz, L. (1840). "Life, Letters and Works of Louis Agassiz," 2 vols. Jules Marcou (Macmillan, New York, 1896).

Ahlquist, C. N., and Gamow, R. I. (1973). *Phycomyces:* Mechanical behavior of stage II and stage IV. *Plant Physiol.* **51,** 586.

Allen, G. E. (1966). A model for the emergence of biochemical novelties. *BioScience* **16,** 325.

Amesz, J. (1973). The function of plastoquinone in photosynthetic electron transport. *Biochim. Biophys. Acta* **301,** 35.

Anderson, C. E., Price, G. B., and Fuller, R. C. (1968). Molecular diversity of the ribulose 1,5-diphosphate carboxylase from photosynthetic microorganisms. *Science* **161,** 482.

Anderson, J. M., and Boardman, N. K. (1966). Fractionation of the photochemical systems of photosynthesis. I. Chlorophyll contents and photochemical activities. *Biochim. Biophys. Acta* **112,** 403.

Anfinsen, C. B. (1963). "The Molecular Basis of Evolution." Wiley, New York.

Arden, G. B., Bridges, C. D. B., Ikeda, H., and Siegel, I. M. (1966). Rapid light-induced potentials common to plant and animal tissues. *Nature (London)* **212,** 1235.

Arnon, D. I. (1965). Ferredoxin and photosynthesis. *Science* **149,** 1460.

Autrum, H. (1950). Die Belichtungspotentiale und das Sehen der Insekten (Untersuchung an *Calliphora* und *Dixippus*). *Z. Vergl. Physiol.* **32,** 176.

Autrum, H. (1958). Electrophysiological analysis of the visual systems in insects. *Exp. Cell Res., Suppl.* **5,** 426.

Autrum, H., and Burkhardt, D. (1961). Spectral sensitivity of single visual cells. *Nature (London)* **190,** 639.

Autrum, H., and von Zwehl, V. (1962). Zur spektralen Empfindlichkeit einzelner Sehzellen der Drone *Apis mellifica*. *Z. Vergl. Physiol.* **46,** 8.

Autrum, H., and von Zwehl, V. (1964). Die spektrale Empfindlichkeit einzelner Sehzellen des Bienenauges. *Z. Vergl. Physiol.* **48,** 357.

Axelrod, J. (1974). The pineal gland: A neurochemical transducer. *Science* **184,** 1341.

Baker, J. T. (1974). Tyrian purple: An ancient dye, a modern problem. *Endeavor* **33,** 11.

Baldwin, E. (1948a). "An Introduction to Comparative Biochemistry." 3rd ed. Cambridge Univ. Press, London and New York.

Baldwin, E. (1948b). "Dynamic Aspects of Biochemistry." Cambridge Univ. Press, London and New York.

Baldwin, E. (1962). "The Nature of Biochemistry." Cambridge Univ. Press, London and New York.

Baldwin, E. (1964). "Introduction to Comparative Biochemistry," 4th ed. Cambridge Univ. Press, London and New York.

Ball, S., Collins, F. D., Dalvi, P. D., and Morton, R. A. (1949). Studies in vitamin A. II. Reactions and retinene$_1$ with amino compounds. *Biochem. J.* **45,** 304.

Bamji, M. S., and Krinsky, N. I. (1965). Carotenoid de-epoxidations in algae epoxidations. Enzymatic conversion of antheraxanthin to zeaxanthin. *J. Biol. Chem.* **240,** 467.

Banbury, G. H. (1952). Physiological studies in the *Mucorales*. Part II. Some observations on growth regulation of *Phycomyces*. *J. Exp. Bot.* **3,** 86.

Barrington, E. J. W. (1965). "The Biology of *Hemichordata* and *Protochordata*." Oliver & Boyd, Edinburgh.

Bartsch, R. G., and Kamen, M. D. (1960). Isolation and properties of two soluble heme proteins in extracts of the photoanaerobe, *Chromatium*. *J. Biol. Chem.* **235,** 825.

Bassham, J. A., and Calvin, M. (1957). "The Path of Carbon in Photosynthesis." Prentice-Hall, Englewood Cliffs, New Jersey.

Bassot, J.-M. (1966). On the comparative morphology of some luminous organs. *In* "Bioluminescence in Progress" (F. H. Johnson and Y. Haneda, eds.), pp. 557–610. Princeton University Press, Princeton, New Jersey.

Bates, H. W. (1864). "The Naturalist on the River Amazon." Murray, London (reprinted by J. M. Dent, London, and Dutton, New York, 1969).

Batra, P., and Tollin, G. (1964). Phototaxis in *Euglena*. I. Isolation of the eyespot granules and identification of the eyespot pigments. *Biochim. Biophys. Acta* **79,** 371.

Beaven, G. H., and Holiday, E. R. (1952). Ultraviolet absorption spectra of proteins and amino acids. *Advan. Protein Chem.* **7,** 319.

Benson, A. A. (1964). Plant membrane lipids. *Annu. Rev. Plant Physiol.* **15,** 1–16.

Benson, A. A. (1966). On the orientation of lipids in chloroplast and cell membranes. *J. Amer. Oil Chem. Soc.* **43,** 265.

Bergman, K., Burke, P. V., Cedrá-Olmedo, E., David, C. N., Delbrück, M., Foster, K. W., Goodell, E. W., Heisenberg, M., Meissner, G., Zalokar, M., Dennison, D. S., and Shropshire, W., Jr. (1969). *Phycomyces*. *Bacteriol. Rev.* **33,** 99.

Berkner, L. V., and Marshall, L. C. (1964). The history of oxygenic concentration in the earth's atmosphere. *Discuss. Faraday Soc.* **37,** 122.

Berl, S., Puszkin, S., and Nicklas, W. J. (1973). Actomyosin-like protein in brain. *Science* **179,** 441.

Bernal, J. D. (1951). "The Physical Basis of Life." Routledge & Kegan Paul Ltd. London.

Bernal, J. D. (1967). "The Origin of Life." Wiedenfeld and Nicholson, London, and World Publ. Co., Cleveland, Ohio.

Bernard, C. (1866). "Leçons sur les propriétés des tissus vivants." Baillière et Fils, Paris.

Bernard, C. (1878). "Leçons sur les phénomènes de la Vie Communs aux animaux et aux végétaux." Baillière et Fils, Paris.

Berns, D. S., and Vaughn, J. R. (1970). Studies on the photopigment system in *Phycomyces. Biochem. Biophys. Res. Commun.* **39,** 1094.

Berrill, N. J. (1955). "The Origin of the Vertebrates." Oxford Univ. Press (Clarendon), London and New York.

Bertholf, L. M. (1931). The distribution of stimulative efficiency in the ultraviolet spectrum of the honeybee. *J. Agr. Res.* **43,** 703.

Blackman, F. F., and Matthaei, G. (1905). Vegetable assimilation and respiration. IV. Carbon dioxide assimilation and leaf temperature. *Proc. Roy. Soc., Ser. B* **76,** 402.

Blasie, J. K. (1972). Location of photopigment molecules in the cross-section of frog retinal disk membranes. *Biophys. J.* **12,** 191.

Blaurock, A. E., and Stoeckenius, W. (1971). Structure of the purple membrane. *Nature (London), New Biol.* **233,** 152.

Blaurock, A. E., and Wilkins, M. H. F. (1969). Structure of frog photoreceptor membranes. *Nature (London)* **233,** 906.

Bliss, A. F. (1943). Derived photosensitive pigments from invertebrate eyes. *J. Gen. Physiol.* **26,** 361.

Bliss, A. F. (1948). The absorption spectra of visual purple of the squid and its bleaching products. *J. Biol. Chem.* **176,** 563.

Bloch, K. (1965). Lipid patterns in the evolution of organisms. *In* "Evolving Genes and Proteins" (V. Bryson and H J. Vogel, eds.), pp. 55–65. Academic Press, New York.

Blum, H. F. (1968). "Time's Arrow and Evolution," 3rd ed. Princeton Univ. Press, Princeton, New Jersey.

Bogorad, L. (1975). Evolution of organelles and eukaryotic genomes. *Science* **188,** 891.

Böll, F. (1876). Zur Anatomie und Physiologie der Retina. *Monatsber. Preuss. Akad. Wiss. Berlin* **41,** p. 783–788.

Bowness, J. M. (1959). Preparation of rhodopsin using columns containing calcium triphosphate. *Biochim. Biophys. Acta* **31,** 305.

Bowness, J. M., and Wolken, J. J. (1959). A light sensitive yellow pigment from the housefly. *J. Gen. Physiol.* **42,** 779.

Boyle, R. (1672). Some observations about shining flesh, both of veal and pullet, and that without any sensible putrefaction in those bodies. *Phil. Trans. Roy. Soc. London* No. 89, 5108.

Brawerman, G., and Eisenstadt, J. M. (1964). DNA from the chloroplasts of *Euglena gracilis. Biochim. Biophys. Acta* **91,** 477.

Bretscher, M. S. (1973). Membrane structure: Some general principles. *Science* **181,** 622.

Briggs, W. R., and Rice, H. V. (1972). Phytochrome: Chemical and physical properties and mechanism of action. *Annu. Rev. Plant Physiol.* **23,** 293.

Brokaw, C. J. (1972). Flagellar movement: A sliding filament model. *Science* **178,** 455.

Brown, F. A., Jr., Hastings, J. W., and Palmer, J. D. (1970). "The Biological Clock: Two Views." Academic Press, New York.

Brown, G., and Doane, J. W. (1974). Liquid crystals and some of their applications. *Appl. Phys.* **4,** 1.

Brown, P. K., and Brown, P. A. (1958). Visual pigments of the octopus and cuttlefish. *Nature (London)* **182,** 1288.

Brown, P. K., and Wald, G. (1964). Visual pigments in single rods and cones of the human retina. *Science* **144,** 45.

Bruce, V. G., and Pittendrigh, C. S. (1956). Temperature independence in unicellular "clock." *Proc. Nat. Acad. Sci. U.S.* **42,** 676.

Bruce, V. G., and Pittendrigh C. S. (1957). Endogenous rhythms in insects and microorganisms. *Amer. Natur.* **71,** 179.

Buder, J. (1918). Die Inversion des Phototropismus bei *Phycomyces*. *Ber. Deut. Bot. Ges.* **36,** 104.

Buetow, D. E., ed. (1968). "The Biology of *Euglena*," Vol. 2. Academic Press, New York.

Buffon, G. L. (1797–1807). "Compte de Buffon's Natural History" (Barr, trans.). J. Johnson, London.

Bünning, E. (1958). Cellular clocks. *Nature (London)* **181,** 1169.

Bünning, E. (1973). "The Physiological Clock," 3rd ed. Springer-Verlag, Berlin and New York.

Bünning, E., and Schneiderhöhn, G. (1956). Über das Aktionspektrum der phototakischen Reaktionen von *Euglena*. *Arch. Mikrobiol.* **24,** 80.

Buriain, H. M., and Ziv, B. (1959). Electric response of the phakic and aphakic human eye to stimulation with near ultraviolet. *Arch. Ophthalmol.* [N.S.] **61,** 347.

Burkhardt, D. (1962). Spectral sensitivity and other response characteristics of single visual cells in the arthropod eye. *In* "Biological Receptor Mechanisms" (J. W. L. Beaument, ed.), pp. 86–109. Academic Press, New York.

Burkhardt, D., and Wendler, L. (1960). Ein direkter Beweis für die Fähigkeit einzelner Sehzellen des Insektenauges, die Schwingungsrichtung polarisierten Lichtes zu analysieren. *Z. Vergl. Physiol.* **43,** 687.

Butenandt, A. (1952). The mode of action of hereditary factors. *Endeavour* **11,** 188.

Butenandt, A., and Neubert, G. (1955). Über Ommochrome. V. Xanthommatins ein Augenfarbstoff der Schmeissfliege. *Hoppe-Seyler's Z. Physiol. Chem.* **301,** 109.

Butenandt, A., Schiedt, V., and Bickert, E. (1954). Über Ommochrome. III. Mitteilung Synthese des Xanthommatins. *Justus Liebig's Ann. Chem.* **588,** 106.

Butler, W. L., Siegelman, H. W., and Miller, C. O. (1964). Denaturation of phytochrome. *Biochemistry* **3,** 851.

Cajal, S. Ramón y (1918). Observaciónes sobre la estructura de los ocelos y vias nerviosas ocelares le algunos insectos. *Trab. Lab. Invest. Biol. Univ. Madrid.* **16,** 109.

Calvin, M. (1959). Energy reception and transfer in photosynthesis. *Rev. Mod. Phys.* **31,** 147.

Calvin, M. (1962). Path of carbon in photosynthesis. *Science* **135,** 879.

Calvin, M. (1969). "Chemical Evolution." Oxford Univ. Press, London and New York.

Cammack, R., Hall, D., and Roa, K. (1971). Ferredoxins: Are they living fossils. *New Sci.* **51,** 696.

Carlile, M. J. (1965). The photobiology of fungi. *Annu. Rev. Plant Physiol.* **16,** 175.

Carlson, S. D. (1972). Microspectrophotometry of the dioptric apparatus and compound rhabdom of the moth (*Manduca sexta*) eye. *J. Insect Physiol.* **18,** 1721.

Carter, H. E., Ohno, K., Nojima, S., Tipton, Z. L., and Stanazev, N. Z. (1961). Wheat flour lipids. II. Isolation and characterization of glycolipids of wheat flour and other plant sources. *J. Lipid Res.* **2,** 215.

Castle, E. S. (1933). The refractive indices of whole cells. *J. Gen. Physiol.* **17,** 41.

Castle, E. S. (1935). The double refraction of chitin. *J. Gen. Physiol.* **19,** 797.

Chalazonitis, N. (1964). Light energy conversion in neuronal membranes. *Photochem. Photobiol.* **3,** 539.

Chappell, R. L., and Dowling, J. E. (1972). Neural organization of the median ocellus of the dragonfly. *J. Gen. Physiol.* **60,** 121.

Clarke, G. L., and Backus, R. H. (1956). Measurements of light penetration in relation to vertical migration and records of luminescence of deep-sea animals. *Deep-Sea Res.* **4,** 1.

Clarke, G. L., and Denton, E. J. (1962). Light and animal life. *In* "The Sea" (M. N. Hill, ed.), pp. 456–468. Wiley (Interscience), New York.

Clayton, R. K. (1953). Studies in the phototaxis of *Rhodospirillum rubrum.* I. Action spectrum, growth in green light, and Weber Law adherence. *Arch. Mikrobiol.* **19,** 107.

Clayton, R. K. (1965). "Molecular Physics in Photosynthesis." Ginn (Blaisdell), Boston, Massachusetts.

Clayton, R. K. (1971). "Light and Living Matter," Vols. I and II. McGraw-Hill, New York.

Cohen, A. I. (1961). The fine structure of the extra-foveal receptors of the Rhesus monkey. *Exp. Eye Res.* **1,** 128.

Cohen, A. I. (1963). Vertebrate retinal cells and their organization. *Biol. Rev. Cambridge Phil. Soc.* **38,** 427.

Cohen, A. I. (1964). Some observations on the fine structure of the retinal receptors of the American gray squirrel. *Invest. Ophthalmol.* **3,** 198.

Cohen, R., and Delbrück, M. (1959). Photoreactions in *Phycomyces.* Growth and tropic responses to the stimulation of narrow test areas. *J. Gen. Physiol.* **42,** 677.

Cohen, S. S. (1970). Are/Were mitochondria and chloroplasts microorganisms? *Amer. Sci.* **58,** 281.

Cohen, S. S. (1973). Mitochondria and chloroplast revisited. *Amer. Sci.* **61,** 437.

Collins, F. D. (1954). Chemistry of vision. *Biol. Rev. Cambridge Phil. Soc.* **29,** 453.

Cormier, M. J., ed. (1973). "Chemiluminescence and Bioluminescence." Plenum, New York.

Crane, F. L. (1959). Internal distribution of coenzyme Q in higher plants. *Plant Physiol.* **34,** 128.

Crescitelli, F. (1972). The visual cells and pigments of the vertebrate. *In* "Handbook of Sensory Physiology" (H. J. A. Dartnall, ed.), Vol. VII, Part 1, pp. 245–363. Springer-Verlag, Berlin and New York.

Crescitelli, F., and Dartnall, H. J. A. (1953). Human visual purple. *Nature (London)* **172,** 195.

Crick, F. H. C., and Orgel, L. E. (1973). Directed panspermia. *Icarus* **19,** 341.

Cundall, R. B., and Gilbert, A. (1970). "Photochemistry." Appleton, New York.

Curry, G. M., and Thimann, K. V. (1961). Phototropism: The nature of the photoreceptor in higher and lower plants. *Progr. Photobiol., Proc. Int. Congr., 3rd, 1960* pp. 127–134.

Danielli, J. F., and Davson, H. (1935). A contribution to the theory of permeability of thin films. *J. Cell Comp. Physiol.* **5,** 495.

Danon, A. and Stoeckenius, W. (1974). Photophosphorylation in *Halobacterium halobium. Proc. Nat. Acad. Sci. U.S.* **71,** 1234.

Dartnall, H. J. A. (1957). "The Visual Pigments." Wiley, New York.

Dartnall, H. J. A. (1962). Photobiology of the visual process. *In* "The Eye" (H. Davson, ed.), Vol. 2, pp. 323–522. Academic Press, New York.

Darwin, C. (1858). On the perpetuation of varieties and species by means of natural selection. *Proc. Linnean Soc. (Zool.)* **III:** 45–62.

Darwin, C. (1859). "On the Origin of Species by Means of Natural Selection, or, The

Preservation of Favored Races in the Struggle for Life." Murray, London (republished by Modern Library, New York, 1936).

Darwin, E. (1794). "Zoonomia," Vol. I, pp. 500–510. J. Johnson, London.

Darwin, F., ed. (1892). "The Autobiography of Charles Darwin and Selected Letters." Appleton, New York (republished by Dover, New York, 1958, p. 220).

Das, S. M. (1948). Nervous mechanism in *Tunicata*. *Nature* (*London*) **162**, 963.

Daumer, K. (1956). Reizmetrische Untersuchung des Farbensehens der Biene. *Z. Vergl. Physiol.* **38**, 413.

Dauvillier, A. (1965). "The Photochemical Origin of Life." Academic Press, New York.

Dayhoff, M. O., ed. (1972). "Atlas of Protein Sequence and Structure," Vol. 5. Nat. Biomed. Res. Found., Silver Springs, Maryland.

De Duve, C., Pressman, B. C., Gianetto, R., Wattiaux, R., and Applemans, F. (1955). Tissue fractionation studies. 6. Intracellular distribution patterns of enzymes in rat-liver tissue. *Biochem. J.* **60**, 604.

Delbrück, M., and Reichardt, W. (1956). System analysis for the light growth reactions in *Phycomyces*. *In* "Cellular Mechanisms in Differentiation and Growth" (D. Rudnick, ed.), pp. 3–44. Princeton Univ. Press, Princeton, New Jersey.

Delbrück, M., and Shropshire, W., Jr. (1960). Action and transmission spectra of *Phycomyces*. *Plant Physiol.* **35**, 194.

Denton, E. J. (1959). The contributions of the oriented photosensitive and other molecules to the absorption of whole retina. *Proc. Roy. Soc., Ser. B* **150**, 78.

Denton, E. J., Gilpen-Brown, J. B., and Wright, P. G. (1970). On the filter in the photophores of mesopelagic fish and on a fish emitting red light and especially sensitive to red light. *J. Physiol.* (*London*) **208**, 72.

Denton, E. J., Gilpen-Brown, J. B., and Wright, P. G. (1972). The angular distribution of the light produced by some mesopelagic fish in relation to their camouflage. *Proc. Roy. Soc., Ser. B* **182**, 145.

De Robertis, E. (1956). Electron microscope observations on the submicroscopic organization of retinal rods. *J. Biophys. Biochem. Cytol.* **2**, 319.

De Robertis, E., and Lassansky, A. (1961). Ultrastructure and chemical organization of photoreceptors. *In* "The Structure of the Eye" (G. K. Smelser, ed.), p. 29. Academic Press, New York.

de Saussure, N. T. (1804). "Recherchés chimiques sur la végétation." Lyon and Paris.

Descartes, R. (1637). "Descartes: Discourse on Method, Optics, Geometry and Meteorology" (P. J. Olscamp, transl.). A Leyde, De l'imprimerie de I. Maine, 1637. (Republished by Bobbs-Merrill, New York, 1965).

Detwiler, S. R. (1943). "Vertebrate Photoreceptors." Macmillan, New York.

De Vries, H. (1956). Physical aspects of the sense organs. *Progr. Biophysics* **6**, 246.

Dickerson, R. E. (1971). Sequence and structure homologies in bacterial and mammalian-type cytochromes. *J. Mol. Biol.* **57**, 1.

Diehn, B. (1973). Phototaxis and sensory transduction in *Euglena*. *Science* **181**, 1009.

Diehn, B., and Tollin, G. (1966a). Phototaxis in *Euglena*. II. Physical factors determining the rate of phototactic response. *Photochem. Photobiol.* **5**, 523.

Diehn, B., and Tollin, G. (1966b). Phototaxis in *Euglena*. III. LAG phenomena and the overall mechanism of the tactic response to light. *Photochem. Photobiol.* **5**, 839.

Dietrich, W. (1909). Die Facettenaugen der *Dipteran*. *Z. Wiss. Zool.* **92**, 465.

Dilly, P. N. (1961). Electron microscope observations of the receptors in the sensory vesicle of the ascidian tadpole. *Nature* (*London*) **191**, 786.

Dilly, P. N. (1964). Studies on the receptors in the cerebral vesicle of the ascidian tadpole. II. The ocellus. *Quart. J. Microsc. Sci.* [N.S.] **105**, 13.

Dilly, P. N. (1969). Studies on the receptors in *Ciona intestinalis*. III. A second type of photoreceptor in the tadpole larva of *Ciona intestinalis*. *Z. Zellforsch. Mikrosk. Anat.* **96,** 63.

Dilly, P. N. (1973). The enigma of colouration and light emission in deep sea animals. *Endeavour* **32,** 25.

Dilly, P. N., and Wolken, J. J. (1973). Studies on the receptors in *Ciona* intestinalis. IV. The ocellus in the adult. *Micron* **4,** 11.

Dobell, C. (1958). "Antony van Leeuwenhoek and His 'Little Animals.'" Russell & Russell Inc. New York.

Dodt, E., and Walther, J. B. (1958). Fluorescence of the crystalline lens and electroretinographic sensitivity determinations. *Nature* (*London*) **181,** 286.

Donner, K. O. (1953). The spectral sensitivity of the pigeon's retinal elements. *J. Physiol.* (*London*) **122,** 524.

Døving, K. B., and Miller, W. H. (1969). Function of insect compound eyes containing crystalline tracts. *J. Gen. Physiol.* **54,** 250.

Dowling, J. E. (1965). Foveal receptors of the monkey retina; fine structure. *Science* **145,** 57.

Dowling, J. E., and Ripps, H. (1970). Visual adaptation in the retina of the skate. *J. Gen. Physiol.* **56,** 491.

Dubois, R. (1885). Note sur la physiologie des pyrophores. *C.R. Soc. Biol.* (Series 8) **2,** 559.

Dubos, R. J. (1960). "Pasteur and Modern Science." Doubleday, Garden City, New York.

Duke-Elder, S. (1958). "System of Ophthamology. The Eye in Evolution," Vol. I. Mosby, St. Louis, Missouri.

Duysens, L. N. M. (1964). Photosynthesis. *Progr. Biophys. Mol. Biol.* **14,** 2–104.

Eakin, R. M. (1963). Lines of evolution of photoreceptors. *In* "General Physiology of Cell Specialization" (D. Mazia and A. Tyler, eds.), pp. 393–425. McGraw-Hill, New York.

Eakin, R. M. (1965). Evolution of photoreceptors. *Cold Spring Harbor Symp. Quant. Biol.* **30,** 363.

Eakin, R. M. (1968). Evolution of photoreceptors. *Evol. Biol.* **2,** 194–242.

Eakin, R. M. (1972). Structure of invertebrate photoreceptors. *In* "Handbook of Sensory Physiology" (H. J. A. Dartnall, ed.), Vol. VII, Part 1, pp. 625–684. Springer-Verlag, Berlin and New York.

Eakin, R. M. (1973). "The Third Eye." Univ. of California Press, Berkeley.

Eakin, R. M., and Kuda, A. (1971). Ultrastructure of sensory receptors in ascidian tadpoles. *Z. Zellforsch. Mikrosk. Anat.* **112,** 287.

Ebrey, T. S. (1967). Fast light-evoked potential from leaves. *Science* **155,** 1556.

Edelman, M., Schiff, J. H., and Epstein, H. T. (1965). Two types of satellite DNA. *J. Mol. Biol.* **11,** 769.

Eguchi, E., and Waterman, T. H. (1966). Fine structure patterns in crustacean rhabdoms. *In* "Functional Organization of the Compound Eye" (C. G. Bernhard, ed.), pp. 105–124. Pergamon, Oxford.

Eichenbaum, D. M., and Goldsmith, T. H. (1968). Properties of intact photoreceptor cells lacking synapses. *J. Exp. Zool.* **169,** 15.

Einstein, A. (1905). On a heuristic viewpoint concerning the production and transformation of light. *Ann. Phys.* (*Leipzig*) **17,** 132.

Elbers, P. F., Minnaert, K., and Thomas, J. B. (1957). Submicroscopic structure of some chloroplasts. *Acta Bot. Neer.* **6,** 345.

Emerson, R. (1956). Effect of temperature on the long-wave limit of photosynthesis. *Science* **123,** 673.

Emerson, R., and Arnold, W. (1932a). A separation of the reactions in photosynthesis by means of intermittent light. *J. Gen. Physiol.* **15,** 391.

Emerson, R., and Arnold, W. (1932b). The photochemical reaction in photosynthesis. *J. Gen. Physiol.* **16,** 191.

Emerson, R., and Lewis, C. M. (1943). The dependence of the quantum yield of *Chlorella* photosynthesis on wave length of light. *Amer. J. Bot.* **30,** 165.

Emerson, R., Chalmers, R. V., and Cederstrand, C. N. (1957). Some factors influencing the long-wave unit of photosynthesis. *Proc. Nat. Acad. Sci. U.S.* **43,** 133.

Englemann, T. W. (1882). Ueber Licht- und Farbenperception niederster Organismen. *Pflüegers Arch. Gesamte Physiol. Menschen Tiere* **29,** 387.

Erwin, J., and Bloch, K. (1962). The linolenic acid content of some photosynthetic microorganisms. *Biochem. Biophys. Res. Commun.* **9,** 103.

Erwin, J., and Bloch, K. (1963). Polyunsaturated fatty acids in some photosynthetic microorganisms. *Biochem. Z.* **338,** 496.

Eversole, R. A., and Wolken, J. J. (1958). Photochemical activity of digitonin extracts of chloroplastin. *Science* **127,** 1287.

Exner, S. (1891). "Die Physiologie der Facettierten Augen von Krebsen und Insekten." Deuticke, Vienna.

Fedele, M. (1923). Sulla organizzazione e le caratteristiche funzionali dell'attività nervosa dei Tunicata. I. Richerche sul sistema nervoso periferico degli Ascidiacea. *Atti. Accad. Naz. Lincei, Mem.* **5,** 98–102, 184–188.

Fernández-Morán, H. (1956). Fine structure of the insect retinula as revealed by electron microscopy. *Nature (London)* **177,** 742.

Fernández-Morán, H. (1958). Fine structure of the light receptors in the eyes of insects. *Exp. Cell Res., Suppl.* **5,** 586.

Fernández-Morán, H. (1961). The fine structure of vertebrate and invertebrate photoreceptors as revealed by low-temperature electron microscopy. *In* "The Structure of the Eye" (G. K. Smelser, ed.), p. 521. Academic Press, New York.

Finean, J. B. (1961). "Chemical Ultrastructure in Living Tissues." Thomas, Springfield, Illinois.

Fingerman, M. (1952). The role of the eye pigments of *Drosophila melanogaster* in photic orientation. *J. Exp. Zool.* **120,** 131.

Fingerman, M. (1960). Tidal rhythmicity in marine organisms. *Cold Spring Harbor Symp. Quant. Biol.* **25,** 481.

Fingerman, M., and Brown, F. A. (1952). A "Purkinje shift" in insect vision. *Science* **116,** 171.

Fingerman, M., and Brown, F. A. (1953). Color discrimination and physiological duplicity of *Drosophila* vision. *Physiol. Zool.* **26,** 59.

Fischer, H., and Stern, A. (1940). "Die Chemie des Pyrrols," Vol. 2, Hälfte, 2. Akad. Verlagsges., Leipzig.

Fitch, W. M., and Margoliash, E. (1967). Construction of phylogenetic trees. *Science* **155,** 279.

Florkin, M. (1949). "Biochemical Evolution." Academic Press, New York.

Forrest, H. S., and Mitchell, H. K. (1954). Pteridines from *Drosophila*. I. Isolation of a yellow pigment. *J. Amer. Chem. Soc.* **76,** 5656.

Forward, R., and Davenport, D. (1968). Red and far-red light effects on a short-term behavioral response of a dinoflagellate. *Science* **161,** 1028.

Fox, D. L. (1953). "Animal Biochromes and Structural Colors." Cambridge Univ. Press, London and New York.

Fox, D. L. (1960). Pigments of plant origin in animal phyla. *In* "Comparative Biochemistry of Photoreactive Systems" (M. B. Allen, ed.), pp. **11–31**. Academic Press, New York.

Fox, S. W. (1965a). A theory of macromolecular and cellular origins. *Nature (London)* **205,** 328.

Fox, S. W., ed. (1965b). "The Origins of Prebiological Systems." Academic Press, New York.

Fox, S. W., and Dose, K. (1972). "Molecular Evolution and the Origin of Life." Freeman, San Francisco.

Fox, S. W., Harada, K., Woods, K. R., and Windsor, C. R. (1963). Amino acid compositions of proteinoids. *Arch. Biochem. Biophys.* **102,** 439.

Fraenkel, G. S., and Gunn, D. L. (1961). "The Orientation of Animals: Kinesis, Taxis, and Compass Reactions." Dover, New York.

Franz, V. (1931). Die Akkomodation des Selachierauges und seine Abblendungsapparate nebst Befunden an der Retina. *Zool. Jahrb., Abt. Allg. Zool. Physiol. Tiere* **49,** 323.

Franz, V. (1934). Vergleichende Anatomie des Wirbeltierauges. *In* "Handbuch der vergleichenden Anatomie der Wirbeltiere" (L. Bolk *et al.,* eds.), Vol. 2, Part 2, pp. 1009–1023. Urban & Schwarzenberg, Berlin.

Frey-Wyssling, A. (1957). "Macromolecules in Cell Structure." Harvard Univ. Press, Cambridge, Massachusetts.

Fujimoto, K., Yanase, T., and Hanaoka, T. (1957). Spectral transmittance of retinal colored oil globules re-examined with microspectrophotometer. *Jap. J. Physiol.* **7,** 339.

Gaffron, H. (1965). The role of light in evolution: The transition from a one quantum to a two quantum mechanism. *In* "The Origins of Prebiological Systems" (S. W. Fox, ed.), pp. 437–460. Academic Press, New York.

Garrison, W. M., Morrison, D. C., Hamilton, J. G., Benson, A. A., and Calvin, M. (1951). The reduction of carbon dioxide in aqueous solutions by ionizing radiation. *Science* **114,** 416.

Garstang, W. (1928). The morphology of the Tunicata, and its bearing on the phylogeny of the Chordata. *Quart. J. Microsc. Sci.* [N.S.] **72,** 51.

Garten, S. (1907). Die Veränderungen der Netzhaut durch Licht. *In* "Graefe Saemisch Handbuch der gesamten Augenheilkunde." sulf. 2, Bd. 3, Kap 12, pp 1–130. Leipzig.

Gascoigne, S. C. B. (1968). The optics of large telescopes. *Quart. J. Roy. Astron. Soc.* **9,** 98.

Gatenby, J. B., and Leslie-Ellis, J. (1951). The vertebrate neuron, the pituitary, and sudan black. *Cellule* **54,** 149.

Gerschler, M. W. (1911). Monographie du *Leptodora kindtii* (Focke). *Arch. Hydrobiol. Planktonk.* **6,** 415.

Gerschler, M. W. (1912). Monographie du *Leptodora kindtii* (Focke). *Arch. Hydrobiol. Planktonk.* **7,** 63.

Gibor, A., and Granick, S. (1964). Plastids and mitochondria: inheritable systems. *Science* **145,** 890.

Gilbert, P. W. (1963). The visual apparatus of sharks. *In* "Sharks and Survival" (P. W. Gilbert, ed.), 283–326. Heath, Lexington, Massachusetts.

Glover, J., Goodwin, T. W., and Morton, R. A. (1948). Conversion *in vivo* of vitamin A aldehyde (retinene$_1$) to vitamin $A_1$. *Biochem. J.* **43,** 109.

Gochnauer, M. B., and Kushner, D. J. (1969). Growth and nutrition of extremely halophilic bacteria. *Can. J. Microbiol.* **15,** 1157.

Gochnauer, M. B., Kushwaha, M. K., and Kushner, D. J. (1972). Nutritional control of pigment and isoprenoid compound formation in extremely halophilic bacteria. *Arch. Mikrobiol.* **84,** 339.

Goldsmith, T. H. (1958a). The visual system of the honeybee. *Proc. Nat. Acad. Sci. U.S.* **44,** 123.

Goldsmith, T. H. (1958b). On the visual system of the bee (*Apis mellifera*). *Ann. N.Y. Acad. Sci.* **74,** 223.

Goldsmith, T. H. (1960). The nature of the retinal action potential and the spectral sensitivities of ultraviolet and green receptor systems of the compound eye of the worker honeybee. *J. Gen. Physiol.* **43,** 775.

Goldsmith, T. H. (1962). Fine structure of the retinulae in the compound eye of the honeybee. *J. Cell Biol.* **14,** 489.

Goldsmith, T. H. (1972). The natural history of invertebrate visual pigments. *In* "Handbook of Sensory Physiology" (H. J. A. Dartnall, ed.), Vol. VII, Part I, pp. 685–719. Springer-Verlag, Berlin, and New York.

Goldsmith, T. H., and Philpott, D. E. (1957). The microstructure of the compound eyes of insects. *J. Biophys. Biochem. Cytol.* **3,** 429.

Goldsmith, T. H., and Warner, L. T. (1964). Vitamin A in the vision of insects. *J. Gen. Physiol.* **47,** 433.

Goldsmith, T. H., Dizon, A. E., and Fernandez, H. R. (1968). Photoreceptor organelles of the prawn *Palaemonetes*. *Science* **161,** 468.

Goodwin, T. W. (1952). "Comparative Biochemistry of Carotenoids." Chapman & Hall, London.

Goodwin, T. W., ed. (1965). "Chemistry and Biochemistry of Plant Pigments." Academic Press, New York.

Goodwin, T. W., and Jamikorn, M. (1954). Studies in carotenogenesis. Some observations on carotenoid synthesis in two varieties of *Euglena gracilis*. *J. Protozool.* **1,** 216.

Gössel, I. (1957). Über das Aktionspektrum der Phototaxis chlorophyllfreier Euglenen und über die Absorption des Augenflecks. *Arch. Mikrobiol.* **27,** 288.

Govindjee and Rabinowitch, E. I. (1960). Two forms of chlorophyll *a in vivo* with distinct photochemical functions. *Science* **132,** 355.

Granick, S. (1948). Magnesium protoporphyrin as a precursor of chlorophyll in *Chlorella*. *J. Biol. Chem.* **175,** 333.

Granick, S. (1950). Magnesium vinyl pheoporphyrin $a_5$, another intermediate in the biological synthesis of chlorophyll. *J. Biol. Chem.* **183,** 713.

Granick, S. (1958). Porphyrin biosynthesis in erythrocytes. I. Formation of ζ-aminolevulinic acid in erythrocytes. *J. Biol. Chem.* **232,** 1101.

Granit, R. (1941). Colour receptors of the frog's retina. *Acta Physiol. Scand.* **3,** 137.

Gregory, R. L. (1966). "Eye and Brain." McGraw-Hill, New York.

Gregory, R. L. (1967). Origin of eyes and brains. *Nature* (*London*) **213,** 369.

Gregory, R. L., Ross, H. E., and Moray, N. (1964). The curious eye of *Copilia*. *Nature* (*London*) **201,** 1166.

Grenacher, H. (1879). "Untersuchungen über das Sehorgan der Arthropoden, insbesondere der Spinnen, Insekten und Crustaceen," p. 195. Vandenhoeck & Ruprecht, Göttingen.

Grenacher, H. (1886). Abhandlungen zur vergleichenden Anatomie des Auges. I. Die Retina der Cephalopoden. *Abh. Naturforsch. Ges. Halle* **16,** 207.

Grossbach, U. (1957). Zur papierchromatographischen Untersuchung von Lepidopteren-Augen. *Z. Naturforsch.* **12b,** 462.

Groth, W. E., and Weyssenhoff, H. V. (1960). Photochemical formation of organic compounds from mixtures of simple gases. *Planet. Space Sci.* **2,** 79.

Gruber, S. H., Hamasaki, D. I., and Bridges, C. D. B. (1963). Cones in the retina of the lemon shark (*Negaprion brevirotris*). *Vision Res.* **3,** 397.

Guzzo, A. V., and Pool, G. L. (1969). Flourescence spectra of the intermediates of rhodopsin bleaching. *Photochem. Photobiol.* **9,** 565.

Haldane, J. B. S. (1928). "Possible World." Harper, New York.

Haldane, J. B. S. (1954). "The Biochemistry of Genetics." Macmillan, New York.

Haldane, J. B. S. (1966). "The Causes of Evolution." Cornell Univ. Press, Ithaca, New York.

Halldal, P. (1964). Phototaxis in protozoa. *In* "Biochemistry and Physiology of Protozoa" (S. H. Hutner, ed.), Vol. 3, pp. 277–296. Academic Press, New York.

Hamasaki, D. I., and Bridges, C. D. B. (1965). Properties of the electroretinogram in three elasmobranch species. *Vision Res.* **5,** 483.

Hamasaki, D. I., and Gruber, S. H. (1965). The photoreceptors of the nurse shark, *Ginglymostoma cirratum* and the sting ray, *Dasyatis sayi. Bull. Mar. Sci.* **15,** 1051.

Hanaoka, T., and Fujimoto, K. (1957). Absorption spectrum of a single cone in the carp retina. *Jap. J. Physiol.* **7,** 276.

Hara, T., and Hara, R. (1972). Cephalopod retinochrome. *In* "Handbook of Sensory Physiology" (H. J. A. Dartnall, ed.), Vol. VII, Part 1, pp. 720–246. Springer-Verlag, Berlin and New York.

Hardy, A. C. (1956). "The Open Ocean." Houghton, Boston, Massachusetts.

Hartmann, K. M. (1966). A general hypothesis to interpret high energy phenomena of photomorphogenesis on the basis of phytochrome. *Photochem. Photobiol.* **5,** 349.

Harvey, E. N. (1952). "Bioluminescence." Academic Press, New York.

Harvey, E. N. (1960). Bioluminescence. *Comp. Biochem.* **2,** 545–591.

Hastings, J. W. (1968). Bioluminescence. *Annu. Rev. Biochem.* **37,** 431.

Hastings, J. W., and Sweeney, B. M. (1957). The mechanisms of temperature independence in a biological clock. *Proc. Nat. Acad. Sci. U.S.* **43,** 804.

Haupt, W. (1973). Role of light in chloroplast movement. *BioScience* **23,** 289.

Hawkins, E. G. E., and Hunter, R. F. (1944). Vitamin A aldehyde. *J. Chem. Soc.* (*London*) **411.**

Hecht, S. (1918). The photic sensitivity of *Ciona intestinalis. J. Gen. Physiol.* **1,** 147.

Hecht, S. (1926). The effect of exposure period and temperature on the photosensory process in *Ciona. J. Gen. Physiol.* **8,** 291.

Heller, J. (1968). Structure of visual pigments. I. Purification, molecular weight, and composition of bovine visual pigment$_{500}$. *Biochemistry* **7,** 2906.

Heller, J. (1969). Comparative study of a membrane protein characterization of bovine, rat and frog visual pigment$_{500}$. *Biochemistry* **8,** 675.

Hendricks, S. B. (1968). How light interacts with living matter. *Sci. Amer.* **219, 174.**

Hendricks, S. B., and Siegelman, H. W. (1967). Phytochrome and photoperiodism in plants. "*Compr. Biochem.*" **27,** 211–235.

Hering, E. (1885). "Ueber Individuelle Verschiedenheiten des Farbensinnes." Lotos, Prague.

Hering, E. (1885). "Outlines of a Theory of the Light Sense" (L. M. Hurvich and D. Jameson, transl.). Harvard Univ. Press, Cambridge, Massachusetts. 1965.

Hertz, M. (1939). New experiments on color vision in bees. *J. Exp. Biol.* **16, 1.**

Hess, W. N. (1943). Visual organs of invertebrate animals. *Sci. Mon.* **57,** 489.

Hill, R., and Bendall, F. (1960). Function of the two cytochrome components in chloroplasts: A working hypothesis. *Nature* (*London*) **186,** 136.

Hill, R., and Scarisbrick, R. (1940a). Production of oxygen by illuminated chloroplasts. *Nature* (*London*) **146,** 61.

Hill, R., and Scarisbrick, R. (1940b). Reduction of ferric oxalate by isolated chloroplasts. *Proc. Roy. Soc., Ser. B* **129,** 238.

Holmes, S. J. (1903). Phototaxis in *Volvox. Biol. Bull.* **4,** 319.

Hooke, R. (1705). "Posthumous works of Robert Hooke." Richard Waller, London.

Hopkins, F. G. (1889). Note on a yellow pigment in butterflies. *Nature* (*London*) **40,** 335.

Horridge, G. A. (1966). The retina of the locust. *In* "The Functional Organization of the Compound Eye" (C. G. Bernhard, ed.), p. 513. Pergamon, Oxford.

Horridge, G. A. (1968). Pigment movement and the crystalline threads of the firefly eye. *Nature (London)* **218,** 778.

Horridge, G. A. (1969). The eye of the firefly *Photuris. Proc. Roy. Soc., Ser.* B **171,** 445.

Hotta, Y., and Benzer, S. (1969). Abnormal electroretinograms in visual mutants of *Drosophila. Nature (London)* **222,** 354.

Hubbard, R. (1954). The molecular weight of rhodopsin and the nature of the rhodopsin-digitonin complex. *J. Gen. Physiol.* **37,** 381.

Hubbard, R., and Kropf, A. (1959). Molecular aspects of visual excitation. *Ann. N.Y. Acad. Sci.* **81,** 442.

Hubbard, R., and St. George, R. C. C. (1958). The rhodopsin system of the squid. *J. Gen. Physiol.* **41,** 501.

Hubbard, R., and Wald, G. (1960). Visual pigment of the horseshoe crab *Limulus polyphemus. Nature (London)* **186,** 212.

Hubbard, R., Brown, P. K., and Bownds, D. (1971). Methodology of vitamin A and visual pigments. *In* "Methods in Enzymology" (D. B. McCormick and L. D. Wright, eds.), Vol. 18C, pp. 615–653. Academic Press, New York.

Hulanicka, D., Erwin, J., and Bloch, K. (1964). Lipid metabolism of *Euglena gracilis. J. Biol. Chem.* **239,** 2778.

Hunter, R. F., and Williams, N. E. (1945). Chemical conversion of β-carotene into vitamin A. *J. Chem. Soc. (London)* pp. 554–555.

Hutner, S. H., ed. (1964). "Biochemistry and Physiology of Protozoa," Vol. 3. Academic Press, New York.

Hutner, S. H., and Lwoff, A., eds. (1955). "Biochemistry and Physiology of Protozoa," Vol. 2. Academic Press, New York.

Hutner, S. H., and Provasoli, L. (1951). The phytoflagellates. *In* "Biochemistry and Physiology of Protozoa" (A. Lwoff, ed.), Vol. 1, pp. 27–128. Academic Press, New York.

Hutner, S. H., and Provasoli, L. (1955). Comparative biochemistry of flagellates. *In* "Biochemistry and Physiology of Protozoa" (S. H. Hutner and A. Lwoff, eds.), Vol. 2, pp. 1–40. Academic Press, New York.

Hutner, S. H., and Provasoli, L. (1965). Comparative physiology: Nutrition. *Annu. Rev. Physiol.* **27,** 19.

Hutner, S. H., Bach, M. K., and Ross, G. I. (1956). A sugar containing basal medium for vitamin $B_{12}$-assay with *Euglena;* application to body fluids. *J. Protozool.* **3,** 101.

Hŭus, J. (1937). Ascidiaceae. *In* "Handbuch der Zoologie" (W. G. Kükenthal and T. Krumbach, eds.), Vol. 5, pp. 545–672. de Gruyter, Berlin.

Huxley, J. (1964). "Evolution; The Modern Synthesis." Wiley, New York.

Hydén, H. (1967). Biochemical and Molecular Aspects of Learning and Memory. *Proc. Amer. Phil. Soc.* **111,** 326.

Hydén, H., and Lange, P. W. (1968). Protein synthesis in the hippocampal pyramidal cells of rats during a behavioral test. *Science* **159,** 1370.

Ingen-Housz, J. (1779). "Experiments upon Vegetables, Discovering their Great Power of Purifying the Common Air in Sunshine and Injuring it in the Shade and at Night." Elmsly & Payne, London.

Ingen-Housz, J. (1798). "Ernährung der Pflanzen und Fruchtbarkeit des Bodens." Leipzig.

Jablonski, A. (1935). Über den Mechanismus der Photoluminescenz von Farbstoffphosphoren. *Z. Phys.* **94,** 38.

Jaffe, M. J. (1970). Evidence for the regulation of phytochrome-mediated processes in bean roots by the neurohormone, acetylcholine. *Plant Physiol.* **46,** 768.

Jagger, J. (1967). "Introduction to Research in Ultraviolet Photobiology." Prentice-Hall, Englewood Cliffs, New Jersey.

Jahn, T. L. (1951). Euglenophyta. *In* "Manual of Phycology" (G. M. Smith, ed.), pp. 69–81. Chronica Botánica, Waltham, Massachusetts.

Jennings, H. S. (1906). "Behavior of the Lower Organisms." Columbia Univ. Press, New York.

Johnson, D. L., and Gamow, R. I. (1972). *Phycomyces:* Growth responses of the sporangium. *Plant Physiol.* **49,** 898.

Johnson, F. H. (1967). Bioluminescence. *Compr. Biochem.* **27,** 79–136.

Johnson, L. P., and Jahn, T. L. (1942). Cause of the green-red color change in *Euglena rubra*. *Physiol. Zool.* **15,** 89.

Jörschke, H. (1914). Die Facettenaugen der Orthopteren und Termiten. *Z. Wiss. Zool.* **111,** 153.

Kamen, M. D. (1956). Hematin compounds in metabolism of photosynthetic tissues. *In* "Enzymes: Units of Biological Structure and Function" (O. H. Gaebler, ed.), pp. 483–504. Academic Press, New York.

Kamen, M. D. (1960). Hematin compounds in photosynthesis. *In* "Comparative Biochemistry of Photoreactive Systems" (M. B. Allen, ed.), pp. 323–327. Academic Press, New York.

Kammerer, P. (1915). "Allegemeine Biologie." Deutsche Verlag-anstalt, Stuttgart and Berlin.

Kampa, E. M. (1955). Euphausiopsin. A new photosensitive pigment from the eyes of the euphausiid crustaceans. *Nature* (*London*) **175,** 996.

Karnaukhov, V. N. (1973). On the nature and function of yellow aging pigment lipofuscin. *Exp. Cell Res.* **80,** 479.

Karrer, P., and Jucker, E. (1950). "Carotenoids" (E. A. Braude, transl.). Amer. Elsevier, New York.

Kästner, A. (1950). Reaktion der Hupfspinnen (Salticidae) auf unbewegte farblose und farbige Gesichtsreige. *Zool. Beitr., Berlin* [N.S.] **1,** 12.

Keeble, F. (1910). "Plant-Animal: A Study in Symbiosis." Cambridge Univ. Press, London and New York.

Keilin, D. (1966). "The History of Cell Respiration and Cytochrome." Cambridge Univ. Press, London and New York.

Kelly, D. (1965). Ultrastructure and development of amphibian pineal organs. *Progr. Brain Res.* **10,** 270.

Kelly, D. (1971). Developmental aspects of amphibian pineal systems. *Pineal Gland, Ciba Found. Symp., 1970* pp. 53–77.

Kelner, A. (1949). Effect of visible light on the recovery of *Streptomyces griseus* conidia from ultraviolet irradiation injury. *Proc. Nat. Acad. Sci. U.S.* **35,** 73.

Kennedy, D., and Bruno, M. S. (1961). The spectral sensitivity of crayfish and lobster vision. *J. Gen. Physiol.* **44,** 1089.

Kenyon, D. H., and Steinman, G. (1969). "Biochemical Predestination." McGraw-Hill, New York.

Keosian, J. (1964). "The Origin of Life." Van Nostrand-Reinhold, Princeton, New Jersey.

Kikkawa, H. (1941). Mechanism of pigment formation in *Bombyx* and *Drosophila*. *Genetics* **26,** 587.

Kikkawa, H., Qgita, Z., and Fujimoto, S. (1955). Nature of pigments derived from tyrosine and tryptophan in animals. *Science* **121,** 43.

King, J. L., and Jukes, T. H. (1969). Non-Darwinian evolution. *Science* **164,** 788.

Koestler, A. (1971). "The Case of the Midwife Toad." Hutchinson, London.

Korn, E. D. (1964). The fatty acids of *Euglena gracilis*. *J. Lipid Res.* **5,** 352.

Krause, W. (1863). Über die Endigung der Muskelnerven. *Z. Ration. Med.* **20,** 1.

Kreger, D. R. (1954). Observations on cell wall of yeasts and some other fungi by X-ray diffraction and solubility tests. *Biochim. Biophys. Acta* **13,** 1.

Krinsky, N. I. (1964). Carotenoid de-epoxidations in algae. I. Photochemical transformation of antheraxanthin to zeaxanthin. *Biochim. Biophys. Acta* **88,** 487.

Krinsky, N. I., and Goldsmith, T. H. (1960). Carotenoids of *Euglena gracilis* (Z strain). *Fed. Proc., Fed. Amer. Soc. Exp. Biol.* **19,** 329.

Krukenberg, C. F. W. (1882). Ueber die Stäbchenfarbe der Cephalopoden. *Untersuch. Physiol. Inst. Heidelberg* **2,** 58.

Kühn, A. (1927). Über den Farbensinn der Bienen. *Z. Vergl. Physiol.* **5,** 762.

Kühn, H. (1968). On possible ways of assembling simple organized systems of molecules. *In* "Structural Chemistry and Molecular Biology" (A. Rich and N. Davidson, eds.), pp. 566–572. Freeman, San Francisco, California.

Kühne, W. (1878). "On the Photochemistry of the Retina and on Visual Purple" (Ed. with notes by M. Foster). Macmillan, New York.

Kuwabara, M. (1957). Bildung des bedingten Reflexes von Paulous Typus bei der Honigbiene *Apis mellifera*. *J. Fac. Sci., Hokkaido Univ., Ser. 6* **13,** 458.

Lamarck, J.-B. (1815). "Système des animaux sans vertèbrates." Paris.

Lamarck, J.-B. (1809). "Philosophie Zoologique," 2 vols. Dentu, Paris.

Langer, H. (1967). Über die Pigmentgranula im Facettenaugen von *Callifora erythrocephala*. *Z. Vergl. Physiol.* **55,** 354.

Langer, H., and Thorell, B. (1966). Microspectrophotometry of single rhabdomeres in the insect eye. *Exp. Cell Res.* **44,** 673.

Lanssansky, A. (1967). Cell junctions in ommatidium of *Limulus*. *J. Cell Biol.* **33,** 365.

Lavoisier, A. L. (1774). "Opuscules Physique et Chimiques." Durand, Paris.

Leedale, G. F. (1967). "Euglenoid Flagellates." Prentice-Hall, Englewood Cliffs, New Jersey.

Lehninger, A. L. (1965). "Bioenergetics," pp. 222–224. Benjamin, New York.

Lemberg, R., and Legge, J. W. (1949). "Hematin Compounds and Bile Pigments." Wiley (Interscience), New York.

Lenard, J., and Singer, S. J. (1966). Protein conformation in cell membrane preparations as studies by optical rotatory dispersion and circular dichroism. *Proc. Nat. Acad. Sci. U.S.* **56,** 1828.

Lentz, T. L. (1968). "Primitive Nervous Systems." Yale Univ. Press, New Haven, Connecticut.

Lerner, A. B., Case, J. D., Takahashi, Y., Lee, T. H., and Mori, W. (1958). Isolation of melatonin, the pineal gland factor that lightens melanocytes. *J. Amer. Chem. Soc.* **80,** 2587.

Lewin, R. A., ed. (1962). "Physiology and Biochemistry of Algae." Academic Press, New York.

Liaanen-Jenson, S., and Andrews, A. G. (1972). Microbial carotenoids. *Annu. Rev. Microbiol.* **26,** 225.

Liebman, P. A. (1962). *In situ* microspectrophotometric studies on the pigments of single retina rod. *Biophys. J.* **2,** 161.

Liebman, P. A., and Entine, G. (1964). Sensitive low-light level of microspectrophotometric detection of photosensitive pigment of retinal cones. *J. Opt. Soc. Amer.* **54,** 1451.

Linzen, B. (1959). Über die Verbreitung der Ommochrome, der dunkeln Augenpigments der Arthropoden. *Zool. Anz., Suppl.* **22,** 154.

Loeb, J. (1918). "Forced Movements, Tropism and Animal Conduct." Lippincott, Philadelphia, Pennsylvania.

Lubbock, J. (1882). "Ants, Bees and Wasps." Appleton, New York.

Lundegårdh, H. (1967). Role of carotenoids in photosynthesis of green plants. *Nature* (*London*) **216,** 981.

Luzzati, V., and Husson, F. (1962). The structure of the liquid-crystalline phases of lipid-water systems. *J. Cell Biol.* **12,** 207.

Lwoff, A., ed. (1951). Introduction to biochemistry of protozoa. *In* "Biochemistry and Physiology of Protozoa" (A. Lwoff, ed.), Vol. 1, pp. 1–27. Academic Press, New York.

Lyell, C. (1830–1833). "Principles of Geology," 3 vols. Murray, London.

Lynch, V. H., and Calvin, M. (1953). $CO_2$ fixation by *Euglena*. *Ann. N.Y. Acad. Sci.* **56,** 890.

Lythgoe, J. N. (1972). List of vertebrate visual pigments. *In* "Handbook of Sensory Physiology" (H. J. A. Dartnall, ed.), Vol. VII, Part 1, pp. 604–624. Springer-Verlag, Berlin and New York.

Macallum, A. B. (1926). The paleochemistry of the body fluids and tissues. *Physiol. Rev.* **6,** 316.

McCapra, F. (1973). The chemistry of bioluminescence. *Endeavor* **32,** 139.

McConnell, J. V. (1962). Memory transfer through cannibalism in planarians. *J. Neuropsychiat.* **3,** Suppl. 1, 42.

McConnell, J. V. (1966). Comparative physiology: Learning in invertebrates. *Annu. Rev. Physiol.* **28,** 107.

McConnell, J. V., and Shelby, J. M. (1970). Memory transfer in invertebrates. *In* "Molecular Mechanisms in Memory and Learning" (G. Ungar, ed.), pp. 71–101. Plenum, New York.

McConnell, J. V., Jacobson, A. L., and Kimble, D. P. (1959). The effects of regeneration upon retention of a conditioned response in the planarian. *J. Comp. Physiol. Psychol.* **52,** 1.

McElroy, W. D., and Seliger, H. H. (1962). Origin and evolution of bioluminescence. *In* "Horizons in Biochemistry" (M. Kasha and B. Pullman, eds.), pp. 91–101. Academic Press, New York.

McElroy, W. D., Seliger, H. H., and White, E. H. (1969). Mechanism of bioluminescence, chemiluminescence and enzyme function in the oxidation of firefly luciferin. *Photochem. Photobiol.* **10,** 153.

MacNichol, E. F. (1964). Three pigment color vision. *Sci. Amer.* **211,** 48.

Manning, J. E., and Richards, O. C. (1972). Synthesis and turnover of *Euglena gracilis* nuclear and chloroplast deoxyribonucleic acid. *Biochemistry* **11,** 2036.

Manten, A. (1948). Phototaxis in the purple bacterium *Rhodospirillum rubrum* and the relation between phototaxis and photosynthesis. *Antonie van Leeuwenhoek; Microbiol. Serol.* **14,** 65.

Marak, G. E., and Wolken, J. J. (1965). An action spectrum of the fire ant: *Solenopsis saevissima*. *Nature* (*London*) **205,** 1328.

Margulis, L. (1970). "Origin of Eucaryotic Cells." Yale Univ. Press, New Haven, Connecticut.

Marks, W. B. (1963). Difference spectra of the visual pigments in single goldfish cones. Thesis, John Hopkins University, Baltimore, Maryland.

Marks, W. B., Dobelle, W. H., and MacNichol, J. R. (1964). Visual pigments of single primate cones. *Science* **143,** 1181.

Marx, K. (1867). "Das Kapital," vol. 1, Hamburg (S. L. Trask, transl.). (Republished by Modern Library, New York, 1932.)

Mast, S. O. (1911). "Light and the Behavior of Organisms." Wiley, New York.

Maupertuis, P. L. (1756). "Système de la Nature," Vol. II, pp. 137–168. Oeuvres de Mr. du Maupertuis à Lyon. Paris.

Maxwell, J. C. (1861). On the theory of compound colours and the relations of the colours of the spectrum. *Phil. Trans. Roy. Soc. London* **150,** 57.

Maxwell, J. C. (1890). "Scientific Papers." On the Theory of Compound Colours and the Relations of the Colours of the Spectrum. Vol. 1, pp. 126–154. Cambridge Univ. Press, London and New York.

Mayer, J. R. (1845). "Die Organische Bewegung in ihrem Zusammenhang mit dem Stoffwechsel." Heilbronn.

Mazokhin-Porshnyakov, G. A. (1969). "Insect Vision." Plenum, New York.

Meissner, G., and Delbrück, M. (1968). Carotenes and retinal in *Phycomyces* mutants. *Plant Physiol.* **43,** 1279.

Mendel, G. (1866). Versuche über Pflanzenhybriden. *In* "Mendel's Principle of Heredity" (W. Bateson, transl.). (Cambridge Univ. Press, London and New York, 1909).

Menger, E. L., ed. (1975). Chemistry of vision. *In* "Accounts of Chemical Research 8," No. 3, pp. 81–112. American Chemical Society, Washington, D.C.

Millar, R. H. (1953). *Ciona,* LMBC Mem. 35. University Press, Liverpool.

Miller, S. L. (1953). A production of amino acids under possible primitive earth conditions. *Science* **117,** 528.

Miller, S. L. (1955). Production of some organic compounds under possible primitive earth conditions. *J. Amer. Chem. Soc.* **77,** 2351.

Miller, S. L. (1957). The mechanism of synthesis of amino acids by electric discharges. *Biochim. Biophys. Acta* **23,** 480.

Miller, S. L., and Orgel, L. E. (1974). "The Origins of Life on the Earth," Prentice-Hall, Englewood Cliffs, New Jersey.

Miller, W. H., and Bernard, G. D. (1968). Butterfly glow. *J. Ultrastruct. Res.* **24,** 286.

Millot, N. (1957). Animal photosensitivity with reference to eyeless forms. *Endeavour* **16,** 19.

Millot, N. (1968). The dermal light sense. *In* "Invertebrate Receptors" (J. D. Carthy and G. E. Newell, eds.), pp. 1–36. Academic Press, New York.

Missotten, L. (1964). L'ultrastructure des tissues oculaire. *Bull. Soc. Belge Ophtalmol.* **135,** 1.

Mitchell, P. (1961). Coupling of phosphorylation to electron and hydrogen transfer by a chemiosmotic type of mechanism. *Nature (London)* **191,** 144.

Mitchell, P. (1966). Chemiosmotic coupling in oxidative and photosynthetic phosphorylation. *Biol. Rev. Cambridge Phil. Soc.* **41,** 445.

Mogus, M. A., and Wolken, J. J. (1974). *Phycomyces:* Electrical response to light stimuli. *Plant Physiol.* **53,** 512.

Molyneux, W. (1709). "Dioptrica Nova," 2nd ed. Printed for B. Tooke, London.

Moody, M. F. (1964). Photoreceptor organelles in animals. *Biol. Rev. Cambridge Phil. Soc.* **39,** 43.

Moore, T. (1953). Vitamin A in the normal individual. *In* "Symposium Nutrition" (R. M. Herriott, ed.), p. 28. John Hopkins Univ. Press, Baltimore, Maryland.

Moray, M. (1972). Visual mechanisms in the copepod *Copilia. Perception* **1,** 193.

Morton, R. A. (1944). Chemical aspects of the visual process. *Nature (London)* **153,** 69.

Morton, R. A., and Collins, F. D. (1950a). Studies on rhodopsin. I. Methods of extraction and the absorption spectrum. *Biochem. J.* **47,** 3.

Morton, R. A., and Collins, F. D. (1950b). Studies on rhodopsin. II. Indicator yellow. *Biochem. J.* **47,** 10.

Morton, R. A., and Collins, F. D. (1950c). Studies on rhodopsin. III. Rhodopsin and transient orange. *Biochem. J.* **47,** 18.

Morton, R. A., and Goodwin, T. W. (1944). Preparation of retinene *in vitro*. *Nature (London)* **153,** 405.

Morton, R. A., and Pitt, G. A. J. (1955). Studies on rhodopsin. IX. pH and the hydrolysis of indicator yellow. *Biochem. J.* **59,** 128.

Morton, R. A., Salah, M. K., and Stubbs, A. L. (1946). Retinene$_2$-vitamin A$_2$ aldehyde. *Biochem. J.* **40,** 49.

Morton, R. A., Salah, M. K., and Stubbs, A. L. (1947). Conversion of retinene$_2$ to vitamin A$_2$ *in vivo*. *Biochem. J.* **41,** 24.

Mote, M. I., and Goldsmith, T. H. (1970). Spectral sensitivities of color receptors in the compound eye of the cockroach *Periplaneta*. *J. Exp. Zool.* **173,** 137.

Murakami, S., and Packer, L. (1970). Protonation and chloroplast membrane structure. *J. Cell Biol.* **47,** 333.

Nagai, R., and Rebhun, L. I. (1966). Cytoplasmic microfilaments in streaming *Nitella* cells. *J. Ultrastruct. Res.* **14,** 571.

Naka, K. I. (1960). Recording of retinal action potentials from single cells in the insect compound eye. *J. Gen. Physiol.* **44,** 571.

Needham, J. (1956). "Science and Civilization in China," Vol. 2, p. 298. Cambridge Univ. Press, London and New York.

Nolte, D. J. (1950). The eye-pigmentary system of *Drosophila:* The pigment cells. *J. Genet.* **50,** 79.

Nowikoff, M. (1932). Über den Bau der Komplexaugen von *Periplaneta (Stylopyga) orientalis*. *Jena. Z. Naturwiss.* **67,** 58.

Oesterhelt, D., and Stoeckenius, W. (1971). Rhodopsin-like protein from the purple membrane of *Halobacterium halobium*. *Nature (London)* **233,** 149.

Oesterhelt, D., and Stoeckenius, W. (1973). Functions of a new photoreceptor membrane. *Proc. Nat. Acad. Sci. U.S.* **70,** 2853.

Ogawa, T., Inoue, Y., Kitajima, M., and Shibata, K. (1973). Action spectra for biosynthesis of chlorophylls a and b and β-carotene. *Photochem. Photobiol.* **18,** 229.

Okada, Y. K. (1966). Observations on rod-like contents in the photogenic tissues of *Watasenia* scintillons through the electron microscope. *In* "Bioluminescence in Progress" (F. H. Johnson and Y. Haneda, eds.), pp. 611–625. Princeton Univ. Press, Princeton, New Jersey.

Oldfield, E. (1973). Are cell membranes fluid? *Science* **180,** 982.

Ootaki, T., and Wolken, J. J. (1973). Octahedral crystals in *Phycomyces. II. J. Cell Biol.* **57,** 278.

Oparin, A. I. (1938). "The Origin of Life" (S. Morgulis, transl.), Macmillan, New York. 2nd ed. (Dover, New York, 1953).

Oparin, A. I. (1968). "Genesis and Evolutionary Development of Life." Academic Press, New York.

Packard, A. (1972). Cephalopods and fish: The limits of convergence. *Biol. Rev. Cambridge Phil. Soc.* **47,** 241.

Packard, A., and Sanders, G. (1969). What the octopus shows the world. *Endeavour* **28,** 92.

Palmer, J. D. (1974). "Biological Clocks in Marine Organisms." Wiley, New York.

Park, R. B., and Biggins, J. (1964). Quantasome: Size and composition. *Science* **144,** 1009.

Pasteur, L. (1857). Mémoire sur la fermentation appelée lactique. *C. R. Acad. Sci.* **45,** 913.

Pasteur, L. (1922–1939). "Oeuvres de Pasteur" (Collected by Vallery-Rodot), Vols. I and II. Masson, Paris.

Patten, W. (1887). Eyes of molluscs and arthropods. *J. Morphol.* **1,** 67.
Peckman, G. W., and Peckman, E. G. (1887). Some observations on the mental powers of spiders. *J. Morphol.* **1,** 383.
Perralet, A., and Baumann, F. (1969). Evidence for extracellular space in the rhabdom of the honeybee drone eye. *J. Cell Biol.* **3,** 825.
Pickett-Heaps, J. D. (1968). Microtubule-like structure in the growing plastids, or chloroplasts of two algae. *Planta* **81,** 193.
Pitelka, D. R. (1963). "Electron-Microscopic Structure of Protozoa." Pergamon, Oxford.
Pitelka, D. R. (1969). Fibrillar systems in protozoa. *In* "Research in Protozoology" (T. T. Chen, ed.), vol. II, pp. 270–388. Pergamon, Oxford.
Pittendrigh, C. S. (1954). On temperature independence in the clock system controlling emergence time in *Drosophila*. *Proc. Nat. Acad. Sci. U.S.* **40,** 1018.
Polyak, S. (1957). "The Vertebrate Visual System" (H. Klüver, ed.). Univ. of Chicago Press, Chicago, Illinois.
Ponnamperuma, C. (1965). A biological synthesis of some nucleic acid constituents. *In* "The Origins of Prebiological Systems" (S. W. Fox, ed.), pp. 221–242. Academic Press, New York.
Ponnamperuma, C. (1972). "The Origins of Life." Dutton, New York.
Porter, G. (1969). Photochemistry of complex molecules. *In* "An Introduction to Photobiology" (C. P. Swanson, ed.), pp. 1–22. Prentice-Hall, Englewood Cliffs, New Jersey.
Pribam, K. H. (1971). "Language of the Brain." Prentice-Hall, Englewood Cliffs, New Jersey.
Price, J. T. (1971). "The Origin and Evolution of Life." English Univ. Press, London.
Priestley, J. (1772). Observations on different kinds of air. *Phil. Trans. Roy. Soc. London,* **62,** 147.
Purkinje, J. (1825). "Beobachtungen und Versuche zur Physiologie der Sinne," Vol. 2. Reimer, Berlin.
Rabinowitch, E. (1945). "Photosynthesis and Related Processes," Vol. I. Wiley (Interscience), New York.
Rabinowitch, E. (1951). "Photosynthesis and Related Processes," Vol. II, Part 1. Wiley (Interscience), New York.
Rabinowitch, E. (1956). "Photosynthesis and Related Processes," Vol. II, Part 2. Wiley (Interscience), New York.
Ramsey, J. A. (1952). "A Physiological Approach to Lower Animals." Cambridge Univ. Press, London and New York.
Ray, D. S., and Hanawalt, P. C. (1964). Properties of the satellite DNA associated with the chloroplasts of *Euglena gracilis*. *J. Molec. Biol.* **9,** 812.
Regan, J. D., and Cook, J. S. (1969). Photoreactivation in an established vertebrate cell line [*Haemulon sciurus*]. *Proc. Nat. Acad. Sci. U.S.* **58,** 2274.
Robertson, J. D. (1966). Granular-fibrillar and globular substructures in unit membranes. *Ann. N.Y. Acad. Sci.* **137,** 421.
Rockwell, R. F., and Seiger, M. B. (1973). Phototaxis in *Drosophila:* A critical evaluation. *Amer. Scient.* **61,** 339.
Röhlich, P. (1966). Sensitivity of regenerating and degenerating planarian photoreceptors to osmium tetraoxide. *Z. Zellforsch. Mikrosk. Anat.* **73,** 165.
Röhlich, P., and Törö, I. (1965). Fine structure of the compound eye of *Daphnia* in normal, dark- and strongly light-adapted state. *In* "The Structure of the Eye, II. Symposium" (J. W. Rohen, ed.), pp. 175–186. Schattauer, Stuttgart.
Röhlich, P., and Török, L. J. (1961). Electronenmikroscopische Untersuchungen des Auges von Planarien. *Z. Zellforsch. Mikrosk. Anat.* **54,** 362.

Rosenberg, A. (1963). A comparison of lipid patterns in photosynthesizing cells of *E. gracilis*. *Biochemistry* **2,** 1148.

Rosenberg, A. (1967). Galactosyl diglycerides: Their possible function in *Euglena* chloroplasts. *Science* **157,** 1191.

Rosenberg, A., and Pecker, M. (1963). Lipid salts in *E. gracilis* during light induced greening. *Biochemistry* **2,** 254.

Rosenberg, B. (1958). Photoconductivity and the visual receptor processes. *J. Opt. Soc. Amer.* **48,** 581.

Rüdiger, W. (1969). Über die struktur des Phytocrom-Chromophors und seine Protein-Bindung. *Justus Liebigs Ann. Chem.* **723,** 208.

Sager, R. (1972). "Cytoplasmic Gene and Organelles." Academic Press, New York.

St. George, R. C. C., and Wald, G. (1949). The photosensitive pigment of the squid retina. *Biol. Bull.* **97,** 248.

Salah, M. K., and Morton, R. A. (1948). Crystalline retinene$_2$. *Biochem. J.* **43,** 6.

Scharrer, E. (1964). A specialized trophospongium in large neurons of *Leptodora* (crustacea). *Z. Zellforsch. Mikrosk. Anat.* **61,** 803.

Schleiden, M. J. (1838). "Beiträge zur Botanik. Gesammelte Auffässa." Engelmann, Leipzig.

Schmidt, W. J. (1934). Dichroismus des Aussengliedes der Stäbchenzellen der Froschnetzhaut, verursachht durch den Sehpurpur. *Naturwissenschaften* **22,** 206.

Schmidt, W. J. (1935). Doppelbrechung, Dichroismus und Feinbau der Aussengleider der Sehzellen vom Frosch. *Z. Zellforsch. Mikrosk. Anat.* **22,** 485.

Schmidt, W. J. (1937). "Die Doppelbrechung von Karoplasma, Zytoplasma, und Metaplasma, Protoplasma," Monogr. II. Borntraeger, Berlin.

Schmidt, W. J. (1938). Polarisationsoptische Analyse eines Eiweisslipoid-systems erlautert am Aussenglied der Sehzellen. *Kolloid-Z.* **85,** 137.

Schmitt, F. O., and Worden, F. G., eds. (1974). "The Neurosciences: Third Study Program." MIT Press, Cambridge, Massachusetts.

Schultze, M. J. (1866). Zur Anatomie und Physiologie der Retina. *Arch. Mikrosk. Anat.* **2,** 175.

Schwann, T. (1839). "Mikroskopische Untersuchungen über die Ueber-stimmung in der Struktur und dem Wachsthum der Tiere und Pflanzen." Sander, Berlin.

Seeliger, O. (1893). Tunicata. *Bronn's* Klassen **3,** Suppl., 319–322.

Seliger, H. H., and McElroy, W. D. (1965). "Light: Physical and Biological Action." Academic Press, New York.

Senebier, J. (1782). "Mémoires physico-chimiques sur l'influence de la lumière solaire pour modifier les êtres de trois règnes, surtout ceux du règne végétal," 3 vols. Chirol, Genève.

Setlow, J. E., and Setlow, R. B. (1963). Nature of the photoreactivable lesion in deoxyribonucleic acid (DNA). *Nature (London)* **197,** 560.

Shemin, D. (1948). The biosynthesis of porphyrins. *Cold Spring Harbor Symp. Quant. Biol.* **13,** 185–192.

Shemin, D. (1955). The succinate-glycine cycle; the role of δ-amino-levulinic acid in porphyrin synthesis. *Porphyrin Biosyn. Metab., Ciba Found. Symp., 1955* pp. 4–26.

Shemin, D. (1956). The biosynthesis of porphyrins. *Harvey Lect.* **50,** 258–284.

Shropshire, W., Jr. (1962). The lens effect and phototropism of *Phycomyces*. *J. Gen. Physiol.* **45,** 949.

Shropshire, W., Jr., (1963). Photoresponses of the fungus *Phycomyces*. *Physiol. Rev.* **43,** 38.

Shropshire, W., Jr. (1972). Phytochrome, a photochromatic sensor. *In* "Photophysiology" (A. L. Geise, ed.), Vol. 7, pp. 31–71. Academic Press, New York.

Sibaoka, T. (1962). Excitable cells in *Mimosa*. *Science* **137,** 226.

Siegelman, H. W., and Firer, E. M. (1964). Purification of phytochrome from oat seedlings. *Biochemistry* **3**, 418.

Singer, S. J., and Nicolson, G. L. (1972). The fluid mosaic model of the structure of cell membranes. *Science* **175**, 720.

Sjöstrand, F. S. (1949). An electron microscope study of the retinal rods of the guinea pig eye. *J. Cell. Comp. Physiol.* **33**, 383.

Sjöstrand, F. S. (1953a). The ultrastructure of the outer segments of rods and cones of the eye as revealed by the electron microscope. *J. Cell. Comp. Physiol.* **42**, 15.

Sjöstrand, F. S. (1953b). The ultrastructure of the inner segments on the retinal rods of the guinea pig as revealed by the electron microscope. *J. Cell. Comp. Physiol.* **42**, 45.

Smith, D. E. (1963). The organization and innervation of the luminescent organ in a firefly, *Photuris pennsylvanica* (*Coleoptera*). *J. Cell. Biol.* **16**, 323.

Smith, J. H. C. (1960). Protochlorophyll transformations. *In* "Comparative Biochemistry of Photoreactive Systems" (M. B. Allen, ed.), pp. 257–277. Academic Press, New York.

Stanier, R. Y. (1959). Formation and function of the photosynthetic pigment system in purple bacteria. *Brookhaven Symp. Biol.* **11**, 43–53.

Stanier, R. Y. (1970). Some aspects of the biology of cells and their possible evolutionary significance. *In* "Prokaryotic and Eukaryotic Cells" (*Symp. Gen. Microbiol.*, No. XX), pp. 1–38.

Steven, D. M. (1963). The dermal light sense. *Biol. Rev. Cambridge Phil. Soc.* **38**, 204.

Stoeckenius, W. (1962). Some electron microscopical observations on liquid-crystalline phases in lipid-water systems. *J. Cell Biol.* **12**, 221.

Strain, H. J. (1951). *In* "Manual of Phycology" (G. M. Smith, ed.), pp. 243–262. Chronica Botanica, Waltham, Massachusetts.

Strehler, B. L. (1962). "Time, Cells and Aging." Academic Press, New York.

Strother, G. K. (1963). Absorption spectra of retinal oil globules in turkey, turtle, and pigeon. *Exp. Cell Res.* **29**, 249.

Strother, G. K., and Casella, A. J. (1972). Microspectrophotometry of arthropod visual screening pigments. *J. Gen. Physiol.* **59**, 616.

Strother, G. K., and Wolken, J. J. (1960). Microspectrophotometry. I. Absorption spectra of colored oil globules in the chicken retina. *Exp. Cell Res.* **21**, 504.

Strother, G. K., and Wolken, J. J. (1961). *In vivo* absorption spectra of *Euglena:* Chloroplast and eyespot. *J. Protozool.* **8**, 261.

Stuermer, W. (1970). Soft parts of cephalopods and trilobites: Some surprising results of X-ray examinations of Devonian slates. *Science* **170**, 1300.

Thornber, J. P., Gregory, R. P. V., Smith, C. A., and Bailey, J. L. (1967a). Studies on the nature of the chloroplast lamella. I. Preparation and some properties of two chlorophyll-protein complexes. *Biochemistry* **6**, 391.

Thornber, J. P., Stewart, J. C., Hatton, M. W. C., and Bailey, J. L. (1967b). Studies on the nature of the chloroplast lamella. II. Chemical composition and further physical properties of two chlorophyll-protein complexes. *Biochemistry* **6**, 2006.

Tischer, J. (1936). Über das Euglenarhodon und andere Carotinoide einer roten Euglene. (Carotinoide der Susswasseralgen, I. Teil.) *Hoppe-Seyler's Z. Physiol. Chem.* **239**, 257.

Tischer, J. (1941). Carotenoids of fresh-water algae. VII. Identity of euglenarhodon with astacin. *Hoppe-Seyler's Z. Physiol. Chem.* **267**, 281.

Tollin, G., and Robinson, M. I. (1969). Phototaxis in *Euglena.* V. Photosuppression of phototactic activity by blue light. *Photochem. Photobiol.* **9**, 411.

Tomita, T. (1970). Electrical activity of vertebrate photoreceptors. *Quart. Rev. Biophys.* **3**, 179.

Towe, K. M. (1973). Trilobite eyes calcified lenses *in vivo*. *Science* **179**, 1007.

Trosper, T., and Sauer, K. (1968). Chlorophyll *a* interactions with chloroplast lipids *in vitro*. *Biochim. Biophys. Acta* **162,** 97.

Ungar, G. (1970). Role of proteins and peptides in learning and memory. *In* "Molecular Mechanisms in Memory and Learning" (G. Ungar, ed.), pp. 149–175. Plenum, New York.

Urey, H. (1952). "The Planets." Yale Univ. Press, New Haven, Connecticut.

"U.S. Standard Atmosphere" (1962). p. 9. Compiled by the National Aeronautics and Space Administration, the United States Air Force, and the United States Weather Bureau, Washington, D.C.

Vaissière, R. (1961). Morphologie et histologie comparées des yeux des Crustacés Copépodes. *Arch. Zool. Exp. Gen.* **100,** 126.

Vallery-Radot, P. (1922–1939). See Pasteur (1922–1939).

Vanderkooi, G., and Green, D. E. (1971). New insights into biological membrane structure. *Bioscience* **21,** 409.

van Dorp, D. A., and Arens, J. F. (1947). Synthesis of vitamin A aldehyde. *Nature (London)* **160,** 189.

van Leeuwenhoek, A. (1674). More observation from Mr. Leeuwenhoek. *Phil. Trans. Roy. Soc. London* **9,** 178.

van Niel, C. B. (1935). Photosynthesis of bacteria. *Cold Spring Harbor Symp. Quant. Biol.* **3,** 158.

van Niel, C. B. (1941). The bacterial photosyntheses and their importance for the general problem of photosynthesis. *Advan. Enzymol.* **1,** 263.

van Niel, C. B. (1943). Biochemical problems of the chemo-autotrophic bacteria. *Physiol. Rev.* **23,** 338.

van Niel, C. B. (1949). The comparative biochemistry of photosynthesis. *Amer. Scientists* **37,** 371.

Varela, F. G., and Wiitanen, W. (1970). The optics of the compound eye of the honeybee (*Apis mellifera*). *J. Gen. Physiol.* **55,** 336.

Velikovsky, I. (1955). "Earth in Upheaval." Doubleday, Garden City, New York.

Virchow, R. (1858). "Die Cellularpathologie in Ihrer Begründung auf Physiologische und Pathologische Gewebelehre." Hirschwald, Berlin.

Vigroux, E. (1953). Contribution to the experimental study of the absorption of ozone. *Ann. Phys.* **8,** 709.

Vigroux, E. (1969). Coefficients d'absorption de l'ozone dans la bande de Hartley. *Ann. Geophys.* **25,** 169.

von Frisch, K. (1914). Der Farbensinn und Formensinn der Beinen. *Zool. Jahrb., Abt. Allg. Zool. Physiol. Tiere* **35,** 1.

von Frisch, K. (1949). Die Polarisation des Himmelslichtes als Faktor der Orientieren bei den Tänzen der Bienen. *Experientia* **5,** 397.

von Frisch, K. (1950). "Bees: Their Vision, Chemical Senses and Language." Cornell Univ. Press, Ithaca, New York, Revised edition 1971.

von Frisch, K. (1953). "The Dancing Bees." Harcourt, New York.

von Frisch, K. (1967). "The Dance Language and Orientation of Bees." Belknap Press, Cambridge, Massachusetts.

von Haffner, K. (1933). Die überzähligen Siphonen und Ocellen von *Ciona intestinalis*. *L. Z. Wiss. Zool.* **143,** 16.

von Helmholtz, H. (1852). Über die Theorie der zusammengesetzten Farben. *Ann. Phys. (Leipzig)* [2] **87,** 45.

von Helmholtz, H. (1867). "Handbuch der Physiologischen Optik," p. 874. Voss, Leipzig.

von Sachs, J. (1864). Wirkungen Farbigen Lichts auf Pflanzen. *Botan. Z.* **22,** 353.

von Sachs, J. (1868). "Lehrbuch der Botanik Nach dem Gegenwärtigen Stand der Wissenschaft." Leipzig (1st Ed.). (English trans., 3rd ed., A. W. Bennett and W. T. Thistleton-Dyer, Oxford, 1875).

von Sachs, J. (1882). "Vorlesungen über Pflanzenphysiologie." Leipzig (English trans., H. M. Ward, Oxford, 1887).

Vowles, D. M. (1954). The orientation of ants. II. Orientation to light, gravity and polarized light. *J. Exp. Biol.* **31**, 356.

Waddington, C. H., and Perry, M. M. (1960). The ultrastructure of the developing eye of *Drosophila*. *Proc. Roy. Soc., Ser. B* **153**, 155.

Wald, G. (1938). On rhodopsin in solution. *J. Gen. Physiol.* **21**, 795.

Wald, G. (1939). On the distribution of vitamins $A_1$ and $A_2$. *J. Gen. Physiol.* **22**, 391.

Wald, G. (1942). The visual system and vitamin A of the sea lamprey. *J. Gen. Physiol.* **25**, 331.

Wald, G. (1945). Human vision and the spectrum. *Science* **101**, 653.

Wald, G. (1948). Galloxanthin, a carotenoid from the chicken retina. *J. Gen. Physiol.* **31**, 377.

Wald, G. (1952). Biochemical evolution. *In* "Modern Trends in Physiology and Biochemistry" (E. S. G. Barrón, ed.), pp. 337–376. Academic Press, New York.

Wald, G. (1953). The biochemistry of vision. *Annu. Rev. Biochem.* **22**, 497.

Wald, G. (1954). On the mechanism of the visual threshold and visual adaptation. *Science* **119**, 887.

Wald, G. (1955). Photoreceptor process in vision. *Amer. J. Ophthalmol.* **40**, 18.

Wald, G. (1956). The biochemistry of visual excitation. *In* "Enzymes: Units of Biological Structure and Function" (O. H. Gaebler, ed.), p. 355. Academic Press, New York.

Wald, G. (1957). The origin of optical activity. *Ann. N.Y. Acad. Sci.* **69**, 352.

Wald, G. (1959). The photoreceptor process in vision. *In* "Handbook of Physiology" (Amer. Physiol. Soc., J. Field, ed.), Sect. 1, Vol. I p. 671. Williams & Wilkins, Baltimore, Maryland.

Wald, G. (1960). The distribution and evolution of visual systems. *Comp. Biochem.* **1**, 311.

Wald, G. (1961a). The molecular organization of visual systems. *In* "Light and Life" (W. D. McElroy and B. Glass, ed.), p. 724. Johns Hopkins Press, Baltimore, Maryland.

Wald, G. (1961b). General discussion of retinal structure in relation to the visual process. *In* "Structure of the Eye" (G. K. Smelser, ed.), pp. 101–115. Academic Press, New York.

Wald, G. (1964a). The origins of life. *Proc. Nat. Acad. Sci. U.S.* **52**, 595.

Wald, G. (1964b). The receptors of human color vision. *Science* **145**, 1007.

Wald, G. (1968). Single and multiple visual systems in arthropods. *J. Gen. Physiol.* **51**, 125.

Wald, G. (1970). Vision and the mansions of life. The first Feodor Lynen Lecture *In* "Miami Winter Symposia," pp. 1–32. North-Holland Publ., Amsterdam.

Wald, G., and Allen, G. (1946). Fractionation of the eye pigments of *Drosophila melanogaster*. *J. Gen. Physiol.* **30**, 41.

Wald, G., and Brown, P. K. (1957). The vitamin A of a euphausiid crustacean. *J. Gen. Physiol.* **40**, 627.

Wald, G., and Brown, P. K. (1958). Human rhodopsin. *Science* **127**, 222.

Wald, G., and Brown, P. K. (1965). Human color vision and color blindness. *Cold Spring Harbor Symp. Quant. Biol.* **30**, 345.

Wald, G., and Burg, S. P. (1957). The vitamin A of the lobster. *J. Gen. Physiol.* **40**, 609.

Wald, G., and Hubbard, R. (1957). Visual pigment of a decapod crustacean: The lobster. *Nature (London)* **180**, 278.

Wald, G., and Seldin, E. B. (1968). Spectral sensitivity of the common prawn, *Palaemonetes vulgaris*. *J. Gen. Physiol.* **51**, 694.

Wald, G., and Zussman, H. (1938). Carotenoids of the chicken retina. *J. Biol. Chem.* **122**, 449.

Wald, G., Brown, P. K., and Gibbons, I. R. (1963). The visual cells and visual pigment of the mudpuppy, *Necturus*. *J. Cell Biol.* **19**, 79.

Wall, E. J. (1925). "History of Three Color Photography." Amer. Phot. Publ., Boston, Massachusetts.

Wallace, A. R. (1858). On the tendency of species to form varieties. *Proc. Linn. Soc. London, Ser. Zool.* **3**, 53–57. See Darwin (1858).

Walls, G. L. (1942). "The Vertebrate Eye." Cranbrook Inst. Sci., Bloomfield Hills, Michigan.

Walther, J. B. (1958). Changes induced in spectral sensitivity and form of retinal act potential of the cockroach eye by selective adaptation. *J. Insect Physiol.* **51**, 694.

Walther, J. B., and Dodt, E. (1957). Electrophysiologische Untersuchungen über die Ultraviollempfindlichkeit von Insektenaugen. *Experientia* **13**, 333.

Walther, J. B., and Dodt, E. (1959). Die Spektralsensitivität von Insekten Komplexaugen in Ultraviolette. *Z. Naturforsch. 3.* **14**, 273.

Warburg, O., and Christian, W. (1938a). Coenzyme of the d-amino acid deaminase. *Biochem. Z.* **295**, 261.

Warburg, O., and Christian, W. (1938b). Coenzyme of the d-alanine-dehydrogenase. *Naturwissenschaften* **26**, 235.

Warburg, O., and Christian, W. (1938c). Coenzyme of d-alanine oxidase. *Biochem. Z.* **296**, 294.

Ward, R. R. (1971). "The Living Clocks." Alfred A. Knopf, Inc., New York.

Watanabe, K. (1959). Ultraviolet absorption process in the upper atmosphere. *Advan. Geophys.* **5**, 153.

Waterman, T. H., Fernandez, H. R., and Goldsmith, T. H. (1969). Dichroism of photosensitive pigment in rhabdoms of the crayfish *Orconectes*. *J. Gen. Physiol.* **54**, 415.

Watson, J. D. (1965). "Molecular Biology of the Gene." Benjamin. New York.

Wigglesworth, V. B. (1964). "The Life of Insects." Weidenfeld & Nicholson, London.

Willmer, E. N., ed. (1960). "Cytology and Evolution," 1st ed. Academic Press, New York.

Winfree, A. T. (1975). Resetting Biological Clocks. *Physics Today* **28**, 34.

Willstäter, R., and Stoll, A. (1913). "Untersuchungen über Chlorophyll. Methoden und Ergebnisse." Springer-Verlag, Berlin.

Witman, C. G., Carlson, K., Berliner, J., and Rosenbaum, J. L. (1972a). *Chlamydomonas* flagella. I. Isolation and electrophoretic analysis of microtubules, matrix, membranes, and mastigonemes. *J. Cell Biol.* **54**, 507.

Witman, C. G., Carlson, K., and Rosenbaum, J. L. (1972b). *Chlamydomonas* flagella. II. The distribution of tubulins 1 and 2 in the outer doublet microtubules. *J. Cell Biol.* **54**, 540.

Wöhler, F. (1828). Über künstliche Bildung des Harmstoffes. *Ann. Phys.* (*Leipzig*) [2] **12**, 253.

Wolken, J. J. (1956). Photoreceptor structures. I. Pigment monolayers and molecular weight. *J. Cell Comp. Physiol.* **48**, 349.

Wolken, J. J. (1958a). The chloroplast structure, pigments and pigment-protein complex. *Brookhaven Symp. Biol.* **11**, 87.

Wolken, J. J. (1958b). Retinal structure. Mollusc cephalopods: *Octopus, Sepia*. *J. Biophys. Biochem. Cytol.* **4**, 835.

Wolken, J. J. (1959). The structure of the chloroplast. *Annu. Rev. Plant Physiol.* **10**, 71.

Wolken, J. J. (1961a). A structural model for a retinal rod. *In* "The Structure of the Eye" (G. K. Smelser, ed.), pp. 173–192. Academic Press, New York.

Wolken, J. J. (1961b). The photoreceptor structure. *Int. Rev. Cytol.* **11,** 195–218.

Wolken, J. J. (1963). The structure and molecular organization of retinal photoreceptors. *J. Opt. Soc. Amer.* **53,** 1.

Wolken, J. J. (1966a). Lipids and the molecular structure of photoreceptors. *J. Amer. Oil Chem. Soc.* **43,** 271.

Wolken, J. J. (1966b). "Vision: Biochemistry and Biophysics of the Retina." Thomas, Springfield, Illinois.

Wolken, J. J. (1967). "Euglena: An Experimental Organism for Biochemical and Biophysical Studies," 2nd ed. Appleton. New York.

Wolken, J. J. (1968a). "Photobiology." Van Nostrand Reinhold, Princeton, New Jersey.

Wolken, J. J. (1968b). The photoreceptors of arthropod eyes. *Symp. Zool. Soc. London* **23,** 113.

Wolken, J. J. (1968c). The cell and the red blood cell. *In* "Hemorheology" (A. L. Copley, ed.), pp. 525–537. Pergamon, Oxford.

Wolken, J. J. (1969). Microspectrophotometry and the photoreceptor of *Phycomyces* I. *J. Cell Biol.* **43,** 354.

Wolken, J. J. (1971). "Invertebrate Photoreceptors: A Comparative Analysis." Academic Press, New York.

Wolken, J. J. (1972). *Phycomyces.* A model photosensory cell. *Int. J. Neurosci.* **3,** 135.

Wolken, J. J. (1973). Comparative visual systems in microorganisms, plants and animals. *In* "Biology Data Book" (P. L. Altman and D. S. Dittmer, eds.), Vol. II, pp. 1268–1271. Fed. Amer. Soc. Exp. Biol., Bethesda, Maryland.

Wolken, J. J. (1974). Comparative structure of invertebrate photoreceptors. *In* "The Eye" (H. Davson and L. T. Graham, Jr., eds), Vol. 6, pp. 111–154. Academic Press, New York.

Wolken, J. J., and Florida, R. (1969). The eye structure and optical system of the crustacean copepod, *Copilia. J. Cell Biol.* **40,** 279.

Wolken, J. J., and Gallik, G. J. (1965). The compound eye of a crustacean: *Leptodora kindtii. J. Cell Biol.* **26,** 968.

Wolken, J. J., and Gross, J. A. (1963). Development and characteristics of the *Euglena* c-type cytochromes. *J. Protozool.* **10,** 189.

Wolken, J. J., and Gupta, P. D. (1961). Photoreceptor structures of the retinal cells of the cockroach eye. IV. *Periplaneta americana* and *Blaberus giganteus. J. Biophys. Biochem. Cytol.* **9,** 720.

Wolken, J. J., and Nakagawa, C. S. (1973). Rhodopsin formed from bacterial retinal and cattle opsin. *Biochem. Biophys. Res. Commun.* **54,** 1262.

Wolken, J. J., and Scheer, I. J. (1963). An eye pigment of the cockroach. *Exp. Eye Res.* **2,** 182.

Wolken, J. J., and Schwertz, F. A. (1953). Chlorophyll monolayers in chloroplasts. *J. Gen. Physiol.* **37,** 111.

Wolken, J. J., and Schwertz, F. A. (1956). The molecular weight of algal chloroplastin. *Nature (London)* **177,** 136.

Wolken, J. J., and Shin, E. (1958). Photomotion in *Euglena gracilis.* I. Photokinesis. II. Phototaxis. *J. Protozool.* **5,** 39.

Wolken, J. J., Mellon, A. D., and Greenblatt, C. L. (1955). Environmental factors affecting growth and chlorophyll synthesis in *Euglena. J. Protozool.* **2,** 89.

Wolken, J. J., Capenos, J., and Turano, A. (1957a). Photoreceptor structures. *Drosophila melanogaster.* III. *J. Biophys. Biochem. Cytol.* **3,** 441.

Wolken, J. J., Mellon, A. D., and Contis, G. (1957b). Photoreceptor structures. *Drosophila melanogaster* II. *J. Exp. Zool.* **134,** 383.

Wolken, J. J., Bowness, J. M., and Scheer, I. J. (1960). The visual complex of the insect: Retinene in the housefly. *Biochim. Biophys. Acta* **43,** 531.

Wolken, J. J., Forsberg, R., Gallik, G. J., and Florida, R. G. (1968). Rapid recording microspectrophotometer. *Rev. Sci. Instrum.* **39,** 1734.

Woodward, R. B. (1961). Total synthesis of chlorophyll. *Pure Appl. Chem.* **2,** 383.

Wulff, V. S. (1956). Physiology of the compound eye. *Physiol. Rev.* **36,** 145.

Wurtman, R. J., Axelrod, J., and Kelly, D. E. (1968). "The Pineal." Academic Press, New York.

Yagi, K. (1962). Chemical determination of flavins. *Methods Biochem. Anal.* 319–354.

Yoshida, M., Ohtsuki, H., and Suguri, S. (1967). Ommochrome from anthomedusan ocelli and its photoreduction. *Photochem. Photobiol.* **6,** 875.

Young, J. Z. (1964). "A Model of the Brain." Oxford Univ. Press, London and New York.

Young, J. Z. (1971). "Introduction to the Study of Man." Oxford Univ. Press, London and New York.

Young, T. (1802). On the theory of light and colors. *Phil. Trans. Roy. Soc. London* **92,** 12.

Young, T. (1807). "Lectures on Natural Philosophy," Vol. I, pp. 315 and 613. W. Savage, London.

Zechmeister, L. (1944). *Cis-trans* isomerization and sterochemistry of carotenoids and diphenylpolyenes. *Chem. Rev.* **34,** 267.

Zechmeister, L. (1962). "Carotenoids: *Cis-Trans* Isomeric Carotenoids, Vitamin A, and Arylpolyenes." Academic Press, New York.

Ziegler, I. (1964). Über naturlich vorkommende Tetrahydropterine. *In* "Pteridine Chemistry" (W. Pfeiderer and E. C. Taylor, eds.), pp. 295–305. Pergamon, Oxford.

Ziegler, I. (1965). Pterine als Wirkstoffe und Pigmente. *Ergeb. Physiol., Biol. Chem. Exp. Pharmakol.* **56,** 1.

Ziegler-Gunder, I. (1956). Pterine: Pigmente und Wirkstoffe im Tierreich. *Biol. Rev. Cambridge Phil. Soc.* **31,** 313.

Zonana, H. V. (1961). Fine structure of the squid retina. *Bull. Johns Hopkins Hosp.* **109,** 185.

# Subject Index

## D



## E

**S**

A 5
B 6
C 7
D 8
E 9
F 0
G 1
H 2
I 3
J 4